VOLUME FORTY EIGHT

ADVANCES IN INSECT PHYSIOLOGY

VOLUME FORTY EIGHT

ADVANCES IN
INSECT PHYSIOLOGY

Genomics, Physiology and Behaviour of Social Insects

Edited by

AMRO ZAYED

Department of Biology, York University, Toronto, Ontario, Canada

CLEMENT F. KENT

Janelia Research Campus, Howard Hughes Medical Institute, Ashburn, VA, USA

ELSEVIER

AMSTERDAM • BOSTON • HEIDELBERG • LONDON
NEW YORK • OXFORD • PARIS • SAN DIEGO
SAN FRANCISCO • SINGAPORE • SYDNEY • TOKYO
Academic Press is an imprint of Elsevier

Academic Press is an imprint of Elsevier
125 London Wall, London, EC2Y 5AS, UK
The Boulevard, Langford Lane, Kidlington, Oxford OX5 1GB, UK
225 Wyman Street, Waltham, MA 02451, USA
525 B Street, Suite 1800, San Diego, CA 92101-4495, USA

First edition 2015

Notices

Knowledge and best practice in this field are constantly changing. As new research and experience broaden our understanding, changes in research methods, professional practices, or medical treatment may become necessary.

Practitioners and researchers must always rely on their own experience and knowledge in evaluating and using any information, methods, compounds, or experiments described herein. In using such information or methods they should be mindful of their own safety and the safety of others, including parties for whom they have a professional responsibility.

To the fullest extent of the law, neither the Publisher nor the authors, contributors, or editors, assume any liability for any injury and/or damage to persons or property as a matter of products liability, negligence or otherwise, or from any use or operation of any methods, products, instructions, or ideas contained in the material herein.

ISBN: 978-0-12-802157-6
ISSN: 0065-2806

CONTENTS

CONTRIBUTORS

Patrick Abbot
Department of Biological Sciences, Vanderbilt University, Nashville, Tennessee, USA

Etya Amsalem
Department of Entomology, Center for Pollinator Research, Center for Chemical Ecology, Huck Institutes of the Life Sciences, The Pennsylvania State University, University Park, Pennsylvania, USA

Andrew B. Barron
Department of Biological Sciences, Macquarie University, Sydney, New South Wales, Australia

Mario S. Cervoni
Departamento de Biologia Celular e Molecular e Bioagentes Patogênicos, Faculdade de Medicina de Ribeirão Preto, Universidade de São Paulo, Ribeirão Preto, Sao Paulo, Brazil

Linh M. Chau
School of Biology, Georgia Institute of Technology, Atlanta, Georgia, USA

Karl M. Glastad
School of Biology, Georgia Institute of Technology, Atlanta, Georgia, USA

Michael A.D. Goodisman
School of Biology, Georgia Institute of Technology, Atlanta, Georgia, USA

Christina M. Grozinger
Department of Entomology, Center for Pollinator Research, Center for Chemical Ecology, Huck Institutes of the Life Sciences, The Pennsylvania State University, University Park, Pennsylvania, USA

Karina R. Guidugli-Lazzarini
Departamento de Biologia Celular e Molecular e Bioagentes Patogênicos, Faculdade de Medicina de Ribeirão Preto, Universidade de São Paulo, Ribeirão Preto, Sao Paulo, Brazil

Klaus Hartfelder
Departamento de Biologia Celular e Molecular e Bioagentes Patogênicos, Faculdade de Medicina de Ribeirão Preto, Universidade de São Paulo, Ribeirão Preto, Sao Paulo, Brazil

Abraham Hefetz
Department of Zoology, George S. Wise Faculty of Life Sciences, Tel Aviv University, Ramat Aviv, Tel Aviv, Israel

Fernanda C. Humann
Departamento de Biologia Celular e Molecular e Bioagentes Patogênicos, Faculdade de Medicina de Ribeirão Preto, Universidade de São Paulo, Ribeirão Preto, and Instituto Educação, Ciência e Tecnologia de São Paulo, Campus do Matão, Matão, Sao Paulo, Brazil

Jennifer M. Jandt
Department of Ecology, Evolution, and Organismal Biology, Iowa State University, Ames, Iowa, USA

Clement F. Kent
Janelia Research Campus, Howard Hughes Medical Institute, Ashburn, VA, USA

Judith Korb
Department of Evolutionary Biology and Ecology, Institute of Biology I, University of Freiburg, Freiburg, Germany

Timothy A. Linksvayer
Department of Biology, University of Pennsylvania, Philadelphia, Pennsylvania, USA

Emma K. Mullen
Department of Biology, Western University, London, Ontario, Canada

Mario Padilla
Department of Entomology, Center for Pollinator Research, Center for Chemical Ecology, Huck Institutes of the Life Sciences, The Pennsylvania State University, University Park, Pennsylvania, USA

Clint J. Perry
School of Chemical and Biological Sciences, Queen Mary University, London, United Kingdom

Douglas E. Santos
Departamento de Biologia Celular e Molecular e Bioagentes Patogênicos, Faculdade de Medicina de Ribeirão Preto, Universidade de São Paulo, Ribeirão Preto, Sao Paulo, Brazil

Eirik Søvik
Department of Biology, Washington University in St. Louis, St. Louis, Missouri, USA

Graham J. Thompson
Department of Biology, Western University, London, Ontario, Canada

Amy L. Toth
Department of Ecology, Evolution, and Organismal Biology, and Department of Entomology, Iowa State University, Ames, Iowa, USA

Amro Zayed
Department of Biology, York University, Toronto, Ontario, Canada

FOREWORD

Social insects have long been prized as exceptionally valuable model organisms in physiology, behaviour, and evolutionary biology. Their remarkably complex and effective systems of communication, behaviour, and social organization, all produced by tiny brains and arising as a result of some of the most extreme forms of self-sacrifice known on the planet, have inspired several of the more important discoveries about the mechanisms and evolution of social life in the twentieth century.

The publication of the honey bee genome sequence in 2006 heralded a new era in which important questions about the mechanisms and evolution of social life could be productively addressed in molecular terms with genomic information and new genome-powered tools. Soon the genomes of a few additional social insects were sequenced, and now there are genomic resources available for about 25 species, and counting.

This special issue highlights many of the advances in social insect sociogenomics that have been made during these past 10 years. The volume is marked by an impressive breadth of study organisms and topics, with chapters authored by some of the best social insect researchers in the world. Judging from the findings covered in these chapters, this has been a decade of significant advancement for social insect research.

In some chapters, genomic information and tools have been used to generate new insights into long-standing problems, such as caste determination and pheromone communication. In other chapters, entirely new analytical approaches are employed, now made possible by the advent of genomics. This includes using DNA methylation to study how environmental information "gets under the skin" to affect diverse aspects of social insect biology, and using genomes as history books to begin to understand which genes and which biological processes have been sculpted by evolution to give rise to the social insects that we know and love. Genomics also can help enhance the development of theory, illustrated nicely by several chapters on indirect genetic effects, inclusive fitness, and the evolution of the reward system.

Sociogenomics is still a young science, and there is much to do. More genome sequencing—more species and more individuals within species—will enhance our ability to use social insects as models to address important questions at all levels of biological organization. Improved tools to manipulate genes and genomes are needed to deepen our understanding of the causal relationships between genotype and phenotype. Early successes with transgenic honey bees from the laboratory of Martin Beye, RNAi, and recently developed genome editing tools provide strong reasons for optimism that we are on the cusp of robust social insect functional genomics. The promise of these new technical advances, together with the rich body of knowledge and insights expertly assembled by the editors of this special issue, makes it easy to conclude that the future of social insect research is very bright.

GENE E. ROBINSON
University of Illinois at Urbana–Champaign

INTRODUCTION TO ADVANCES IN INSECT PHYSIOLOGY: GENOMICS, PHYSIOLOGY AND BEHAVIOUR OF SOCIAL INSECTS

Amro Zayed*, Clement F. Kent†
*Department of Biology, York University, Toronto, Ontario, Canada
†Howard Hughes Medical Institute, Chevy Chase, Maryland, USA

Sociality evolved multiple times in insects, and social insects dominate many terrestrial ecosystems. They are the 'little things that run the world' (Wilson, 1987). Reference to social insects adorns our cultural and scientific heritage—from cave paintings and ancient Egyptian temples (Harpur and Zayed, 2014) to Darwin's 'special difficulty' in *The Origins of Species* (Darwin, 1859). Social insects exhibit some of the most charismatic behaviours and traits found in the animal world—from the agriculturally inclined ants to the dancing honey bees. Research on social insect biology is just as rich, diverse and wonderful as the social insects themselves. The research questions are seemingly very simple—how and why did sociality evolve. But answering these questions requires integrative and interdisciplinary research that is arguably unrivalled in the field of Biology. Unlike other disciplines where reduction is the path to discovery, social insect researchers embrace complexity and diversity—as you will find by reading this volume.

We are happy to present this special volume of *Advances in Insect Physiology* that reviews and synthesizes some of the best integrative research on social insects carried out over the past decade. It was just a little under a decade ago when the first social insect genome was sequenced (The Honeybee Genome Sequencing Consortium, 2006). We have witnessed an explosion of social insect genome projects since then, and more importantly, an explosion of insightful studies that have used genomic tools to revisit long-standing questions in the field. The impetus of this volume was to broadly summarize and synthesize genomics-empowered research on the physiology, behaviour and evolution of social insects, written by experts in the field.

We start our volume with juvenile hormone and caste development in honey bees, but Chapter 1 'Old Threads Make New Tapestry—Rewiring of Signalling Pathways Underlies Caste Phenotypic Plasticity in the Honey Bee, *Apis mellifera* L.' by Hartfelder and colleagues is far from a retelling of a classic story. The authors weave an amazing amount of physiological and evolutionary insight to show us how 'old threads', such as conserved nutrient-sensing pathways and hormones, can be rewired to give us the 'new tapestry' of caste-specific phenotypes in the bee.

Chapter 2 'The Physiological and Genomic Bases of Bumble Bee Social Behaviour' by Amsalem and colleagues provides a comprehensive review of the relationship between genes, physiology and behaviour in the primitively social Bumblebees. The chapter is a tome of knowledge and an excellent starting point for students and researchers interested in bumblebee sociobiology.

In Chapter 3 'Physiological and Genomic Mechanisms of Social Organization in Wasps (Family: Vespidae)', Jandt and Toth discuss how wasps in the Vespidae family provide a great platform for studying the evolution of sociality. They neatly review the physiological and genomic mechanisms underlying caste determination in this group and introduce key mechanistically explicit hypotheses for how caste differences arise.

Termites are social insects that differ from ants, bees and wasps in many interesting ways. In Chapter 4 'Juvenile Hormone: A Central Regulator of Termites' Caste Polyphenism', Korb provides a wonderful introduction to the social biology of lower termites and summarizes how the environment, genetics and juvenile hormone underlie termite division of labour.

Research on non-hymenopteran social insects will surely add many interesting insights on how sociality evolves. In Chapter 5 'The Physiology and Genomics of Social Transitions in Aphids', Abbot provides a captivating and exciting review of the biology of social aphids and how the recent availability of the pea aphid genome offers new opportunities for studying the genetics and evolution of the soldier caste.

We change themes in Chapter 6 'Insect Reward Systems: Comparing Flies and Bees' by Søvik and colleagues from the physiology and genes underlying caste determination to the neurobiology of the reward system in a model social insect, the honey bee, and a model solitary insect, the fruit fly. In addition to providing a comprehensive review of the similarities and differences between the reward system in the bee and the fly, the authors provide an intriguing discussion of how modifications of the reward system could have contributed to the evolution of sociality.

It is increasingly apparent that epigenetics play an important role in influencing the remarkable levels of phenotypic plasticity found in social insects. In Chapter 7 'Epigenetics in Social Insects', Glastad and colleagues provide an authoritative review of the major epigenetic mechanisms found in social insects and how these epigenetic mechanisms influence social traits.

Chapter 8 'The Molecular and Evolutionary Genetic Implications of Being Truly Social for the Social Insects' by Linksvayer provides an important 'call to arms' for studying the genetics of social insects within the framework of 'indirect genetic effects'. Social insects obviously interact with related kin, creating ripe conditions for the evolution of genes with indirect effects—where an individual's phenotype is controlled by a social partner's genotype. Linksvayer provides a succinct review of how indirect genetic effects influence behaviour and physiology of individuals within colonies and discusses prospects for investigating how genes with indirect genetic effects contribute to social evolution.

In Chapter 9 'Population and Phylogenomic Insights into the Evolution of Physiology and Behaviour in Social Insects', we (Kent and Zayed) discuss the importance of using studies of molecular evolution—on a genomic scale—as a way to 'evolutionarily' annotate the genomes of social insects. We review studies on the evolution of genes and regulatory sequences in social insects as revealed by comparative genomics and population genomics and illustrate how these data can be used to link mechanistic and ultimate hypotheses for the evolution of sociality.

The field of sociobiology is theory-rich, but recognizing the 'genes for altruism' (Hamilton, 1964a,b) among the hundreds of interesting candidate genes revealed by genomic studies is not so obvious. In Chapter 10 'Understanding Honey Bee Worker Self-Sacrifice: A Conceptual–Empirical Framework', Mullen and Thompson synthesize the properties of genes controlling worker self-sacrifice and provide a set of genetic, genomic and molecular predictions to guide the discovery of these 'genes for altruism' in the age of sociogenomics.

We hope that this volume will provide an excellent summary of the state of the field as well as a road map for the future of social insect research.

REFERENCES

Darwin, C., 1859. The Origin of Species. P. F. Collier & Son, New York.

Hamilton, W.D., 1964a. The genetical evolution of social behavior I. J. Theor. Biol. 7, 1–16.

Hamilton, W.D., 1964b. The genetical evolution of social behavior II. J. Theor. Biol. 7, 17–52.

Harpur, B.A., Zayed, A., 2014. Honey bee: management. In: Smith, C. (Ed.), Encyclopedia of Global Archaeology. Springer, New York.

The Honeybee Genome Sequencing Consortium, 2006. Insights into social insects from the genome of the honeybee Apis mellifera. Nature 443, 931–949.

Wilson, E.O., 1987. The little things that run the world (the importance and conservation of invertebrates). Conserv. Biol. 1, 344–346.

CHAPTER ONE

Old Threads Make New Tapestry—Rewiring of Signalling Pathways Underlies Caste Phenotypic Plasticity in the Honey Bee, *Apis mellifera* L.

Klaus Hartfelder*[,1], Karina R. Guidugli-Lazzarini*, Mario S. Cervoni*, Douglas E. Santos*, Fernanda C. Humann*[,†]

*Departamento de Biologia Celular e Molecular e Bioagentes Patogênicos, Faculdade de Medicina de Ribeirão Preto, Universidade de São Paulo, Ribeirão Preto, Sao Paulo, Brazil
[†]Instituto Educação, Ciência e Tecnologia de São Paulo, Campus do Matão, Matão, Sao Paulo, Brazil
[1]Corresponding author: e-mail address: klaus@fmrp.usp.br

Contents

Abstract

Caste development in social Hymenoptera represents a prime example of phenotypic plasticity and variation in life histories and is best understood in honey bees. We review the larval feeding regimes and major signalling pathways involved in the divergent developmental trajectories. The role of juvenile hormone (JH) is discussed with an emphasis on the regulation of JH synthesis, degradation and the molecular mode of JH action. As there are apparent inconsistencies between the expression of insulin, TOR and Egfr pathway genes and larval growth rates, we propose that the circuitries and integration of these nutrient-sensing pathways have been remodelled in social evolution. A potential

Advances in Insect Physiology, Volume 48
ISSN 0065-2806
http://dx.doi.org/10.1016/bs.aiip.2014.12.001

integrator of nutrient sensing with growth regulation is the hypoxia response, indicating that regulation of oxidative metabolism, possibly through epigenetic effects, is a major determinant of caste fate. Finally, we compare signalling pathways acting in larvae and adult females, with a special emphasis on the role(s) of vitellogenin.

1. INTRODUCTION

Long-lived, large and highly reproductive queens and short-lived, small and facultatively or anatomically sterile workers are the most prominent and characteristic features of reproductive division of labour among females in the highly eusocial bees, wasps and ants. This astounding association of high reproductivity with long lifespan obviously represents a challenge to life history theory for sexually reproducing metazoans, where a trade-off between reproduction (propagation through the germline) and longevity (somatic maintenance) is the rule. As Flatt et al. (2013) emphasise, this trade-off can be conceptually interpreted as plasticity in life history responses to environmental conditions. Under favourable conditions, the allocation of resources should go towards reproduction at the expense of somatic maintenance, and *vice versa* under unfavourable ones. In highly eusocial insects, this plasticity in resource allocation at the individual level has been split into two morphologically and functionally different life histories, the queen and the worker, where the queen caste has, by and large, a monopoly on reproduction, and the worker caste maintains the colony or, in other words, represents the extended soma.

A crucial decision in the reproduction/longevity trade-off in insects and other metazoans with fixed adult size, such as insects and mammals, is when to stop growing. In holometabolous insects, this decision is set by three factors, minimal viable weight, critical weight for metamorphosis and post-critical weight growth until pupation (Mirth and Riddiford, 2007; Nijhout, 2003; Nijhout and Williams, 1974). Within this framework, the critical weight threshold is a developmental switch that enables a larva to enter metamorphosis. Physiologically, this involves an integration of nutrient sensing and growth regulation on the one hand and synchronisation of moulting events on the other, or, in other terms, an integration of the insulin–insulin-like signalling (IIS) and target-of-rapamycin (TOR) pathways with the secretion of juvenile hormone (JH) from the corpora allata (CA) and of moult-inducing ecdysteroids from the prothoracic gland. The scenario of this integration is best worked out in *Drosophila melanogaster*, where the reduction of haemolymph JH levels and the production and release of insulin-like peptides (ILPs) by brain or fat body cells stimulate ecdysone synthesis in the prothoracic gland (Colombani et al., 2005; Mirth et al., 2005, 2014), thus determining the onset of metamorphosis and, consequently, fixing adult size.

While adult size can differ between the two sexes in insects, likely due to their differences in genetic make up, the caste systems of highly eusocial insects introduce a new facet into this scenario, namely phenotypic plasticity based on a divergence in developmental trajectories. When speaking of castes, it is worthy of note to distinguish between the caste systems of highly eusocial Hymenoptera (ants, bees and wasps) and those seen in termites. The former are built on reproductive division of labour among the adult queen(s) and workers present in a colony, whereas in the latter, workers are larvae that can still show great flexibility on their caste options, especially so in the basal termite clades (Hoffmann et al., 2014; Korb and Hartfelder, 2008; Korb et al., 2009; Korb, 2015, this volume).

The question of how a system could evolve where a female bee, wasp or ant forgoes reproduction and opts to help the queen raise her offspring has been addressed and explained within the framework inclusive fitness theory (Hamilton, 1964; Ratnieks and Helanterä, 2009) and the formation of a super-organism structure, the colony phenotype (Wilson and Hölldobler, 2005). But what is the ontogenetic programme that allows for the flexibility to generate such two distinct phenotypes, the queen and the worker? Strikingly, the ontogenetic programme underlying the development of the two caste phenotypes is, with few exceptions (see Schwander et al. (2008)), not built on genetic (allelic) differences between queens and workers, but is environmentally driven, primarily by the differential feeding of female larvae. So, the question becomes as to how signalling pathways involved in nutrient sensing are integrated with the metamorphosis-regulating endocrine system. Phrased in other terms the question is, whether and how the regulatory circuits that govern metamorphosis and set adult size in *D. melanogaster*, *Manduca sexta* or *Tribolium castaneum*, the three major model systems for holometabolous insects, and *Blattella germanica*, as the major model in hemimetabolans (Belles and Santos, 2014), have been co-opted and rewired to generate the caste phenotypic plasticity.

2. DIFFERENTIAL FEEDING OF HONEY BEE LARVAE

The question as to how a dietary regime can trigger the development of the alternative queen or worker phenotypes is best studied in the honey bee, *Apis mellifera*. This is not only due to the ecological, economic and cultural importance of this species, but also because its biology is favourable to investigation and experimental manipulation of postembryonic development. The queen and worker larvae receive well-defined diets, and their developmental fate is predetermined by the locality where they are raised, either in a small worker cell or in a large queen cell. This is the advantage of the honey bee over social wasps, which rear their larvae on diverse, predigested prey items, or in ants, where larvae are constantly carried around the brood nest, making it difficult to track the developmental fate of individual larvae.

Honey bee larvae are fed a mix of two secretions, a milky-white and a clear one, produced by the hypopharyngeal and mandibular glands of nurse bees (Hoffmann, 1960; Jung-Hoffmann, 1966). Queen larvae receive such glandular products, called royal jelly, throughout the larval feeding period, whereas worker larvae undergo a switch in diet as they grow. Worker larvae first also receive a glandular secretion, but this differs from that fed to queens in the relative proportions of the milky-white to clear components (Haydak, 1970; Jung-Hoffmann, 1966). Then, approximately from the fourth larval instar onwards, a yellowish component is added to the two of secretory origin. The yellowish component contains pollen grains and it is thought to come from the honey crop of the nurse bees (Haydak, 1970). So, there are essentially two different types of worker jelly: one of purely secretory origin fed to young larvae and a mixed diet that is provisioned to older larvae.

Larvae of both castes are fed progressively by nurse bees, which are young adult workers performing brood care tasks inside the nest, but the frequency at which they tend queen and worker larvae differs by a factor 10, as shown by Lindauer (1952), who observed that queen larvae are fed approximately 1500 times during their development, whereas typical worker larvae only receive about 150 such visits. About one-third into the final fifth instar (approximately 120 h after hatching from the egg), feeding stops and nurse bees seal the brood cells with a wax cap. Once a cell is sealed, the larva spins a fine cocoon lining the inside of the brood cell and then enters the pre-pupal stage. This initiates the larval-to-pupal transition, i.e., first metamorphic moult. Pupal development then takes place within the sealed brood cell and, after the second metamorphic moult representing the pupal–adult transition, the adult bee emerges from the brood cell. Based on estimates of larval food conversion, a queen larva would typically receive about 360 mg of royal jelly, this resulting in an adult queen at emergence of about 200 mg body weight (Sasaki and Okada, 1972).

3. THE SECRETS OF ROYAL JELLY

Judging from royal jelly sales in health food stores, customers connect it with mystery properties, such as the promise for long and healthy life. Hence, analyses on its composition started already in the late nineteenth century (von Planta, 1888) so as to bring scientific insight into its queen promoting properties. Such studies are ongoing until today and are not only driven by scientific but also economic interests, the latter to set and assure quality standards, especially so for the global market for royal jelly.

In general terms, royal jelly contains more sugar and lipids than worker jelly. Royal jelly consists of approximately 60–70% water (Dietz and Haydak, 1971), and the dry matter contains around 30% of crude protein (around 15% for fresh weight) and 20–30% of reducible sugars (typically 12% for fresh weight; Asencot and Lensky, 1988; Shuel and Dixon, 1968; Takenaka and Takahashi, 1980). In contrast, worker jelly fed to young larvae has a water content of approximately 75% and a much lower sugar content of only 12% relative to dry matter (or around 4% for fresh weight). The diet fed to older worker larvae, the mixed diet, has approximately the same water content as worker jelly fed to young larvae, but the sugar content is increased to close to 50% dry matter (Asencot and Lensky, 1976, 1988). So, while the dietary composition in terms of crude matter is kept relatively constant for queen larvae, which are also fed at very high frequency, there is a clear switch in diet composition in the case of worker larvae. Interestingly, a similar switch is also seen in the food provided to drone larvae (Matsuka et al., 1973), an aspect which is rarely taken into account when effects of diet on caste development in honey bees are discussed.

An important approach to understand the role of diet composition and of its individual compounds was the attempt to develop a diet that could sustain queen development *in vitro*. Early experiments, starting in the 1950s with royal jelly diets, had low rates of success (reviewed in Kaftanoglu et al., 2011). This only changed once diets were developed that were mixes of commercial royal jelly, water, sugars and yeast (Rembold and Lackner, 1981; Vandenberg and Shimanuki, 1987). Subsequently, most attempts towards more efficient *in vitro* diets essentially varied and adjusted the sugar content according to the age of the larvae (Aupinel et al., 2005; Brodschneider et al., 2009; Kaftanoglu et al., 2011).

In vitro rearing experiments of honey bee larvae on such diets done between 1950 and 1980 essentially had the goal to identify individual compounds in royal jelly that may determine caste fate. But since at the time no single caste-determining factor could be fully purified (Rembold et al., 1974b), this gave weight to an alternative hypothesis, stating that not a single compound in royal jelly would be the critical factor for determining caste, but the balance of nutrients (Weaver, 1974). Coincidently, it was exactly at this time that synthetic JH became commercially available, which, when applied to honey bee larvae, showed a high potential of transforming the caste fate of worker-destined larvae into that of a queen-like phenotype (Rembold et al., 1974a; Wirtz and Beetsma, 1972). Further experiments then reared honey bee larvae on stored royal jelly—it is important to emphasise here that the experiments were done on stored and not fresh royal jelly, as we will see later—and treated the larvae with JH (Asencot and Lensky, 1984) or supplemented stored royal jelly with soluble sugars (Asencot and Lensky, 1988). These experiments led to the hypothesis that the higher sugar content in royal jelly, when compared to worker jelly, would act as a phagostimulant that, on the one hand would increase the growth rate of queen-destined larvae, and on the other, induce an enhanced JH synthesis in the CA.

In essence, these experiments essentially brought together a growth-stimulating effect of royal jelly, already inferred through analyses of the growth rates of queen and worker larvae (Rembold et al., 1980; Stabe, 1930), and the endocrine system, especially JH. Nonetheless, the way sugars may actually act as phagostimulants has never been fully elucidated, and to our knowledge, it was so far only experimentally addressed in a PhD thesis via electric recordings from labial taste receptors (Goewie, 1978). Furthermore, it is still unclear whether such phagostimulatory activity may act on the endocrine system via the central nervous system through neurosecretion from the pars intercerebralis (Dogra et al., 1977) or via the stomatogastric nervous system (Boleli et al., 1998). In a different approach, mathematical modelling was recently employed to analyse how the sugar concentrations in the respective larval diets would be reflected in the larval JH titres and, consequently, in the adult caste phenotypes (Leimar et al., 2012). The strength of this model is that it explains not only the divergence in the naturally produced caste phenotypes but also explores the whole range of intermediary ones (frequently referred to as inter-castes), which can be produced in *in vitro* rearing protocols.

The other main point of interest concerning royal jelly composition is its proteins. Secreted by the hypopharyngeal glands, these constitute a special class, the major royal jelly proteins (MRJPs). These proteins, first sequenced by Schmitzova et al. (1998), are the major components of the royal jelly and the hypopharyngeal gland proteomes (Santos et al., 2005; Scarselli et al.,

2005). Once the honey bee genome had been fully sequenced, the MRJPs were fully annotated and found to be members of the Yellow protein family (The Honeybee Genome Sequencing Consortium, 2006; Drapeau et al., 2006). Recent comparative genomics analyses then confirmed that MRJPs represent a gene family that was specifically expanded in the genus *Apis*, from an MRJP-like precursor gene, and is genomically flanked by the same two *yellow* genes across all different bee and ant species investigated so far (Buttstedt et al., 2014). While some of the MRJPs are indeed the dominant proteins in royal jelly, not all *mrjp* genes are actually expressed in the hypopharyngeal gland of worker bees and, even more strikingly, some are expressed in other tissues, including the brain, and during different life cycle stages of workers and even in the male sex (Buttstedt et al., 2013). Thus, even though they are an expanded gene family in the genus *Apis* and an important fraction of the royal jelly proteome, they probably did not evolve with the single and unique function of serving as larval food proteins.

Their peculiar evolutionary history in the genus *Apis*, however, makes them a novelty in terms of caste development, and the clearly most exciting recent finding on the queen-promoting properties of royal jelly was the discovery of Royalactin function (Kamakura, 2011). Royalactin is a temperature- and storage-sensitive 57-kDa monomeric form of MRJP1 (Kamakura, 2011). After storage of royal jelly for 30–40 days at elevated temperatures, the content of Royalactin was found to be decreased, this leading to low success in queen rearing. However, when purified or heterologously expressed, Royalactin was added to such stored royal jelly, it fully restored the queen-inducing effects of fresh royal jelly, which was not the case when other proteins were added (Kamakura, 2011). Furthermore, Royalactin did not only stimulate larval growth and the queen phenotype in honey bee larvae, but it also had a growth-promoting effect when added to the diet of *Drosophila* larvae (Kamakura, 2011), and it was this finding which ultimately facilitated the identification of its mode of action via the EGF receptor (Egfr). As mentioned, these recent findings now re-opened the discussion on which royal jelly factors may play a crucial and singular role in honey bee caste development and on how these may ultimately affect endocrine system functions.

4. THE ROLE OF JH IN HONEY BEE CASTE DEVELOPMENT

Like in all insects, the moulting and metamorphosis processes in honey bees are induced and synchronised by the interplay between ecdysteroids and JH synthesised in and released from the prothoracic gland and the CA, respectively. The caste phenotypic plasticity in the highly eusocial insects, however, adds an additional layer of complexity to this already intricate hormonal symphony (Hartfelder and Emlen, 2012).

As mentioned above, a prominent role of JH in honey bee caste development was firmly established through topical applications of synthetic JH to fourth and early fifth instar worker larvae. Such treatment led to the induction of queen characters in the emerging adults, independent of whether these larvae were reared *in vitro* on a royal-jelly-based diet or whether they were treated in their natural brood cells then further tended by nurse bees in their hives of origin (Asencot and Lensky, 1984; Dietz et al., 1979; Rembold et al., 1974a; Wirtz and Beetsma, 1972). Subsequently, measurements of endogenous JH-III levels in whole-body extracts (Rembold, 1987) or those actually circulating in haemolymph (Rachinsky et al., 1990) revealed that JH levels reach a peak titre in the third larval instar of queens, before they gradually decline during the fourth instar and go through a minimum during the feeding phase of the fifth instar (Fig. 1A). In worker larvae, the JH levels during the corresponding developmental period were found to be much lower, but they also undergo a gradual decline in the fifth instar. As this is the period with the steepest growth rates in larvae of both castes, it is not surprising that the rates of JH synthesis in the CA remain high until the end of the larval feeding phase of queen larvae, and only decline once the brood cells are sealed and the larvae begin to spin their cocoons inside their brood cells (Rachinsky and Hartfelder, 1990). In fact, the differences between queen and worker larvae are even more pronounced when it comes to JH synthesis rates in the CA than with respect to the actual JH titres (Rachinsky and Hartfelder, 1990).

As the haemolymph JH titre is a snapshot picture reflecting JH synthesis and release from the CA on the one hand and JH metabolism by JH degrading enzymes in target tissues on the other, investigating these two sides underlying the dynamics of the physiologically relevant hormone levels and their difference among the castes was important. Upon challenging the CA of honey bee larvae cultured *in vitro* with an excess of the JH precursor farnesoic acid, it became evident that queen larvae were competent to convert this precursor at a high rate into JH, whereas glands from worker larvae were not and, instead, accumulated methyl farnesoic acid (Rachinsky and Hartfelder, 1991). As this was indicative of a restricted activity of a methyl farnesoate epoxidase in honey bee larvae, the question was as to what may be causing this difference in the activity of an enzyme that catalyses the terminal

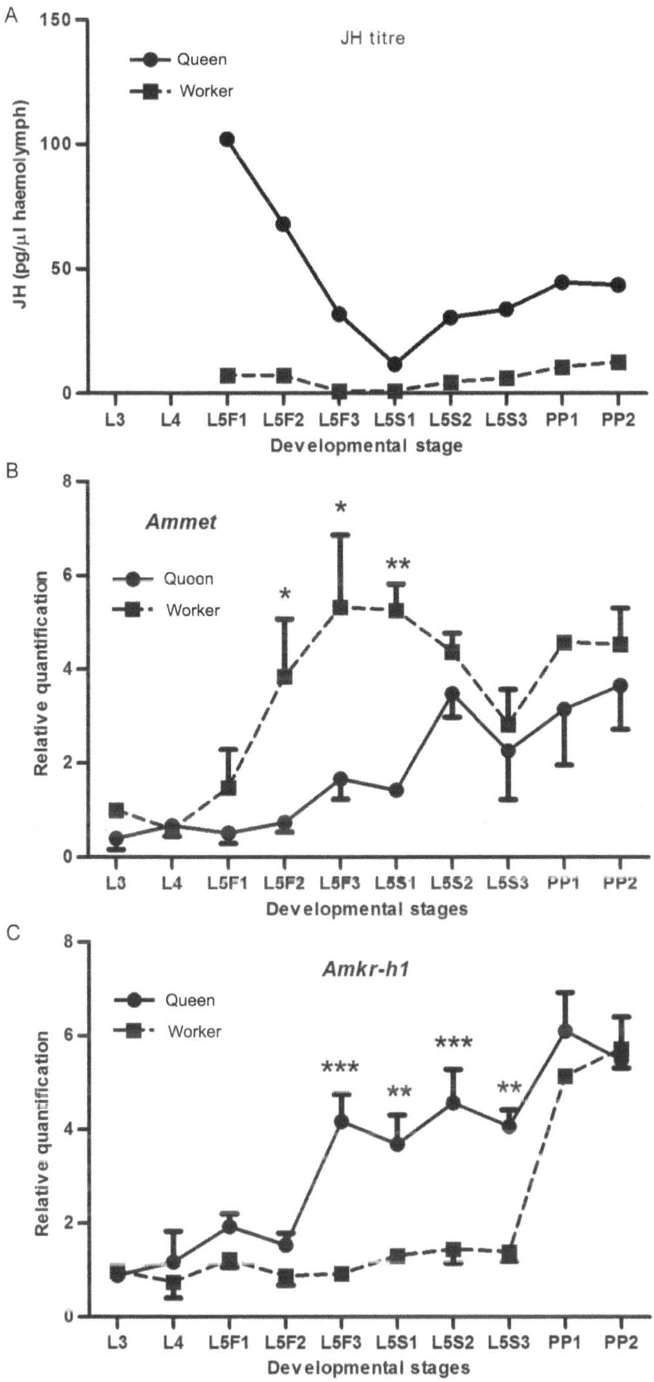

Figure 1 Juvenile hormone (JH) signalling in honey bee queen and worker larvae. (A) JH titre in the haemolymph of fifth instar larvae (L5F1–PP2); this graph was generated based on the radioimmunoassay data reported by Rachinsky and Hartfelder (1990). (*Continued*)

step of JH biosynthesis. *In vitro* experiments testing the response of CA to allatotropic peptides soon revealed that none were capable of breaking the block on JH synthesis in worker larval glands (Rachinsky and Feldlaufer, 2000). Among the biogenic amines that are also known to stimulate JH synthesis in insects, octopamine and serotonin were capable of stimulating JH synthesis *in vitro* in honey bee CA, but they also did not fully remove the block on the terminal JH synthesis step (Rachinsky, 1994).

With respect to JH metabolism in haemolymph, a JH esterase and a JH epoxy hydrolase encoding gene were recently annotated in the honey bee genome (Mackert et al., 2008, 2010). Functional assays indicated that the JH esterase is more important in JH metabolism than the JH epoxy hydrolase. Nonetheless, expression of the JH esterase encoding gene is only increased in fifth instar larvae (Mackert et al., 2008), but not in the earlier instars when the JH titre differences are maximal between queens and workers. Thus, current evidence indicates that the low JH haemolymph titre seen in third and fourth instar worker larvae is likely due to a repression in the JH biosynthesis pathway via a block on the terminal step catalysed by a methyl farnesoate epoxidase.

One of the major effects of the high JH titre in queen larvae is that it prevents the onset of autophagic cell death in the larval ovaries (Schmidt Capella and Hartfelder, 1998). Autophagic cell death, which leads to the degradation and removal of over 95% of the ovariole primordia in the worker ovaries, is the developmental mechanism through which worker fertility is so markedly reduced in the honey bee. Molecularly, the mode of action of JH on the larval honey bee ovary seems to take two distinct routes.

Figure 1—Cont'd (B) Relative expression levels of the putative honey bee JH receptor gene *methoprene-tolerant* (*Ammet*). (C) Relative expression levels of the putative honey bee JH response gene *krüppel-homolog 1* (*Amkr-h1*). Gene expression levels were measured in whole-body RNA extracts of queen and worker larvae by means of SYBR Green (Applied Biosystems) methodology in a StepOne Plus system (Applied Biosystems). Relative expression levels were calculated by the $2^{-\Delta\Delta Ct}$ method (Livak and Schmittgen, 2001) using a honey bee ribosomal protein gene (*rpl32*, formerly named *rp49*) as an endogenous control and a third instar worker sample as a calibrator. The developmental stages were characterised according to Michelette and Soares (1993). These were third instar (L3) and fourth instar (L4) larvae and a series of substages of the fifth instar, starting with the feeding stages (L5F1–L5F3), the cocoon-spinning stages (L5S1–L5S3) and pre-pupal stages (PP1 and PP2). All developmental stages were analysed as biological triplicates, each run as technical triplicate. The gene expression data shown as means $\pm$ SEM were analysed by two-way ANOVA using Bonferroni correction. Statistical differences are denoted as * ($P \leq 0.05$), ** ($P \leq 0.01$) and *** ($P \leq 0.0001$).

One goes through the maintenance of cytoskeletal elements that constitute the polyfusomes of germ cell clusters within the ovarioles (Schmidt Capella and Hartfelder, 2002), while the other route implies differential gene expression (Humann and Hartfelder, 2011), including a possible role for two long non-coding RNAs in the divergence of the queen/worker ovary phenotypes (Humann et al., 2013).

Despite its fundamental role in insect development and reproduction, the mode of action of insect JH has long been an enigma, due to the fact that its receptor had not been identified. Due to genomic information accumulated over the last decade, however, the JH signalling pathway has now been worked out (for review, see Jindra et al. (2013). The genuine JH receptor is the Methoprene-tolerant (Met) protein, an intra-cellular receptor molecule belonging to the bHLH-PAS family. In the honey bee genome, Met is represented as a single copy gene, like in other insects, and not as a duplicated gene (*met* and *gce*), as is the case in *Drosophila* (Baumann et al., 2010). The JH/JH-receptor complex binds to the promoter of the transcription factor Krüppel homolog 1 (Kr-h1), which is a primary response gene and, thus, a direct read-out of the JH response in hemi- and holometabolous insects (Belles and Santos, 2014; Minakuchi et al., 2009).

When analysing the expression profiles of the honey bee *Ammet* gene (GB44365, XM_395005) starting from the third larval instar, we found that *Ammet* transcript levels were higher in fifth instar worker larvae, especially so in the fifth instar feeding and spinning stage, whereas in queen larvae, *Ammet* transcript levels were lower and reached a peak only later, way into the spinning phase (Fig. 1B). These findings were surprising, because based on the JH haemolymph titres we had expected that the JH receptor transcript levels would be higher in queens. Possibly, JH may be functioning here as a negative regulator on the expression of its own receptor, thus desensitising the systemic response to the high JH levels seen in young queen larvae.

To a certain extent, this inverted relationship between the JH titre and *Ammet* transcript levels in honey bee larvae is reminiscent of their relationship in the cockroach *Blattella germanica*, where an elevated *met* expression coincides with basal JH levels, and this was shown to be crucial for wing development (Lozano and Belles, 2014). While this interpretation involving a negative transcriptional effect of JH on the expression of its own receptor obviously needs experimental validation, for instance through JH application experiments, it is nonetheless in accordance with other recent findings, for example, on the expression of the ecdysone receptor isoform A (EcR-A), which is the predominant EcR isoform in honey bee post-embryonic development (Mello et al., 2014). EcR-A expression shows a peak in the spinning phase of both queen and worker larvae. But this peak appears earlier and is higher in spinning-stage worker larvae than in queens, contrasting with the developmental profile of the ecdysteroid haemolymph titre, which is higher in queens than in workers (Rachinsky et al., 1990). Both hormone systems, thus, apparently show opposite trends considering caste-specific responses in the hormone production sites (CA and prothoracic gland) versus the hormone target sites where the respective receptors are expressed. This tight balance between hormone levels and their receptors may actually be a key to understand how divergent phenotypes can evolve without causing major disruptions in the general metamorphosis programme of insects.

With respect to *Amkr-h1* (GB45427, XM_001123084), which, as mentioned above, is a primary JH response gene and, thus, the genomic read-out response to JH in insects, including honeybees (Fussnecker and Grozinger, 2008; Kayukawa et al., 2012), we could see that its transcript levels start to differ between queen and worker larvae from the fourth instar onwards. Furthermore, *Amkr-h1* is strongly expressed in queen larvae all throughout the fifth larval instar (Fig. 1C). These significantly higher *Amkr-h1* transcript levels in queen larvae seen throughout the last larval instar likely correspond to tissue differentiation responses to the elevated JH titre in spinning-stage queen larvae (Rachinsky et al., 1990) on the one hand and to ecdysteroid-activated gene expression on the other, particularly in imaginal discs and the larval ovary (Hepperle and Hartfelder, 2001). A positive interaction effect of JH on the ecdysteroid titre in honey bee larvae has already been shown (Rachinsky and Engels, 1995), as well as effects of these hormones on the expression of the *ultraspiracle* (*usp*) gene, which encodes a dimerisation partner for the ecdysone receptor (EcR) in insects and which is up-regulated by JH in honey bee development (Barchuk et al., 2004). USP, together with the competence factor protein Broad Complex, may thus be a link in the JH and ecdysteroid responses in honey bee metamorphosis, but this requires further investigation.

In conclusion, role of the endocrine system in honey bee caste development can be envisaged as setting and synchronising the timing of the moulting and metamorphosis events at a general scale, and tissue-specific differentiation at a more refined one. Such tissue-specific differentiation events would, for instance, include differential gene expression underlying the induction of autophagy in the larval ovary (Humann and Hartfelder, 2011), or the differentiation of the corbicula bristle pattern on the worker hind leg, which is likely to involve patterning by Hox genes (Bomtorin et al., 2012). On a more general level, gene regulatory networks for caste development have been revealed through microarray analysis of queen and worker larval transcriptomes, showing major differences in connectivity and a separation among the two castes into growth-related physio-metabolic genes being over-expressed in queen larvae and differentiation-related genes in worker larvae (Barchuk et al., 2007). So, while differentiation events can most likely be linked to JH and ecdysteroid functions, nutrient-sensing/growth factor signalling pathways are favourite candidates for driving the differential expression of physio-metabolic genes in caste development.

5. IIS, TOR AND EGFR PATHWAY ACTIVITIES IN HONEY BEE CASTE DEVELOPMENT

With TOR and IIS being two of the most ancient growth-regulating pathways in eukaryotes and bilaterians, respectively (Oldham, 2011), these were the first pathways receiving close attention once the honey bee genome had been sequenced, and most genes in these pathways were annotated. Within the IIS pathway, the components that received most attention were the two predicted insulin-like peptides (AmILP1 and AmILP2) and the two predicted insulin receptors (AmInR-1 and AmInR-2). The temporal expression patterns of these were investigated by quantitative RT-PCR in two independent studies with slightly different focus. One set of studies (Wheeler et al., 2006, 2014) investigated their expression in the early larval feeding stages in queen and worker larvae and in larvae that had been shifted from a queen to a worker feeding regime and *vice versa*. So, it had its main focus on the influence of early larval diets on IIS-mediated nutrient sensing. The second study (Azevedo and Hartfelder, 2008) was designed to investigate IIS gene expression in the major larval growth phase and during preparation for metamorphosis in the fifth larval instar.

The general conclusions that could be drawn with respect to the expression of the ILP genes is that *Amilp1* is consistently over-expressed in queen larvae, whereas *Amilp2* is so in worker larvae, thus indicating possibly different functions for the two ILPs in nutrient sensing and growth regulation among the castes. The experiments shifting queen larvae to a worker feeding regime and worker larvae to a queen environment provided results that are highly consistent with the transcript level differences seen for the two ILPs in naturally raised queens and workers, with *Amilp1* expression being up-regulated when a worker larvae came to receive royal treatment, and *Amilp2* being up-regulated upon transferring a queen larvae into a worker brood environment (Wheeler et al., 2006). With respect to insulin receptor gene expression, the results were even more surprising. First of all, the transcript levels for the two insulin receptors are highly similar within each caste, indicating caste-specific co-regulation, but they are very different between queens and workers. Even more so, the most surprising results was that the transcript levels of the two insulin receptors drop sharply in queen larvae (Azevedo and Hartfelder, 2008) just as these enter their major growth phase (Stabe, 1930; Wang, 1965).

As transcript levels are only proxies for pathway functionality, especially in a pathway as complex as the IIS pathway, it is worthy of note that most of the intra-cellular pathway components are functionally regulated by phosphorylation and, accordingly, transcript level analyses for these (AKT, PDK1, PTEN and others) showed little variation in relation to prospective caste fate (Wheeler et al., 2006, 2014). Not surprisingly, it was the entry point components of IIS, namely the release of ILPs and their binding to the respective InRs, and less so the downstream elements of this pathway that showed a closer link with caste in the gene expression analyses. Accordingly, assays using single or double knock-down of *Amilp1* and *Amilp2* gene function by RNAi provide evidence for functional differences among the two ILPs in honey bee caste development (Wang et al., 2013). While none of the single or double ILP knock-down experiments had any effects on haemolymph glucose or trehalose levels, Amilp1 RNAi caused a reduction in JH levels, and Amilp2 RNAi affected adult ovary size (ovariole number) and body mass. Interestingly, in *Drosophila*, JH and IIS are linked through FoxO, which had its activity increased in allatectomised larvae (Mirth et al., 2014), and FoxO activity is a known regulator of insulin receptor expression acting in a negative feedback loop (Puig and Tijan, 2005). The honey bee ILP RNAi experiments, thus, led us to infer that in honey bee larvae AmILP1 is expressed in the brain, as is the case with ILPs of many insect species across orders (Wu and Brown, 2006), where it positively regulates JH synthesis in the CA, thus, possibly affecting developmental timing. AmILP2 on the other hand is thought to act in a nutrient-sensing pathway in the fat body, affecting general growth of the larvae. This is plausible since *Amilp2* has been shown to be expressed in honey bee fat body, in both oenocytes and trophocytes (Nilsen et al., 2011). Nonetheless, none of the honey bee ILP RNAi experiments had a major effect on caste fate (Wang et al., 2013), thus indicating that the IIS pathway may actually not be a major switch point in caste development, but may play a role as a modulator, acting in concert with other signalling pathways, thus canalising developmental decisions into clearly divergent paths.

IIS interacts with other growth-regulating signalling pathways in a complex structural network, but its strongest link is clearly with TOR. In honey bees, TOR activity has been shown to strongly affect caste fate, as the RNAi-mediated knock-down of *Amtor* gene function in larvae reared in a queen-promoting environment caused all these to express a worker phenotype, with a prolonged developmental period, lower body mass and strongly reduced ovariole number (Patel et al., 2007). In a subsequent experiment using an RNAi approach to knock-down insulin receptor substrate (IRS) and TOR function, the major effect of TOR knock-down was further put in evidence, but also its close link with IIS and JH (Mutti et al., 2011). TOR knock-down led to a decrease in AmIRS mRNA levels (but not *vice versa*) and caused an increase in *Amilp2* and a decrease in *Amilp1* expression. Furthermore, TOR RNAi had a strong effect on specific DNA methylation, and both TOR RNAi and IRS RNAi caused a major reduction in JH content (Mutti et al., 2011).

This suite of effects, common on the one hand and distinct on the other, for IIS and TOR in honey bee caste development may be explained by the convergence of the two pathways at the level of AKT in the setting of adult size in insects (Edgar, 2006; Mirth and Riddiford, 2007). But distinct from the IIS pathway, which is a module with homeostatic properties composed of a series of positive and negative regulators upstream of AKT, TOR functions as an intra-cellular nutrient sensor and is directly linked to AKT via tuberosis sclerosis complex (TSC). Thus, we expected to see higher *Amtor* transcript levels in queen larvae and lower ones in workers, as originally postulated from the measurement of relative *Amtor* expression at two developmental time points (Patel et al., 2007). Surprisingly though, when measuring *Amtor* (GB44905, XM_625127) transcript levels for the developmental period from the third larval instar until the pre-pupal stage, thus covering both the nutritional switch for worker larvae (in L3 L4) and the major growth phase in the fifth larval instar, we found that *Amtor* is significantly more expressed in feeding stage fifth instar worker larvae (Fig. 2A), and it is exactly in this period when several morphological caste characters are determined (Dedej et al., 1998).

The question as to whether and how other signalling pathways may intersect and integrate with IIS/TOR in honey bee caste development has gained force through the investigation of the mode of action by which Royalactin primes queen development (Kamakura, 2011). In this study, the Egfr pathway was identified as the one that most powerfully translates the Royalactin stimulus into a growth and differentiation-relevant signal.

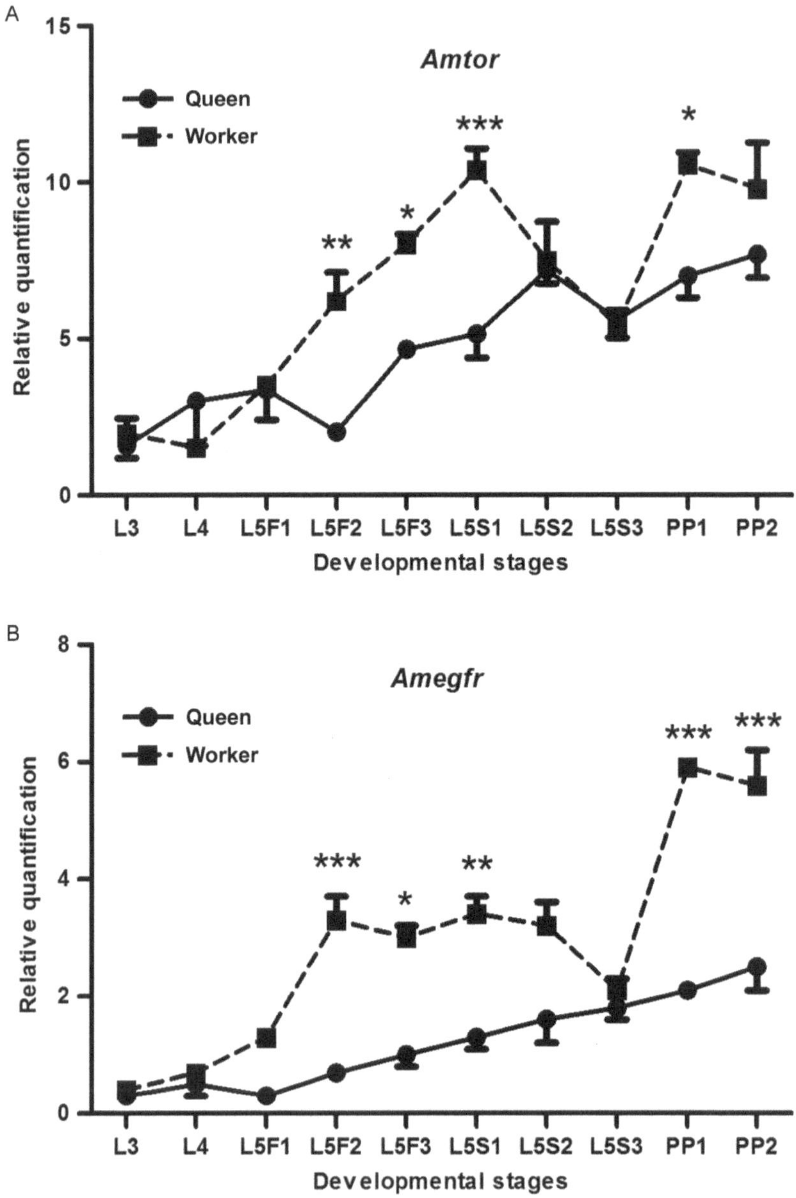

Figure 2 Relative expression levels of (A) a honey bee TOR (*Amtor*) and (B) a honey bee Egfr (*Amegfr*) homolog. The developmental stages and samples and the analysis method were the same as those described in the legend of Fig. 1. The gene expression data shown as means $\pm$ SEM were analysed by two-way ANOVA using Bonferroni correction. Statistical differences are denoted as * ($P \leq 0.05$), ** ($P \leq 0.01$) and *** $P \leq 0.0001$).

This was shown through feeding Royalactin to a *Drosophila* dEgfr-RNAi line and by over-expressing Royalactin in fat body using a UAS-GAL system, and in the honey bee by feeding larvae with a combination of Royalactin and Egfr-dsRNA (Kamakura, 2011). Furthermore, Amegfr RNAi strongly reduced the larval JH titre, leading to infer that Egfr signalling should be upstream of JH. Also, the effect of Amegfr RNAi on the expression of caste characters was much stronger than those of Aminr RNAi, leading to the hypothesis that provisioning honey bee larvae with Royalactin-rich royal jelly maybe a master caste-determining signal acting via Egfr signalling, and downstream of Egfr this would involve a MAPK and S6K signal in the larval fat body (Kamakura, 2011). Furthermore, the interaction with IIS and the TOR pathway is thought to occur via a MAPK/ERK cascade, as also evidenced for eye development and glucose metabolism in *D. melanogaster* (McNeill et al., 2008; Zhang et al., 2011). But there are two conceptual problems for the Royalactin-Egfr hypothesis, one is the fact that both queen and worker larvae are constantly fed fresh hypopharyngeal gland secretions, which would be expected to contain Royalactin, and the other is the ligand function of Royalactin to Egfr. The canonical Egfr ligand in *Drosophila* is the TGFα homolog Spitz, and there is no significant protein-level similarity between Royalactin and Spitz or other *Drosophila* Egfr ligands, such as Gurken and Argos. Nonetheless, not taking into account these conceptual problems, one would expect that the Royalactin-activated Egfr signal in queen larvae be reflected in an up-regulation of *Amegfr* gene expression. Strikingly though, when running RT-qPCR assays for honey bee *Amegfr* (GB54477, XM_003249562) transcript levels in queen and worker larvae, we noted that this gene is significantly over-expressed in fifth instar worker larva and not in queens (Fig. 2B). Furthermore, this over-expression occurs not only in the "wrong" caste, but also at the "wrong" time, as it does not coincide with the switch in the larval feeding regime that worker larvae experience in the third to fourth instar, but at the end of the larval feeding stage and throughout the metamorphosis process.

So as it stands, there are currently more open questions than answers concerning the role of nutrient-sensing and growth-regulating pathways in honey bee development, and it is only for the endocrine system factors, JH and ecdysteroids, that the observations match up and are consistent. An explanation for the apparent contradictions with expected system functions described in model organisms, such as *Drosophila*, may simply be due to the fact that the interactions in these ancient and ancestral pathways have

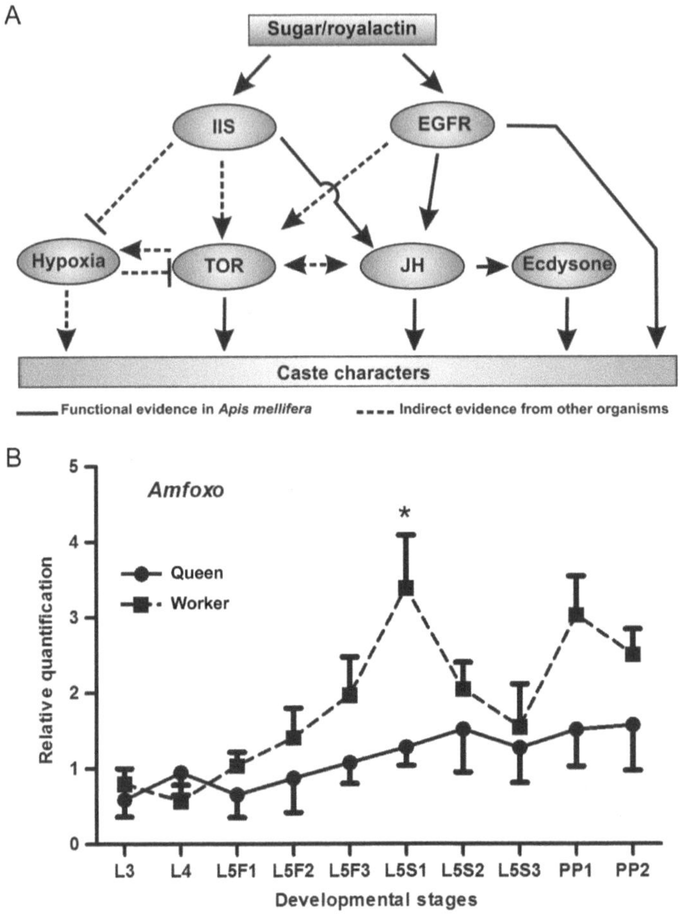

Figure 3 Signalling pathways in honey bee development and their cross-talk. (A) Larval food with its inducing factors is represented as sugar and Royalactin. These interact with the insulin–insulin-like signalling (IIS) and the Egfr pathways. And these affect downstream pathways, such as TOR and the morphogenetic hormones JH and ecdysteroids. Hypoxia signalling is placed at the same level, but its exact position is still unclear. All together, these pathways influence caste fate and affect the expression of the respective morphological and anatomical characters in queen and worker larvae. Connections supported by direct experimental evidence in honey bees are shown as full lines, whereas dotted lines represent those based on evidence from other organisms, mainly *Drosophila melanogaster*. For details and references, see text. (B) Relative expression of a honey bee FoxO gene (*Amfoxo*) in queen and worker larvae. FoxO is a convergence point of the IIS and TOR pathways and a key factor regulating cell proliferation and body and organ growth. The developmental stages and samples and the analysis method were the same as those described in the legend of Fig. 1. The gene expression data shown as means ± SEM were analysed by two-way ANOVA using Bonferroni correction. Statistical differences are denoted as * with $P \leq 0.05$.

been remodelled to some extent in the evolution of the morphology-based caste systems and highly efficient division of labour seen in social insects. In Fig. 3A, we tried to give a brief and highly simplified overview on the network structure inferred from actual experimental evidence in honey bees and from putatively conserved elements composing the respective pathways.

The graphical representation for the three nutrient-sensing and growth-regulating pathways indicates that these intersect at the level of AKT and MAPK/ERK. Nonetheless, these are cross-roads for a plethora of intracellular signalling pathways. Their read-outs in the form of growth-promoting kinase activity, especially S6K, and metabolic repressors such as 4EBP are only touched upon in honey bees, given as inferences from RT-qPCR analyses (Wheeler et al., 2014). Most integrative insights that are potentially relevant for honey bee caste development are still drawn from *Drosophila* and lack experimental confirmation in bees, for example, a negative regulator effect of FoxO activity (nuclear localisation of dephosphorylated FoxO) on insulin receptor expression (Puig and Tijan, 2005), that could explain the down-regulation of *Aminr-1* and *Aminr-2* expression in queen larvae.

So as to ascertain at least initial insights into FoxO function in honey bee development, we analysed the expression levels of a honey bee FoxO homolog (GB48301, XM_01122804) in the same developmental stages for which we already had information on insulin receptor, TOR and Egfr expression. The developmental expression profile of *Amfoxo* was strikingly similar to that seen for the *Amtor* gene (Fig. 3B). This is a result that is clearly difficult to explain, as FoxO, which is the key downstream integrator of IIS and TOR activity, is a negative regulator of growth. On the one hand, such a negative regulation of growth would be expected for worker larvae, as these are fed on a less nutritive diet than queen larvae, and on the other, the elevated *Amfoxo* transcript levels seen in worker larvae are difficult to reconcile with TOR and IIS activity, as these negative regulators of FoxO activity are over-expressed in worker larvae. Nonetheless, and in line with our results, Wheeler et al. (2014) also found that developing queen larvae that were switched to the worker feeding regime showed an increased expression of IIS/TOR pathway genes, including *Amfoxo*, but worker larvae that were switched to the larval queen diet did not. At this point, a possible explanation is that certain elements in the circuitry of these major signalling pathways have apparently been rewired so as to make possible a certain degree of plasticity in the responses to nutrient quality and quantity. Another point of interest is that in the cockroach *Blattella germanica*, BgFoxO inhibits JH biosynthesis in the CA in a situation of nutrient deprivation

(Süren-Castillo et al., 2012). This actually fits with the findings in honey bee larvae, where *Amfoxo* expression shows a peak exactly when JH titres drop to basal at the end of the fifth instar feeding stage, and with the interpretation that honey bee worker larvae are to a certain extent reared under a sub-optimal dietary regime.

6. ENDOGENOUS HYPOXIA IN WORKER LARVAE—CONVERGENCE OF NUTRIENT SIGNALLING AND OXIDATIVE METABOLISM

The rather unexpected down-regulation of insulin receptor expression in queen larvae made us search for a possibly overarching explanation, i.e., a signalling pathway other than TOR, Egfr or JH that could connect IIS activity with growth and developmental regulation. This led us to look at the hypoxia response pathway in honey bee larvae. The reason was that in *Drosophila* larvae the hypoxia response is closely linked to IIS (Dekanty et al., 2005; Romero et al., 2007). The core module of the hypoxia response comprises three proteins, two are members of the bHLH-PAS family: Sima, which is the homolog of vertebrate HIFα, and Tango, the vertebrate homolog of HIFβ. The third is Fatiga, a homolog of the vertebrate HIF prolyl hydroxylase PHD. PHD is a *bona fide* oxygen sensor which hydroxylates specific prolyl residues in the oxygen-sensitive domain of HIFα (Bruick and McKnight, 2001). This causes the ubiquitination and subsequent proteasomal degradation of HIFα (Ivan et al., 2001). PHD is active under normoxic conditions but inhibited under hypoxia, thus causing HIFα to be stabilised, enter the nucleus, and form a dimeric transcription factor with HIFβ, which then drives the expression of hypoxia responsive genes (Wang et al., 1995). In *Drosophila* larvae, this response can be seen as a branching response in tracheolar trees (Centanin et al., 2010).

As oxygen consumption is dependent both on exogenous oxygen levels as well as on metabolic state, the activity of PHD is also connected to nutrient-sensing systems. In this context, the nucleolar localisation of Sima is promoted by AKT and TOR function (Dekanty et al., 2005), whereas the TSC proteins down-regulate the hypoxia response (Fig. 4). This link between the nutrient- and oxygen-sensing pathways, which is well established in the fruit fly, made us investigate the hypoxia response in honey bee caste development. To do so, we annotated the three hypoxia response core genes in the honey bee genome and quantified their transcript levels in queen and worker larvae (Azevedo et al., 2011). For all three genes, we observed a gradual increase in expression from the third to the early fifth

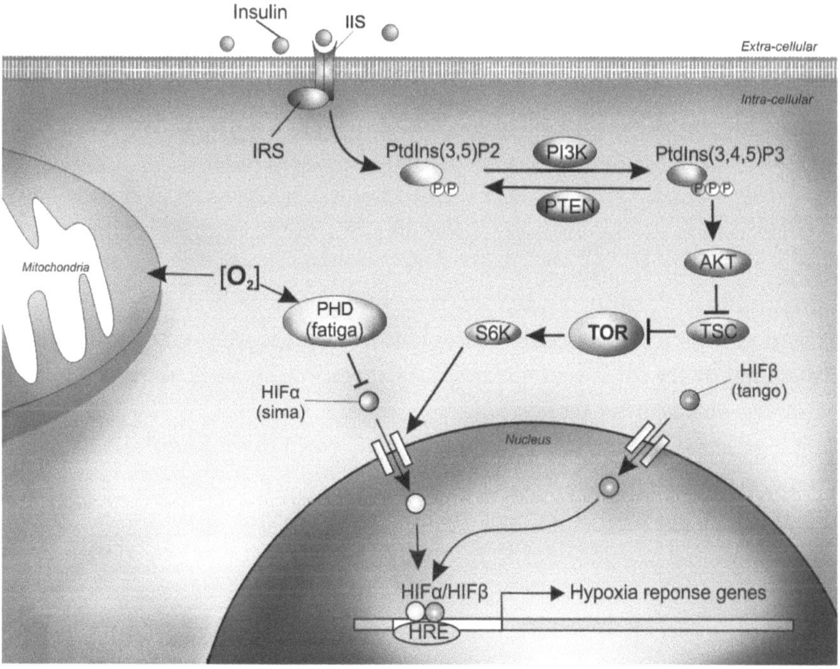

Figure 4 The hypoxia response and its connection with insulin and TOR signalling. The graph is based on a model described for *Drosophila melanogaster* (Dekanty et al., 2005). The hypoxia response in honey bee larvae, put in evidence as an over-expression of the three hypoxia core genes *Amfatiga, Amsima* and *Amtango* (Azevedo et al., 2011), is interpreted as the result of an endogenous hypoxia. The latter is thought to be related to differences in oxidative metabolism in queen and worker larvae. Activated TOR is thought to stabilise the honey bee HIFα homolog and to facilitate its translocation into the nucleus, where it can dimerise with HIFβ and form an active transcription factor driving the expression of hypoxia response genes.

larval instar, i.e., throughout the larval feeding period, and their temporal expression profiles indicated that the three genes are co-regulated. Even more strikingly, all three genes were significantly over-expressed in worker larvae throughout the fifth instar (Azevedo et al., 2011).

The reason for this hypoxia response in workers is not obvious, as there was no reason to assume a difference for oxygen levels in the rearing environment of queen and worker larvae. Both are reared next to each other in unsealed brood cells. Consequently, the assumption is that worker larvae could be in a state of endogenous hypoxia, and this assumption receives strong support from classical physiology and biochemistry analyses done in the 1930s–1980s, i.e., prior to the availability of synthetic JH, which essentially lead to a paradigm change in insect caste development and research focus (Nijhout and Wheeler, 1982). Classical studies on oxygen consumption in honey bee larvae (Melampy and Willis, 1939) and on mitochondrial enzyme activity (Eder et al., 1983; Osanai and Rembold, 1969) had already put in evidence that queen larvae have higher oxygen consumption rates, higher respiratory rates and a higher cytochrome *c* content. These results also indicated that worker larvae have a mitochondrial deficit (Osanai and Rembold, 1968). To investigate whether the hypothesised mitochondrial deficit may be due to lower mitochondria number or dysfunctional mitochondria, we prepared larval fat body samples for ultrastructure analysis. As seen in Fig. 5, mitochondrial numbers in trophocytes, which are the primary cell type for the insect intermediary metabolism, are strikingly different for queen and worker larvae.

There are three obvious questions arising from these findings, which are actually not all that new. The first question is when do caste differences in mitochondria numbers first become established during larval development? Answering this question will require serial analyses of fat body ultrastructure and/or mitochondria counts via specific labelling of mitochondria in laser confocal microscopy. The second question concerns mechanisms underlying the divergence in mitochondria numbers. These could either be due to differences in mitochondrial biogenesis on the one hand or be due to mitochondrial fusion or autophagy on the other. The overarching, third question, however is, how may the mechanisms driving mitochondrial biogenesis or degradation be coupled to nutrient-sensing signalling pathways? A recent study has shown that a genetic activation of Sima function in *Drosophila* larval fat body can promote autophagy through impairment of the proteasomal system (Löw et al., 2013). Another connection could be through TOR signalling which, in mammalian systems, has been shown to regulate many aspects of mitochondrial function, including mitochondrial biogenesis and mitochondrial degradation (Wei et al., 2014). In a much wider sense, this eventually leads to the question as to whether and how nutrition may act as an epigenetic modifier and become integrated with signalling pathways and transcriptional regulatory networks.

7. EPIGENETICS AND CASTE DEVELOPMENT

The question of epigenetics became a key point for understanding honey bee caste development with the annotation of the honey bee genome (The Honeybee Genome Sequencing Consortium, 2006) where, for the first time, a full DNA methylation toolkit was described in an insect, with components and characteristics similar to that found in mammals. The

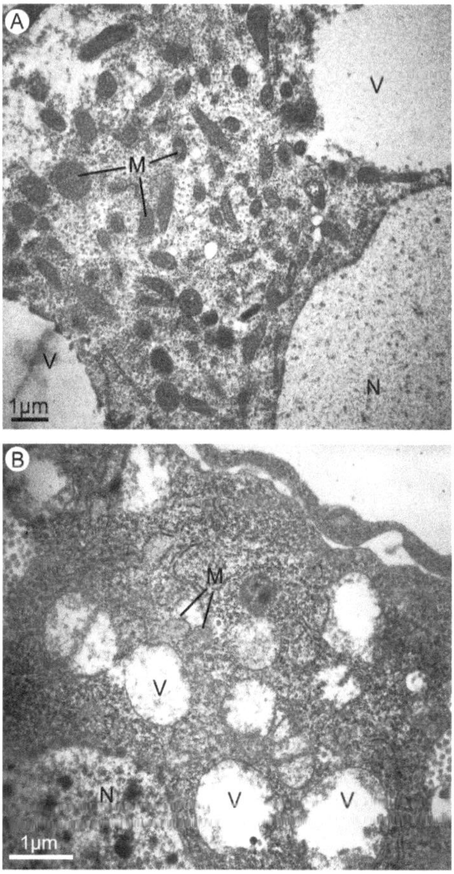

Figure 5 Ultra-structure of fat body trophocytes of fourth instar queen (A) and worker larvae (B). Queen trophocytes are rich in mitochondria (M), especially so in the perinuclear (N) space. Lipid vacuoles (V) are found in trophocytes of both castes. Space bars represent 1 μm.

breakthrough, however, came through an RNA interference experiment designed to functionally silence a honey bee DNA methyltransferase 3 (*Amdnmt3*) gene (Kucharski et al., 2008). Larva fed with dsdnmt3 RNA showed a strong bias to develop into queen-like females, whereas controls developed as workers. The caste character that was most prominently affected was ovary size, leading to infer that down-regulation of *de novo* DNA methylation apparently led to a widespread epigenetic modification across caste-related genes. Subsequent analyses of differential DNA methylation in adult queens and workers (Lyko et al., 2010) and, especially so, in queen and worker larvae (Foret et al., 2011) then revealed enrichment in differentially methylated genes for three central metabolic pathways:

(i) enzymes of the Krebs cycle and associated entry and output reactions, (ii) genes involved in IIS/TOR signalling, especially so the regulation of alternative splicing of a honey bee *alk* gene homolog and (iii) genes involved in ubiquitination and proteasomal degradation. Interestingly, several of the genes showing differential methylation between queen and worker larvae were also differentially methylated at another important stage in honey bee life history, the transition from within-nest to foraging tasks seen in adult workers (Foret et al., 2011).

The Dnmt3 silencing and methylation status analyses not only were the first ones to actually demonstrate how important epigenetic modification is to caste development, but the diet context-dependent expression of caste characters in social insects also nicely fits the epigenetic landscape concept of Waddington (for review, see West-Eberhard (2003)). This classical concept of evolutionary developmental biology considers developmental pathways as sequential bifurcation decisions, each shaped by the preceding one, and the probability landscape of the subsequent genetic/epigenetic constellation. Furthermore, epigenetic marks *sensu stricto* revealed in these studies may also be linked to epigenetic marks *sensu lato*, i.e., chromatin modification through histone modification and/or long non-coding RNAs. Such chromatin modifiers have barely been touched upon in honey bees. Histone post-translational modifications have only recently been investigated in queen ovaries and in fifth instar larvae (Dickman et al., 2013), and these may provide a link to an earlier observation that a fatty acid component of royal jelly, (*E*)-10-hydroxy-2-decenoic acid, actually has properties as an inhibitor of histone deacetylation (Spannhoff et al., 2011). Long non-coding RNAs, for which the X-chromosome inactivating gene *Xist* is the certainly best known member, have recently called attention as the black box of the genome, consisting of thousands of expressed genes involved in the fine tuning of developmental processes (Mercer et al., 2009). In the honey bee, only three long non-coding RNA genes have been identified so far, two of these involved in caste-specific ovary development (Humann et al., 2013).

8. ARE THERE COMMONALITIES BETWEEN CONTEXT-DEPENDENT SIGNALLING PATHWAYS IN LARVAL CASTE DEVELOPMENT AND ADULT CASTE FUNCTION?

As already mentioned above, the behavioural maturation of a worker from performing tasks within the hive to foraging tasks outside the hive is a major life history transition. In fact, this behavioural transition implicitly determines the worker's lifespan. This means that the effective lifespan of a worker is not only determined by external factors associated with the high risk tasks of foraging outside the safe and well-defended hive environment but also governed by intrinsic factors. These are a switch from a protein-rich diet distributed among in-hive bees (Crailsheim, 1990) to a practically pure carbohydrate diet and an concomitant progressive decline of the innate immune response (immunosenescence) in foragers (Amdam et al., 2005).

Physiologically, the two prominent regulators associated with task and lifespan are the haemolymph JH and vitellogenin titres, which are intertwined in a mutual repressor circuitry (Amdam and Omholt, 2002; Page and Amdam, 2007). Young, in-hive bees have high vitellogenin and low JH titres, whereas foragers have high JH and low vitellogenin titres (Hartfelder and Engels, 1998; Robinson et al., 1992). But even though the switch from an in-hive (nurse) bee to a forager is frequently referred to as age polyethism, it is not strictly dependent on a bee's age but on the colony environment, including colony age demography (Huang and Robinson, 1992), amount of brood and nutrient stores. Also, while in temperate climates most adult summer bees generally do not live more than 35–40 days, there is a special generation of bees reared in autumn, the so-called winter or *diutinus* bees. These survive within the hive for several months and only gradually die off once they have started to forage for a few days in early spring (Maurizio, 1950; Seehuus et al., 2006), and they provide the colony with new pollen and honey stores for the new growth period. Furthermore, the signs of immunosenescence could be reverted in old workers that were forced to re-assume brood caring functions, and consistent with this reverted function in division of labour, they also showed decreased JH and increased vitellogenin levels (Amdam et al., 2005).

An even higher degree of plasticity in lifespan separates queens from workers in highly eusocial Hymenoptera, and this, of course, makes these insects an attractive model for gerontology studies. Similar to worker lifespan, the long-lived queens are characterised by low JH and high vitellogenin titres in the haemolymph, and it is actually this inverted relationship between JH and vitellogenin which sets the highly eusocial honey bees apart from the other insects, including the primitively eusocial bumble bees (Bloch et al., 2000). The dissociation of JH from its gonadotropic function during adult life—it only induces vitellogenin synthesis in the late pharate adult stage (Barchuk et al., 2002)—and its conversion into a positive regulator of division of labour in the sterile worker caste and, implicitly, a factor limiting lifespan, has been discussed in terms of switch-function or split-function hypotheses (Giray et al., 2005; Robinson and Vargo, 1997; West-Eberhard, 1996).

Most clearly, this inverted relationship of JH and vitellogenin has been brought to light by Corona et al. (2007) in their study on the relationship among JH, vitellogenin and IIS in adult queens and workers. They could show that vitellogenin expression by fat body cells in the head is a characteristic of queens and absent in workers. This was associated with an inverted relationship for *Amilp1* expression, which decreases with age in queens and increases with age in workers. Furthermore, treatment of queens with the JH-mimic methoprene caused a decrease in vitellogenin and an increase in *Amilp1* expression in the heads of queens, thus emphasising an intricate and body compartment-specific interaction among JH, vitellogenin and IIS.

The remodelling of the classical trade-off relationship between reproduction and lifespan seen in most model organisms, thus clearly comes to light in the physiology of adult queens and workers, and here it is vitellogenin which assumes a truly novel role, that of an antioxidant, in addition to its ancestral reproductive function as an egg yolk protein. This antioxidant function is due to its zinc-binding properties which guarantee the survival of haemocytes and suppresses immunosenescence (Amdam et al., 2004). Evolutionarily, this new role of vitellogenin may be associated with the replacement of traditional antioxidant systems and the reduction of genes encoding functional extracellular peroxidases (Corona and Robinson, 2006).

An interesting question then arises as to what may be the possible function(s) of vitellogenin expression seen in honey bee larvae, where this gene is highly expressed in second to fourth instar larvae before it declines in the early fifth instar and remains undetectable until the late pharate adult stage (Guidugli et al., 2005). Strikingly, despite the considerable levels of vitellogenin mRNA, there were no or only trace amounts of vitellogenin protein found in larval haemolymph, thus indicating that vitellogenin may be playing an intra-cellular role in the larval stage, possibly related to protection against oxidative damage, like in the adult females. The question then becomes whether this could be a direct link to the striking caste differences in the hypoxia response and mitochondrial content in these early developmental stages.

9. CONCLUSIONS

Honey bee queens and workers are alternative phenotypes expressed from the same genotype. These alternative phenotypes in the female sex co-operate in a colony context within which they are functionally united and where one cannot exist without the other. Morphologically and physiologically differentiated castes are considered terminal steps in the evolution of insect societies, and the inter-dependence of the two castes on one another is considered to represent an evolutionary point of no return (Wilson, 1971; Wilson and Hölldobler, 2005). This poses an interesting question to evolutionary developmental biology, as the caste phenotypes, once determined by extrinsic (diet, temperature, etc.) or intrinsic (genetic) factors, develop as two different types of individuals, which are then also independently subject to selection. An important aspect herein is that the developmental programmes should be structured so as to guarantee that only clearly distinct caste phenotypes are formed, but not inter-castes, i.e., individuals with a mix of caste characters, as these would decrease colony fitness. In other words, this would mean that selection should act to guarantee that the developmental pathways into divergent phenotypes are canalised.

In this review, we discuss current evidence on how a context-dependent nutritional regime possibly became integrated with conserved modules of nutrient-sensing pathways, body size regulation and insect metamorphic hormones. Upon analysing temporal expression profiles of several key components of these signalling pathways, we came to the conclusion that the discrepancies seen when comparing honey bee castes with major model organisms, especially *D. melanogaster*, maybe the result of considerable rewiring within and among certain pathways, which made possible the divergence in the queen/worker developmental pathways. Interestingly, this divergence is only seen in the female sex, whereas in the male sex, drones may be larger or smaller, depending on the larval rearing conditions, but they show no essential differences in morphology. The only difference except for general body size is in sperm number, fitness so to speak, which is lower in smaller drones (Berg et al., 1997; Koeniger et al., 2014).

As an overarching argument for all these "incompatibilities" in the signalling pathways, we found that the hypoxia response core genes are significantly over-expressed in worker larvae, indicating an endogenous hypoxia response. Upon analysing the ultrastructure of the fat body cells of honey bee larvae, it became evident that queens have much higher mitochondrial numbers than workers, indicating that queens have much higher oxygen consumption and, thus, a higher oxidative metabolism. This has already been indicated by several studies done between the 1930s and 1980s (Eder et al., 1983; Melampy and Willis, 1939; Osanai and Rembold, 1969), but these were apparently over-shadowed by findings arising between 1970 and 1980 that established JH as a major endocrine regulator of caste phenotype.

Interestingly, the question of oxidative metabolism and its regulation takes us, on the one hand, back to basic principles of physiology of the insect intermediary metabolism, which of course has a direct link to nutrient sensing. On the other hand, the question of oxidative metabolism is also a link with more recent findings on epigenetic mechanisms (see Galstad et al., 2015, this volume) and their role in caste development, as revealed by differential methylation of tricarboxylic acid cycle and IIS/TOR pathway genes (Foret et al., 2011).

And finally, we argue that there may be parallels between signalling pathways and their integration in the larval stage related to caste development, and in the adult stage, where they are related to caste function, especially the differences in reproduction and lifespan. One of the major transitions, or novelties so to speak, is the dissociation of the intricate link between JH and vitellogenin seen in adult honey bees. Whereas in most insect clades, this link underlies the reproductive physiology, in honey bees JH and vitellogenin were recombined in a mutually negative regulatory circuitry, and this apparently occurred in combination with a rewiring in their association with IIS. Interestingly, vitellogenin has been shown to be expressed not only in adult females but also in honey bee larvae (Guidugli et al., 2005), i.e., at a time when it is normally not expressed in other insects, and we currently have no explanation for this, except that it occurs exactly at a time when queen and worker larvae also show major differences in their haemolymph JH titres. It may thus be worthwhile to keep an eye on these apparent commonalities shared between the adult castes and their pre-imaginal stages in a wider sense, thinking about possible evolutionary steps that may be underlying and have driven the remodelling of nutrient sensing, signalling pathways and their endocrine connections in the social insects.

The obvious difference between the physiology of these highly eusocial insects and the common models of life histories and their physiologies lies in the separation of reproduction and somatic maintenance into two separate castes. These are literally united and function as a superorganism, not only in terms of behavioural interactions, which is the more explored side of this concept, but also in their physiology, genomic architecture and epigenetics. While the latter is the side where we have barely scratched the surface, conclusions coming out of a comparative genome analysis of 10 bee species over all levels of sociality (Kapheim et al., submitted manuscript) are in support of the view that major signalling pathways, especially those related to nutrient sensing, have apparently undergone rewiring in their connectivity, and that epigenetics has gained a major role in phenotypic plasticity of the castes. Such rewiring may have become possible because of relaxed selection acting during the early stages of eusocial evolution in Hymenoptera (Hunt et al., 2011), and this could have broken up the constraints embedded in the reproduction/longevity trade-off seen in most metazoan clades. On the other hand, once such rewiring had occurred in basal eusocial species, it may have actually opened the path for positive selection in the highly eusocial ones, especially so on worker phenotypic traits. Such selection has recently been revealed through a population genomic analysis of honey bees (Harpur et al., 2014), showing evidence for positive selection on genes in pathways that were particularly relevant in caste development, such as IIS, JH and vitellogenin.

ACKNOWLEDGEMENTS

We thank Zilá Luz Paulino Simões and Marcia M. Gentile Bitondi for ongoing discussions on the physiology and genomics of honey bees. Carolina Gonçalves Santos provided valuable assistance in the qPCR analyses. Funding was provided by grants from Fundação de Amparo à Pesquisa doe Estado de São Paulo (FAPESP 2012/01808-9) and Conselho Nacional de Desenvolvimento Científico e Tecnológico (CNPq (470335/2009-2 and 305378/2007-4)).

REFERENCES

Amdam, G.V., Omholt, S.W., 2002. The regulatory anatomy of honeybee lifespan. J. Theor. Biol. 21, 209–228.

Amdam, G.V., Simões, Z.L.P., Hagen, A., Norberg, K., Schroder, K., Mikkelsen, O., Kirkwood, T.B., Omholt, S.W., 2004. Hormonal control of the yolk precursor vitellogenin regulates immune function and longevity in honeybees. Exp. Gerontol. 39, 767–773.

Amdam, G.V., Aase, A.L., Seehuus, S.C., Fondrk, M.K., Norberg, K., Hartfelder, K., 2005. Social reversion of immunosenescence in honey bee workers. Exp. Gerontol. 40, 939–947.

Asencot, M., Lensky, Y., 1976. The effect of sugar and juvenile hormone on the differentiation of the female honeybee larvae (*Apis mellifera* L.). Life Sci. 18, 693–700.

Asencot, M., Lensky, Y., 1984. Juvenile hormone induction of 'queenliness' on female honeybee (*Apis mellifera* L.) larvae reared on worker jelly and on stored royal jelly. Comp. Biochem. Physiol. B 78, 109–117.

Asencot, M., Lensky, Y., 1988. The effect of soluble sugars in stored royal jelly on the differentiation of female honeybee (*Apis mellifera* L.) larvae to queens. Insect Biochem. 18, 127–133.

Aupinel, P., Fortini, D., Dufour, H., Tasei, J.N., Michaud, B., Odoux, J.F., Delegue, M.H. P., 2005. Improvement of artificial feeding in a standard in vitro method for rearing *Apis mellifera* larvae. Bull. Insectol. 58, 107–111.

Azevedo, S.V., Hartfelder, K., 2008. The insulin signaling pathway in honey bee (*Apis mellifera*) caste development—differential expression of insulin-like peptides and insulin receptors in queen and worker larvae. J. Insect Physiol. 54, 1064–1071.

Azevedo, S.V., Carantona, O.A.M., Oliveira, T.L., Hartfelder, K., 2011. Differential expression of hypoxia pathway genes in honey bee (*Apis mellifera* L.) caste development. J. Insect Physiol. 57, 38–45.
Barchuk, A.R., Bitondi, M.M.G., Simões, Z.L.P., 2002. Effects of juvenile hormone and ecdysone on the timing of vitellogenin appearance in hemolymph of queen and worker pupae of *Apis mellifera*. J. Insect Sci. 2, e8.
Barchuk, A.R., Maleszka, R., Simoes, Z.L.P., 2004. *Apis mellifera ultraspiracle*: cDNA sequence and rapid up-regulation by juvenile hormone. Insect Mol. Biol. 13, 459–467.
Barchuk, A.R., Cristino, A.S., Kucharski, R., Costa, L.F., Simoes, Z.L.P., Maleszka, R., 2007. Molecular determinants of caste differentiation in the highly eusocial honeybee *Apis mellifera*. BMC Dev. Biol. 7, e70.
Baumann, A., Barry, J., Wang, S., Fiujiwara, Y., Wilson, T.G., 2010. Paralogous genes involved in juvenile hormone action in Drosophila melanogaster. Genetics 185, 1327–1336.
Belles, X., Santos, C.G., 2014. The MEKRE93 (Methoprene tolerant-Krüppel homolog1-E93) pathway in the regulation of insect metamorphosis and the homology of the pupal stage. Insect Biochem. Mol. Biol. 52, 60–68.
Berg, S., Koeniger, N., Koeniger, G., Fuchs, S., 1997. Body size and reproductive success of drones (*Apis mellifera* L.). Apidologie 28, 449–460.
Bloch, G., Borst, D.W., Hyang, Z., Robinson, G.E., Cnaani, J., Hefetz, A., 2000. Juvenile hormone titers, juvenile hormone biosynthesis, ovarian development and social environment in *Bombus terrestris*. J. Insect Physiol. 46, 47–57.
Boleli, I.C., Simões, Z.L.P., Hartfelder, K., 1998. The stomatogastric nervous system of the honey bee (*Apis mellifera*) in a critical phase of caste development. J. Morphol. 236, 139–148.
Bomtorin, A.D., Barchuk, A.R., Moda, L.M., Simoes, Z.L.P., 2012. Hox gene expression leads to differential hind leg development between honeybee castes. PLoS One 7, e40111.
Brodschneider, R., Riessberger-Galle, U., Crailsheim, K., 2009. Flight performance of artificially reared honeybees (*Apis mellifera*). Apidologie 40, 441–449.
Bruick, R.K., McKnight, S.L., 2001. A conserved family of prolyl-4-hydroxylases that modify HIF. Science 294, 1337–1340.
Buttstedt, A., Moritz, R.F.A., Erler, S., 2013. More than royal food—*major royal jelly protein* genes in sexuals and workers of the honeybee *Apis mellifera*. Front. Zool. 10, e72.
Buttstedt, A., Moritz, R.F.A., Erler, S., 2014. Origin and function of the major royal jelly proteins of the honeybee (*Apis mellifera*) as members of the *yellow* gene family. Biol. Rev. 89, 255–269.
Centanin, L., Gorr, T.A., Wappner, P., 2010. Tracheal remodelling in response to hypoxia. J. Insect Physiol. 56, 447–454.
Colombani, J., Bianchini, L., Layalle, S., Pondeville, E., Dauphin-Villemant, C., Antoniewski, C., Carre, C., Noselli, S., Leopold, P., 2005. Antagonistic actions of ecdysone and insulins determine final size in *Drosophila*. Science 310, 667–670.
Corona, M., Robinson, G.E., 2006. Genes of the antioxidant system of the honey bee: annotation and phylogeny. Insect Mol. Biol. 15, 687–701.
Corona, M., Velarde, R.A., Remolina, S., Moran-Lauter, A., Wang, Y., Hughes, K.A., Robinson, G.E., 2007. Vitellogenin, juvenile hormone, insulin signaling, and queen honey bee longevity. Proc. Natl. Acad. Sci. U.S.A. 104, 7128–7133.
Crailsheim, K., 1990. The protein balance of the honey bee worker. Apidologie 21, 417–429.
Dedej, S., Hartfelder, K., Aumeier, P., Rosenkranz, P., Engels, W., 1998. Caste determination is a sequential process: effect of larval age at grafting on ovariole number, hind leg size and cephalic volatiles in the honey bee (*Apis mellifera carnica*). J. Apic. Res. 37, 183–190.

Dekanty, A., Lavista-Llanos, S., Irisarri, M., Oldham, S., Wappner, P., 2005. The insulin-PI3K/TOR pathway induces a HIF-dependent transcriptional response in *Drosophila* by promoting nuclear localization of HIF-alpha/Sima. J. Cell Sci. 118, 5431–5441.

Dickman, M.J., Kucharski, R., Maleszka, R., Hurd, P.J., 2013. Extensive histone post-translational modification in honey bees. Insect Biochem. Mol. Biol. 43, 125–137.

Dietz, A., Haydak, M.H., 1971. Caste determination in honey bees I. The significance of moisture in larval food. J. Exp. Zool. 177, 353–358.

Dietz, A., Hermann, H.R., Blum, M., 1979. The role of exogenous JH I, JH II, JH III and anti-JH (Precocene II) on queen induction of 4-5-day-old worker honeybee larvae. J. Insect Physiol. 25, 503–512.

Dogra, G.S., Ulrich, G.M., Rembold, H., 1977. A comparative study of the endocrine system of the honey bee larvae under normal and experimental conditions. Z. Naturforsch. 32, 637–642.

Drapeau, M.D., Albert, S., Kucharski, R., Prusko, C., Maleszka, R., 2006. Evolution of the Yellow/Major Royal Jelly Protein family and the emergence of social behavior in honey bees. Genome Res. 16, 1385–1394.

Eder, J., Kremer, J.P., Rembold, H., 1983. Correlation of cytochrome c titer and respiration in *Apis mellifera*: adaptative response to caste determination defines workers, intercastes and queens. Comp. Biochem. Physiol. 76B, 703–716.

Edgar, B.A., 2006. How flies get their size: genetics meets physiology. Nat. Rev. Genet. 7, 907–916.

Flatt, T., Amdam, G.V., Kirkwood, B.L., Omholt, S.W., 2013. Life-history evolution and the polyphenic regulation of somatic maintenance and survival. Q. Rev. Biol. 88, 185–218.

Foret, S., Kucharski, R., Pellegrini, M., Feng, S., Jacobsen, S.E., Robinson, G.E., Maleszka, R., 2011. DNA methylation dynamics, metabolic fluxes, gene splicing, and alternative phenotypes in honey bees. Proc. Natl. Acad. Sci. U.S.A. 109, 4968–4973.

Fussnecker, B., Grozinger, C.M., 2008. Dissection of the role of Kr-h1 brain gene expression in foraging behavior in honey bees (Apis mellifera). Insect Mol. Biol. 17, 515–522.

Giray, T., Giovanetti, M., West-Eberhard, M.J., 2005. Juvenile hormone, reproduction, and worker behavior in the neotropical social wasp *Polistes canadensis*. Proc. Natl. Acad. Sci. U.S.A. 102, 3330–3335.

Glastad, K.M., Chau, L.M., Goodisman, M.A.D., 2015. Epigenetics in Social Insects. In: Zayed, A., Kent, C.F. (Eds.), Genomics, Physiology and Behavior of Social Insects Vol. 48. London, Academic Press, pp. 227–270.

Goewie, E.A., 1978. Regulation of caste differentiation in the honey bee (Apis mellifera L.). Meded. Landb. Wageningen 78 (15), 1–55.

Guidugli, K.R., Piulachs, M.-D., Bellés, X., Lourenço, A.P., Simões, Z.L.P., 2005. Vitellogenin expression in queen ovaries and in larvae of both sexes of *Apis mellifera*. Arch. Insect Biochem. Physiol. 59, 211–218

Hamilton, W.D., 1964. The genetical theory of social behaviour I & II. J. Theor. Biol. 7, 1–52.

Harpur, B.A., Kent, C.F., Molodtsova, D., Lebon, J.M., Alqarni, A.S., Owayss, A.A., Zayed, A., 2014. Population genomics of the honey bee reveals strong positive signatures of positive selection on worker traits. Proc. Natl. Acad. Sci. U.S.A. 111, 2614–2619.

Hartfelder, K., Emlen, D.J., 2012. Endocrine control of insect polyphenism. In: Gilbert, L.I. (Ed.), Insect Endocrinology. Academic Press, London, pp. 464–522.

Hartfelder, K., Engels, W., 1998. Social insect polymorphism: hormonal regulation of plasticity in development and reproduction in the honeybee. Curr. Top. Dev. Biol. 40, 45–77.

Haydak, H.M., 1970. Honey bee nutrition. Annu. Rev. Entomol. 15, 143–156.

Hepperle, C., Hartfelder, K., 2001. Differentially expressed regulatory genes in honey bee caste development. Naturwissenschaften 88, 113–116.

Hoffmann, I., 1960. Untersuchungen über die Herkunft von Komponenten des Königinnenfuttersaftes der Honigbiene. Z. Bienenforsch. 5, 101–111.

Hoffmann, K., Gowin, J., Hartfelder, K., Korb, J., 2014. The scent of royalty: a P450 gene signals reproductive status in a social insect. Mol. Biol. Evol. 31, 2689–2696.

Huang, Z.Y., Robinson, G.E., 1992. Honeybee colony integration: worker-worker interactions mediate hormonally regulated plasticity in division of labor. Proc. Natl. Acad. Sci. U.S.A. 89, 11726–11729.

Humann, F.C., Hartfelder, K., 2011. Representational difference analysis (RDA) reveals differential expression of conserved as well as novel genes during caste-specific development of the honey bee (*Apis mellifera* L.) ovary. Insect Biochem. Mol. Biol. 41, 602–612.

Humann, F.C., Tiberio, G.J., Hartfelder, K., 2013. Sequence and expression characteristics of long noncoding RNAs in honey bee caste development—poterntial regulatirs of transgresssive ovary size. PLoS One 8, e78915.

Hunt, B.G., Ometto, L., Wurm, Y., Shoemaker, D., Yi, S.V., Keller, L., Goodisman, M.A. D., 2011. Relaxed selection is a precursor to the evolution of phenotypic plasticity. Proc. Natl. Acad. Sci. U.S.A. 108, 15936–15941.

Ivan, M., Kondo, K., Yang, H.F., Kim, W., Valiando, J., Ohh, M., Salic, A., Asara, J.M., Lane, W.S., Kaelin, W.G., 2001. HIF alpha targeted for VHL-mediated destruction by proline hydroxylation: implications for O_2 sensing. Science 292, 464–468.

Jindra, M., Palli, S.R., Riddiford, L.M., 2013. The juvenile hormone signaling pathway in insect development. Annu. Rev. Entomol. 58, 181–204.

Jung-Hoffmann, I., 1966. Die determination von königin und arbeiterin der honigbiene. Z. Bienenforsch. 8, 296–322.

Kaftanoglu, O., Linksvayer, T.A., Page, R.E., 2011. Rearing honey bees (*Apis mellifera* L.) in vitro I: effects of sugar concentrations on survival and development. J. Insect Sci. 11, e96.

Kamakura, M., 2011. Royalactin induces queen differentiation in honeybees. Nature 473, 478–483.

Kayukawa, T., Minakuchi, C., Namiki, T., Togawa, T., Yoshiyama, M., Kamimura, M., Mita, K., Imanishi, S., Kiuchi, M., Ishikawa, Y., Shinoda, T., 2012. Transcriptional regulation of juvenile hormone-mediated induction of Krüppel homolog 1, a repressor of insect metamorphosis. Proc. Natl. Acad. Sci. U.S.A. 109, 11729–11734.

Koeniger, G., Koeniger, N., Ellis, J., Connor, L., 2014. Mating Biology of Honey Bees (*Apis mellifera*). Wicwas Press, Kalamazoo, MI.

Korb, J., 2015. Juvenile hormone: A central regulator of termites' caste polyphenism. In: Zayed, A., Kent, C.F. (Eds.), Genomics, Physiology and Behavior of Social Insects Vol. 48. London, Academic Press, pp. 131–162.

Korb, J., Hartfelder, K., 2008. Life history and development—a framework for understanding developmental plasticity in lower termites. Biol. Rev. 83, 295–313.

Korb, J., Hoffmann, K., Hartfelder, K., 2009. Endocrine signatures underlying plasticity in postembryonic development of a lower termite, *Cryptotermes secundus* (Kalotermitidae). Evol. Dev. 6, 342–352.

Kucharski, R., Maleszka, J., Foret, S., Maleszka, R., 2008. Nutritional control of reproductive status in honeybees via DNA methylation. Science 319, 1827–1830.

Leimar, O., Hartfelder, K., Laubichler, M.D., Page Jr., R.E., 2012. Development and evolution of caste dimorphism in honeybees—a modeling approach. Ecol. Evol. 2, 3098–3109.

Lindauer, M., 1952. Ein Beitrag zur Frage der Arbeitsteilung im Bienenstaat. Z. Vergl. Physiol. 34, 49–53.

Livak, K.J., Schmittgen, T.D., 2001. Analysis of relative gene expression data using real-time quantitative PCR and the $2^{-\Delta\Delta CT}$ method. Methods 25, 402–408.

Löw, P., Varga, A., Pircs, K., Nagy, P., Szatmári, Z., Juhász, G., 2013. Impaired proteasomal degradation enhances autophagy via hypoxia signaling in Drosophila. BMC Cell Biol. 14, e29.

Lozano, J., Belles, X., 2014. Role of methoprene-tolerant (Met) in adult morphogenesis and in adult ecdysis of *Blattella germanica*. PLoS One 9, e103614.

Lyko, F., Foret, S., Kucharski, R., Wolf, S., Falkenhayn, C., Maleszka, R., 2010. The honey bee epigenomes: differential methylation of brain DNA in queens and workers. PLoS Biol. 8, e10000506.

Mackert, A., Donascimento, A.M., Bitondi, M.M.G., Hartfelder, K., Simões, Z.L.P., 2008. Identification of a juvenile hormone esterase-like gene in the honey bee. Apis mellifera L.—expression analysis and functional assays. Comp. Biochem. Physiol. B 150, 33–44.

Mackert, A., Hartfelder, K., Bitondi, M.M.G., Simões, Z.L.P., 2010. The juvenile hormone (JH) epoxide hydrolase gene in the honey bee (*Apis mellifera)* genome encodes a protein which has negligible participation in JH degradation. J. Insect Physiol. 56, 1139–1146.

Matsuka, M., Watanabe, N., Takeuchi, K., 1973. Analysis of the food of larval drone honeybees. J. Apic. Res. 12, 1–5.

Maurizio, A., 1950. The influence of pollen feeding and brood rearing on the length of life and physiological conditions of the honeybee. Bee World 31, 9–12.

McNeill, H., Craig, G.M., Bateman, J.M., 2008. Regulation of neurogenesis and epidermal growth factor receptor signaling by the insulin receptor/target of rapamycin pathway in Drosophila. Genetics 179, 843–853.

Melampy, R.M., Willis, E.R., 1939. Respiratory metabolism during larval and pupal development of the female honeybee (*Apis mellifica* L.). Physiol. Zool. 12, 302–311.

Mello, T.R.P., Aleixo, A.C., Pinheiro, D.G., Nunes, F.M.F., Bitondi, M.M.G., Hartfelder, K., Barchuk, A.R., Simões, Z.L.P., 2014. Developmental regulation of ecdysone receptor (EcR) gene expression and its targets in the honeybee, *Apis mellifera*. Front. Genet. 4, e445.

Mercer, T.R., Dinger, M.E., Mattick, J.S., 2009. Long noncoding RNAs: insights into functions. Nat. Rev. Genet. 10, 155–159.

Michelette, E.R.F., Soares, A.E.E., 1993. Characterization of preimaginal developmental stages in Africanized honey bee workers (*Apis mellifera* L.). Apidologie 24, 431–440.

Minakuchi, C., Namiki, T., Shinoda, T., 2009. Krüppel homolog 1, an early juvenile hormone-response gene downstream of Methoprene-tolerant, mediates its anti-metamorphic action in the red flour beetle *Tribolium castaneum*. Dev. Biol. 325, 341–350.

Mirth, C.K., Riddiford, L.M., 2007. Size assessment and growth control: how adult size is determined in insects. BioEssays 29, 344–355.

Mirth, C., Truman, J.W., Riddiford, L.M., 2005. The role of the prothoracic gland in determining critical weight to metamorphosis in *Drosophila melanogaster*. Curr. Biol. 15, 1796–1807.

Mirth, C.K., Tang, H.Y., Makohon-Moore, S.C., Salhader, S., Gokhale, R.H., Warner, R.D., Koyama, T., Riddiford, L.M., Shingleton, A.W., 2014. Juvenile hormone regulates body size and perturbs insulin signaling in *Drosophila*. Proc. Natl. Acad. Sci. U.S.A. 111, 7018–7023.

Mutti, N.S., Dolezal, A.G., Wolschin, F., Mutti, J.S., Gill, K.S., Amdam, G.V., 2011. IRS and TOR nutrient-signaling pathways act via juvenile hormone to influence honey bee caste fate. J. Exp. Biol. 214, 3977–3984.

Nijhout, H.F., 2003. The control of body size in insects. Dev. Biol. 261, 1–9.

Nijhout, H.F., Wheeler, D.E., 1982. Juvenile hormone and the physiological basis of insect polymorphism. Q. Rev. Biol. 57, 109–133.

Nijhout, H.F., Williams, C.M., 1974. Control of moulting and metamorphosis in the tobacco hornworm, *Manduca sexta* (L.): cessation of juvenile hormone secretion as a trigger for pupation. J. Exp. Biol. 61, 493–501.

Nilsen, K.-A., Ihle, K.E., Frederick, K., Fondrk, M.K., Smedal, B., Hartfelder, K., Amdam, G.V., 2011. Insulin-like peptide genes in honey bee fat body respond differentially to manipulation of social behavioral physiology. J. Exp. Biol. 214, 1488–1497.

Oldham, S., 2011. Obesity and nutrient sensing TOR pathway in flies and vertebrates: functional conservation of genetic mechanisms. Trends Endocrinol. Metab. 22, 45–52.

Osanai, M., Rembold, H., 1968. Entwicklungsabhängige Enzymaktivitäten in den Mitochondrien von Königinnen und Arbeiterinnen der Honigbiene. Biochim. Biophys. Acta 162, 22–31.

Osanai, M., Rembold, H., 1969. Caste-specific differences in mitochondrial cytochrome systems of the honey bee (Apis mellifica L.). Hoppe-Seylers, Z. Physiol. Chem. 350, 1160.

Page, R.E.J., Amdam, G.V., 2007. The making of a social insect: developmental architectures of social design. BioEssays 29, 334–343.

Patel, A., Fondrk, M.K., Kaftanoglu, O., Emore, C., Hunt, G., Frederick, K., Amdam, G.V., 2007. The making of a queen: TOR pathway is a key player in diphenic caste development. PLoS One 2, e509.

Puig, O., Tijan, R., 2005. Transcriptional feedback control of insulin receptor by dFOXO/FOXO1. Genes Dev. 19, 2435–2446.

Rachinsky, A., 1994. Octopamine and serotonin influence on corpora allata activity in honey bee (*Apis mellifera*) Larvae. J. Insect Physiol. 40, 549–554.

Rachinsky, A., Engels, W., 1995. Caste development in honeybees (*Apis mellifera*)—juvenile hormone turns on ecdysteroids. Naturwissenschaften 82, 378–379.

Rachinsky, A., Feldlaufer, M.F., 2000. Responsiveness of honey bee (*Apis mellifera* L.) corpora allata to allatoregulatory peptides from four insect species. J. Insect Physiol. 46, 41–46.

Rachinsky, A., Hartfelder, K., 1990. Corpora allata activity, a prime regulating element for caste-specific juvenile hormone titre in honey bee larvae (*Apis mellifera carnica*). J. Insect Physiol. 36, 189–194.

Rachinsky, A., Hartfelder, K., 1991. Differential production of juvenile hormone and its desoxy precursor by corpora allata of honeybees during a critical period of caste development. Naturwissenschaften 78, 270–272.

Rachinsky, A., Strambi, C., Strambi, A., Hartfelder, K., 1990. Caste and metamorphosis—hemolymph titers of juvenile hormone and ecdysteroids in last instar honeybee larvae. Gen. Comp. Endocrinol. 79, 31–38.

Ratnieks, F.L., Helanterä, H., 2009. The evolution of extreme altruism and inequality in insect societies. Philos. Trans. R. Soc. Lond. B Biol. Sci. 364, 3169–3179.

Rembold, H., 1987. Caste specific modulation of juvenile hormone titers in *Apis mellifera*. Insect Biochem. 17, 1003–1006.

Rembold, H., Lackner, B., 1981. Rearing of honeybee larvae in vitro: effect of yeast extract on queen differentiation. J. Apic. Res. 20, 165–171.

Rembold, H., Czoppelt, C., Rao, P.J., 1974a. Effect of juvenile hormone on caste differentiation in the honey bee. Apis mellifera. J. Insect Physiol. 20, 1193–1202.

Rembold, H., Lackner, B., Geistbeck, J., 1974b. The chemical basis of queen bee determinator from royal jelly. J. Insect Physiol. 20, 307–314.

Rembold, H., Kremer, J.-P., Ulrich, G.M., 1980. Characerization of the postembryonic stages of the female castes of the honey bee, *Apis mellifera* L. Apidologie 11, 29–38.

Robinson, G.E., Vargo, E.L., 1997. Juvenile hormone in adult eusocial Hymenoptera: gonadotropin and behavioral pacemaker. Arch. Insect Biochem. Physiol. 35, 559–583.

Robinson, G.E., Strambi, C., Strambi, A., Huang, Z.Y., 1992. Reproduction in worker honey bees is associated with low juvenile hormone titers and rates of biosynthesis. Gen. Comp. Endocrinol. 87, 471–480.

Romero, N.M., Dekanty, A., Wappner, P., 2007. Cellular and developmental adaptations to hypoxia: a *Drosophila* perspective. Methods Enzymol. 435, 123–144.

Santos, K.S., dos Santos, L.D., Mendes, M.A., de Souza, B.M., Malaspina, O., Palma, M.S., 2005. Profiling the proteome complement of the secretion from hypopharyngeal gland of Africanized nurse-honeybees (*Apis mellifera* L.). Insect Biochem. Mol. Biol. 35, 85–91.

Sasaki, M., Okada, I., 1972. Efficiency of conversion of royal jelly during the development of the queen honeybee. J. Apic. Res. 11, 135–140.

Scarselli, R., Donadio, E., Giuffrida, M.G., Fortunato, D., Conti, A., Balesteri, E., Felicioli, R., Pinzauti, M., Sabatini, A.G., Felicioli, A., 2005. Towards the royal jelly proteome. Proteomics 5, 769–776.

Schmidt Capella, I.C., Hartfelder, K., 1998. Juvenile hormone effect on DNA synthesis and apoptosis in caste-specific differentiation of the larval honey bee (*Apis mellifera* L.) ovary. J. Insect Physiol. 44, 385–391.

Schmidt Capella, I.C., Hartfelder, K., 2002. Juvenile-hormone-dependent interaction of actin and spectrin is crucial for polymorphic differentiation of the larval honey bee ovary. Cell Tissue Res. 307, 265–272.

Schmitzova, J., Klaudiny, J., Albert, S., Schroeder, W., Schneckengost, W., Hanes, J., Judova, J., Simuth, J., 1998. A family of major royal jelly proteins of the honeybee *Apis mellifera*. Cell. Mol. Life Sci. 54, 1020–1030.

Schwander, T., Humbert, J.Y., Brent, C.S., Cahan, S.H., Chapuis, L., Renai, E., Keller, L., 2008. Maternal effect on female caste determination in a social insect. Curr. Biol. 18, 265–269.

Seehuus, S.C., Krekling, T., Amdam, G.V., 2006. Cellular senescence in honey bee brain is largely independent of chronological age. Exp. Gerontol. 41, 1117–1125.

Shuel, R.W., Dixon, S.E., 1968. The importance of sugar for the population of the worker honeybee. J. Apic. Res. 7, 109–112.

Spannhoff, A., Kim, Y.K., Raynal, N.J.-M., Ghaibyan, V., Su, M.-B., Zhou, Y.-Y., Li, J., Castellano, S., Sbardella, G., Issa, J.-P.J., Bedford, M.T., 2011. Histone deacetylase inhibitor activity in royal jelly might facilitate caste switching in bees. EMBO Rep. 12, 238–243.

Stabe, H.A., 1930. The growth rate of worker, drone and queen larvae of the honeybee, *Apis mellifera* Linn. J. Econ. Entomol. 23, 447–453.

Süren-Castillo, S., Abrisqueta, M., Maestro, J.L., 2012. FoxO inhibits juvenile hormone biosynthesis and vitellogenin production in the German cockroach. Insect Biochem. Mol. Biol. 42, 491–498.

Takenaka, T., Takahashi, E., 1980. General composition of the royal jelly. Bull. Fac. Agric. Tamagawa Univ. Tokyo 20, 71–78.

The Honeybee Genome Sequencing Consortium, 2006. Insights into social insects from the genome of the honeybee *Apis mellifera*. Nature 443, 931–949.

Vandenberg, J.D., Shimanuki, H., 1987. Technique for rearing honeybees in the laboratory. J. Apic. Res. 26, 90–97.

von Planta, A., 1888. Über den Futtersaft der Bienen. Hoppe-Seylers Z. Physiol. Chem. 12, 327–354.

Wang, I.D., 1965. Growth rates of young queen and worker honeybee larvae. J. Apic. Res. 4, 3–5.

Wang, G.L., Jiang, B.H., Rue, E.A., Semenza, G.L., 1995. Hypoxia-inducible factor-1 is a basic-helix-loop-helix-PAS heterodimer regulated by cellular O_2 tension. Proc. Natl. Acad. Sci. U.S.A. 92, 5510–5514.

Wang, Y., Azevedo, S.V., Hartfelder, K., Amdam, G.V., 2013. Insulin-like peptides (AmILP1 and AmILP2) differentially affect female caste development in the honey bee (*Apis mellifera* L.). J. Exp. Biol. 216, 4347–4357.

Weaver, N., 1974. Control of dimorphism in hte female honeybee. 3. The balance of nutrients. J. Apic. Res. 13, 93–101.
Wei, Y., Zhang, Y.-J., Cai, Y., Xu, M.-H., 2014. The role of mitochondria in mTOR-regulated longevity. Biol. Rev.. http://dx.doi.org/10.1111/brv.12103 (Epub).
West-Eberhard, M.J., 1996. Wasp societies as microcosms for the study of development and evolution. In: Turillazzi, S., West-Eberhard, M.J. (Eds.), Natural History and Evolution of Paper Wasps. Oxford University Press, London, pp. 290–317.
West-Eberhard, M.J., 2003. Developmental Plasticity and Evolution. Oxford University Press, Oxford.
Wheeler, D.E., Buck, N., Evans, J.D., 2006. Expression of insulin pathway genes during the period of caste determination in the honey bee. Apis mellifera. Insect Mol. Biol. 15, 597–602.
Wheeler, D.E., Buck, N., Evans, J.D., 2014. Expression of insulin/insulin-like signaling and TOR pathway genes in honey bee caste determination. Insect Mol. Biol. 23, 113–121.
Wilson, E.O., 1971. The Insect Societies. Belknapp Press of Harvard University Press, Cambridge, MA.
Wilson, E.O., Hölldobler, B., 2005. Eusociality: origins and consequences. Proc. Natl. Acad. Sci. U.S.A. 102, 13367–13371.
Wirtz, P., Beetsma, J., 1972. Induction of caste differentiation in the honey bee (*A. mellifera*) by juvenile hormone. Entomol. Exp. Appl. 15, 517–520.
Wu, Q., Brown, M.R., 2006. Signaling and function of insulin-like peptides in insects. Annu. Rev. Entomol. 51, 1–24.
Zhang, W., Thompson, B.J., Hietakangas, V., Cohen, S.M., 2011. MAP/ERK signaling regulates insulin sensitivity to control glucose metabolism in *Drosophila*. PLoS Genet. 7, e1002429.

CHAPTER TWO

The Physiological and Genomic Bases of Bumble Bee Social Behaviour

Etya Amsalem*, Christina M. Grozinger*, Mario Padilla*, Abraham Hefetz†,1

*Department of Entomology, Center for Pollinator Research, Center for Chemical Ecology, Huck Institutes of the Life Sciences, The Pennsylvania State University, University Park, Pennsylvania, USA
†Department of Zoology, George S. Wise Faculty of Life Sciences, Tel Aviv University, Ramat Aviv, Tel Aviv, Israel
[1]Corresponding author: e-mail address: hefetz@post.tau.ac.il

Contents

Advances in Insect Physiology, Volume 48
ISSN 0065-2806
http://dx.doi.org/10.1016/bs.aiip.2015.01.001

Abstract

Bumble bees are an outstanding model system in which to study the organization and evolution of complex social behaviour. Bumble bees pass through several distinct phases during their annual life cycle, including solitary and eusocial phases, and the final stage of the colony cycle is marked by intense competition among the queen and workers over the production of males. Furthermore, there are approximately 250 species of bumble bees, and of the few species that have been examined, it is clear that multiple life history strategies are possible. Thus, the ultimate and proximate mechanisms underlying co-operation, conflict, and behavioural plasticity can readily be examined in bumble bees. Here, we describe the current state of knowledge about the evolutionary, ecological, behavioural, physiological, chemical, and genomic mechanisms and factors underpinning bumble bee social behaviour throughout the colony cycle. We highlight long-standing questions in the field and discuss how advances in genomics and comparative approaches across bumble bee species can provide profound insights into this fascinating system and the evolution of social behaviour.

ABBREVIATIONS

CA corpora allata
CP competition phase
JH juvenile hormone
Kr-h1 *Krüppel-homolog 1*
QLC queenless compartment
Pre-CP pre-competition phase
QEC queen-excluded compartment
QL queenless
QR queenright
QRC queenright compartment
Vg/*vg* vitellogenin (protein/mRNA)

1. INTRODUCTION

Bumble bees are one of the most conspicuous bees in meadows of the temperate zones. They comprise about 250 species, all of which belong to the genus *Bombus*, except for the social parasitic species that belong to the genus *Psythirus*. Most bumble bees live in the temperate zones of the Northern hemisphere, but some species also occupy temperate zones of the Southern hemisphere and some are even present in tropical zones (Cameron et al., 2007; Michener, 1974; Sakagami, 1976; Williams et al., 2008). All species in the genus *Bombus* are social but are misleadingly considered 'primitively' eusocial due to their solitary mode of colony foundation, their annual colony life cycle, and because the queens possess pollen-collecting morphological characters. However, this classification is not completely accurate because in most species, castes are clearly recognizable by size and thus are morphologically distinct, and females also store food in cells different from the brood cells, both characteristics of highly eusocial bees (Goulson, 2010; Michener, 1974; Wilson, 1971).

Bumble bees are both economically important pollinators and have served as a principal model system for a variety of studies in social behaviour and ecology (Goulson, 2010). Moreover, with the development of new genomic tools and resources (Sadd et al., 2015), bumble bees are emerging as an outstanding model system to study the sociogenomic mechanisms mediating social behaviours. With few exceptions, bumble bee colonies have an annual life cycle, which encompasses first a solitary phase in which the queen cares for the first-generation brood, then a co-operative eusocial phase with a clear reproductive division of labour, and finally transitions to a period of intense competition and conflict. Thus, bumble bees can be used to study the proximate mechanisms that regulate complex social behaviour, as well as the ultimate mechanisms underlying the evolution of these behaviours.

Though there are hundreds of bumble bee species, studies have focused on only a handful of species, particularly *Bombus terrestris*. In this chapter, we will focus on the physiology and genomic mechanisms underlying social behaviour in *B. terrestris* and refer to other species whenever information is available.

1.1 The *B. terrestris* colony life history

The general descriptions of bumble bee life histories, nesting, and social behaviours were published before (Alford, 1978; Duchateau and Velthuis, 1988; Goulson, 2003; Michener, 1974). Here, we briefly outline their life cycle with emphasis on traits that might have an impact on their social behaviour and social physiology (Fig. 1). Since most of the studies pertaining to the physiology and genomics of bumble bees were performed with *B. terrestris*, we will first describe its life cycle, and thereafter allude to deviations from this general scheme with references to how it may affect social

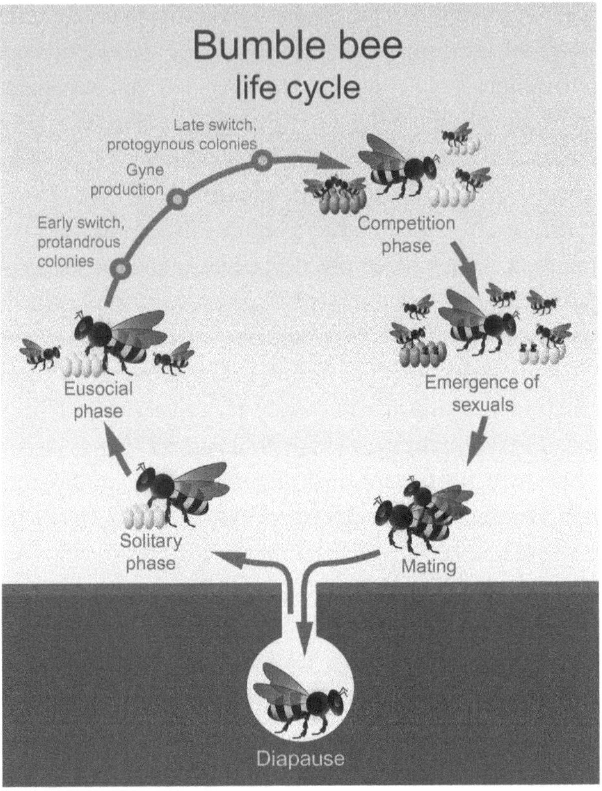

Figure 1 *Bombus terrestris* life cycle. Colonies rear one generation per year. New queens emerge near the end of colony cycle and leave the colony shortly after. During the solitary phase, the queens mate and enter a winter diapause that can last 6–9 months. Upon completion of diapause, queens forage, activate their ovaries, and lay their first brood (solitary phase). The colony's social phase begins with the emergence of the first worker (eusocial phase). The queen is the sole reproducer until the onset of the competition phase where females aggressively compete over male production. Males are produced at the 'switch point', where the queen switches from diploid egg to haploid egg production. The timing of the switch point varies between colonies and is likely influenced by ecological conditions. Gynes are produced towards the end of life cycle. *Graphical design by Nick Sloff, Department of Entomology, Pennsylvania State University.*

structure. It is also noteworthy that the geographical distribution of *B. terrestris* is very large, from Northern Europe to the Middle East, and probably encompasses several cryptic species or at least sub-species (Lecocq et al., 2013). This can explain the occasional differences in traits investigated by different laboratories that may have used different populations (e.g. Central Europe and Israel).

The colony cycle in *B. terrestris* starts with the diapause of mated queens, which can be initiated in the fall and lasts for 6–9 months (Alford, 1969b). Mated queens search for a subterranean cavity where they overwinter relying on their food reserves for survival (see Section 3.2, for more information). While in nature diapause is required to survive the winter, laboratory-reared queens can be induced to bypass diapause and are still able to produce colonies. These colonies, however, produce a greater number of gynes (future queens) (Beekman and van Stratum, 2000; Gosterit and Gurel, 2009). In nature, queens that skip diapause are bivoltine and are therefore selected to produce reproductives earlier in order for the next-generation queens to survive the approaching winter (see Section 1.3), and thus these laboratory-reared colonies may be reflecting a bivoltine life cycle.

After emerging from diapause, the founder queen searches for a nesting site (in *B. terrestris* nests are underground, usually in abandoned rodent burrows). Here, she builds several wax cells, provisions them with pollen, and lays the first batch of eggs (8–16 eggs). As the first batch of larvae develops, the queen continues foraging to provide them with pollen and nectar. When the larvae pupate, the queen typically builds additional egg cells on top of them to rear the second cohort of workers, which she continuously provides with pollen and nectar. This constitutes the *solitary phase* of the colony cycle.

In the next, *eusocial phase* of the colony, the queen stops foraging and is engaged solely in egg laying, while the workers take over all nest duties. During this phase, the nest population grows exponentially. After several generations of worker production and towards the end of the foraging season, reproductives (males and female gynes) are produced. Gyne production follows the 'Bang-Bang' strategy (Macevicz and Oster, 1976; Oster and Wilson, 1978), e.g., an abrupt shift from producing workers to producing gynes, consistent with *B. terrestris* having annual colonies and being semelparous (i.e. reproducing only once in a lifetime). Timing of male production in *B. terrestris* is more complex: depending on the colony, male production either precedes that of gyne production (protandry), or gyne production precedes that of male production (protogyny). There are also incidences of split sex ratios, i.e., within a population some colonies predominately produce gynes, whereas other colonies predominately produce males (Beekman and van Stratum, 1998; Bourke, 1997; Duchateau and Velthuis, 1988; Muller et al., 1992). The ecological and sociobiological factors underpinning these strategies will be discussed below. Males generally do not stay in the nest but disperse a few days post-emergence to establish mating territories, while gynes reside for a period of time in their maternal nest before dispersing, mating, and seeking an appropriate location to diapause (Goulson, 2003).

1.2 The social history of the *B. terrestris* colony

Within the colony's *eusocial phase*, there are several distinct periods and events that have been revealed by behavioural observation of laboratory-reared colonies of *B. terrestris* (Duchateau and Velthuis, 1988). The first is the *harmonious period* during which the queen is the sole egg-layer and the workers assume all other nest duties. Behaviourally, this period is characterized by no observable conflict between the queens and workers, and an exponential ergonomic colony growth through the production of workers. It can also be considered the *pre-competition phase* (pre-CP, see below). The next event occurs when the queen switches her oviposition strategy, from laying diploid, female-destined, to haploid, male-destined eggs. Accordingly, the initiation of this phase was termed the *switch point*. Colonies that show protandry are described as 'early switch' colonies, and those exhibiting protogyny as 'late switch' colonies (Duchateau and Velthuis, 1988). Towards the end of the colony life cycle, nest harmony is disrupted because workers that possess activated ovaries concomitantly lay eggs and become aggressive not only towards the queen but also towards other reproductive workers. This period is termed the *competition phase* (CP). The CP can be readily recognized by the presence of several open egg cells. Typically, egg cells are sealed immediately after oviposition, but during the CP, competing workers or the queen may open each other's egg cells to cannibalize the eggs. The CP is also highly correlated with the switch in the developmental fate of the diploid larvae from worker-destined to queen-destined development (caste determination) (Cnaani et al., 2000a; Duchateau and Velthuis, 1988). The CP and emergence of reproductives mark the end of colony cycle, and after males and gynes disperse, the colony deteriorates rapidly.

1.3 Alternative life history strategies

The above described life and social histories are probably characteristic of most temperate bumble bee species, but deviations from the strict annual life cycle have been reported for several bumble bee species (Buttermore, 1997; Plowright and Lavert, 1984 and references therein; Sakagami, 1976; Skyrm et al., 2012; Stelzer et al., 2010). Indeed, even *B. terrestris* may be occasionally bivoltine under lab conditions (Beekman et al., 1999). How these alternative life history strategies impact the various social phases in the colony cycle remains to be determined. The social structure of colonies of the tropical species, *B. atratus*, for example, is quite distinct (Cameron and Jost, 1998; Garofalo et al., 1986; cited in Sakagami, 1976; Zucci, 1973). Colonies of this species seem to be perennial and present year round. One colony survived in the laboratory for 10 years, producing reproductives two to three times a year. While some of the mated young queens dispersed, other returned to the nest (in which the old queen died) creating a temporary polygyne nest. Polygyny however did not last long as the queens fought until a single reproductive queen remained. There is no mention of either worker aggression or worker reproduction. Unfortunately, these two detailed studies with *B. atratus* comprised only a single colony each, preventing any generalization.

1.4 Ultimate mechanisms shaping bumble bee social behaviour

Kin selection and inclusive fitness theory are considered the primary selection forces that shaped sociality, particularly in eusocial insect societies (Hamilton, 1964, 1972). However, this was recently challenged (Allen et al., 2013; Nowak et al., 2010; Wilson and Nowak, 2014), with the argument that relatedness among members of a eusocial colony is the consequence rather than the cause for sociality. Subsequent analyses demonstrated various flaws in this argument and re-validated the robust theoretical and empirical data supporting Hamilton's inclusive fitness theory for the evolution of eusociality (Abbot et al., 2011; Bourke, 2014). Inclusive fitness theory takes into account not only the degree of relatedness between the co-operating individuals but also group (or colony) productivity that results from such co-operation. Thus, for social insect colonies, reproductive output in terms of male and gyne production is a major factor (colony level selection), in addition to genetic relatedness, which affects the selection of social traits (Hammond and Keller, 2004). Thus, though relatedness may favour traits such as worker reproduction or queen nepotism, these traits cannot prevail if they hamper colony reproductive output because such colonies have lower fitness (reproductive output) than other conspecific colonies that do not express this trait. The evolution of social traits in social insects therefore reflects a delicate balance between individual and colony level selection. Below, we will discuss the fundamental role inclusive fitness and the resulting kin selection theory plays in the evolution of bumble bee social behaviour.

Most bumble bee species that have been studied are monogyne (a single queen per colony), the queen of which is monandrous (singly inseminated)

(Estoup et al., 1995; Owen and Whidden, 2013; Schmid-Hempel and Schmid-Hempel, 2000). As in all Hymenoptera, bumble bees have a haplodiploid sex determination system, where males are haploid and females are diploid. The queen is thus equally related to her male and female progeny ($r=0.5$) but is less related to worker-born males ($r=0.25$). She is therefore selected to monopolize the production of all progeny, males and females, as well as develop mechanisms that suppress worker reproduction. Workers, on the other hand, have more complex genetic gain. They are more related to their sister gynes ($r=0.75$) than to their would-be daughter gynes ($r=0.5$). It is therefore advantageous for the workers to rear sister gynes, which aligns with the queen's interest, resulting in harmony for gyne (and worker) production. In the case of male production, however, a worker is more related to her sons ($r=0.5$) than to her nephews (another nestmate worker's son $r=0.375$) or brothers (queen-born sons $r=0.25$) (Hamilton, 1972; Trivers and Hare, 1976). Therefore, kin selection predicts that workers are selected to compete with the queen and their sisters over male parentage. However, there are two limitations for worker reproduction: (1) workers must wait for the production of sufficient worker force before attempting to reproduce (according to the efficiency theory; Hammond and Keller, 2004) and (2) since workers gain more by rearing sister gynes than their own sons, they should be selected to defer reproduction until at least several diploid larvae became committed to gyne development (Alaux et al., 2005; Bourke and Ratnieks, 2001; Duchateau et al., 2004). Accordingly, queens can dominate male production in one of two ways: either coercively inhibit worker reproduction (queen control hypothesis; Keller and Nonacs, 1993) or control larval developmental fate and defer gyne production as late as possible in the colony cycle (given resource availability), thereby reducing the time period during which worker reproduction is favoured.

2. KIN CONFLICT OVER SEX RATIOS IN BUMBLE BEES

Inclusive fitness theory also provides a framework for explaining variation in male versus gyne production among bumble bee colonies. According to Trivers and Hare (1976) model of the sex ratio for the social Hymenoptera, a single, outbred, singly mated queen that is the sole reproducer (as in most bumble bee species) should equally invest in males and females (1:1), while workers are predicted to prefer females over males in a ratio of 3:1, consistent with the ratio of their relatedness to the reproductives ($r=0.75$:0.25 sister gynes:brothers). While sex investment ratios should be either even or female biased (and accordingly closer to the queen or worker optimum), it is in practice clearly male biased (Bourke, 1997; Duchateau et al., 2004; Owen et al., 1980; Yeninar et al., 2000), with the early switching, protandrous colonies producing most of the males (Duchateau and Velthuis, 1988), posing an evolutionary question of both the proximate and ultimate reasons of this phenomenon. It was suggested that male bias is the result of worker reproduction in orphaned colonies (Owen et al., 1980), but male bias is present in both orphaned and queenright (QR) colonies (Paxton et al., 2001) and most of the males are queen-born (Alaux et al., 2004b; Brown et al., 2003a; Owen and Plowright, 1982; Paxton et al., 2001). It was further suggested that sexual selection for protandry in bumble bees leads to male-biased sex investment ratios (Bourke, 1997), and since both protandrous and protogynous colonies may exist within a protandrous population, it may result in bimodal sex variation with colonies producing either mostly males or mostly females, also termed as 'split sex ratio' (Beekman and van Stratum, 1998; Duchateau et al., 2004). Producing a queen is 1.69–2.11 times as costly as producing a male (Duchateau et al., 2004). Thus, a split sex ratio is hypothesized to be the result of optimal allocation of resources, with colonies under low resource conditions showing a male-biased investment ratio, whereas colonies under high resource conditions allocate more resources towards females (Beekman and van Stratum, 1998).

Thus, ecological and social constraints interact to shape the annual life cycle of *B. terrestris*. Since this species cannot store resources for long-term survival, the colony is unable to survive the harsh winter and must have an annual life cycle. Thus, neither the queen nor the workers have interest in deferring the conflict to later years (Bulmer, 1981), which allows aggression to dominate the final phase of the colony cycle. Finally, resource availability can determine the timing of the switch point and CP within the colony and the resulting sex ratio. Thus, this malleable annual colony cycle provides the researcher with an excellent opportunity to study the proximate mechanisms pertaining to queen–worker conflict.

3. BEHAVIOURAL, PHYSIOLOGICAL, AND GENETIC TRAITS ASSOCIATED WITH THE SOLITARY PHASE

3.1 Pre-mating and mating behaviour

The bumble bee queen undergoes several behavioural and physiological changes in the period after emergence and before diapause. The newly emerged adult queens do not participate in colony tasks (Roseler and Roseler, 1988). Instead, they spend their first days accumulating fats and carbohydrates that are crucial for their ability to survive the winter diapause (Alford, 1969a; Fliszkiewicz and Wilkaniec, 2007). When they are 6–7 days old, the gynes leave their mother colony (Roseler and Van Honk, 1990) and are attracted to a territorial pheromone produced by the labial glands of conspecific males who patrol in relatively fixed flight routes (Ayasse and Jarau, 2014; Bergman and Bergström, 1997; Kullenberg et al., 1970). Copulation is triggered by sex pheromones produced by the gynes' mandibular glands (Krieger et al., 2006; van Honk et al., 1978). In most of the studied bumble bee species, queens mate once (Estoup et al., 1995; Paxton et al., 2001; Schmid-Hempel and Schmid-Hempel, 2000). Among eight European species that were examined (including *B. terrestris*), only *Bombus hypnorum* was reported to be facultatively polyandrous (Brown et al., 2003a; Schmid-Hempel and Schmid-Hempel, 2000). Among the 11 North American species studied, 5 were reported to mate with multiple males (*Bombus bimaculatus*, *Bombus impatiens*, *Bombus mixtus*, *Bombus citrinus*, and *Bombus ternarius*) (Payne et al., 2003). *Bombus ignitus*, the species most studied in East Asia, was found to be monandrous (Takahashi et al., 2008), and in the tropical species *Bombus wilmattae*, seven of nine queens were singly mated and two were doubly mated (Huth-Schwarz et al., 2011b). Multiple mating can be advantageous to males (increased reproductive success), to queens (increase fertility), and to the colonies they produce (increased genetic diversity of the workers can improve efficiency and decrease pathogen load) (Arnqvist and Nilsson, 2000). Multiple mating can also have negative effects, and indeed, multiply mated queens had lower performance (diapause survival, female longevity, and fitness) as compared to singly mated queens (Baer and Schmid-Hempel, 2004). Indeed, although some bumble bee species are multiply mated, the effective mating number in bumble bees is close to one (Strassmann, 2001), suggesting post-mating mechanisms by which the sperm of one male is favoured. An inefficient mating plug that the male transfers to the queen during copulation and a trade-off between mating duration and the likelihood of re-mating were suggested as proximate factors underlying multiple mating in bumble bees (Brown et al., 2002; Payne et al., 2003).

Mating causes extensive short- and long-term physiological as well as behavioural modifications in females (Chapman et al., 1995; Gillott, 2003; Kocher et al., 2008). However, the mechanisms underlying post-mating changes in bumble bee queens have not been explored in great detail.

Ovarian activation (demonstrated by the average of the terminal oocytes in the ovaries, Fig. 2) is not induced by mating events (unlike in honey bee queens; Kocher et al., 2008); rather, ovaries remain inactive until diapause terminates, several months after mating. Furthermore, when kept separated from their natal colonies under laboratory conditions, virgin queens can lay haploid eggs that develop into males (Amsalem, E., personal observations). Levels of juvenile hormone (JH), the apparent gonadotropin in *B. terrestris*, are low in virgin queens and remain low in mated and diapause queens, increasing only several days after the termination of diapause (Larrere et al., 1993). The same is true for ecdysteroids (the exact identity of the ecdysteroid has not yet been elucidated for bumble bees), which remain low before and after mating (Geva et al., 2005). Although most of the queens that survive diapause are mated (Alford, 1969b), mating is not obligatory for entering diapause, and unmated queens are able to survive diapause (Greeff and Schmid-Hempel, 2008). Indeed, mating may have negative effects, since mated queens have reduced longevity and increased melanization

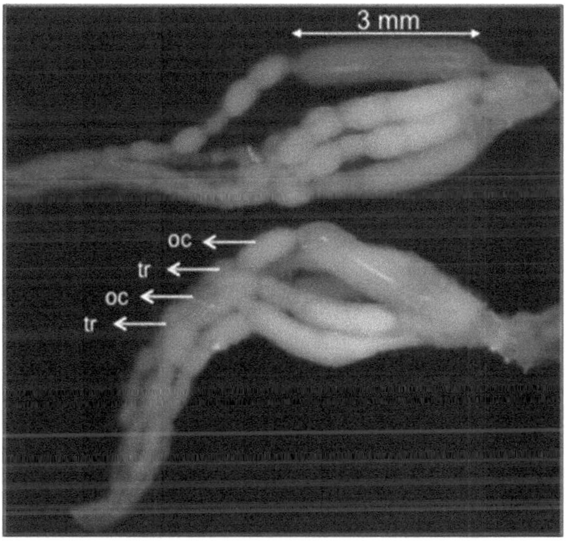

Figure 2 Ovaries of *Bombus terrestris* females. Workers and queens possess four ovarioles per ovary, for a total of eight ovarioles per individual. The length of the terminal oocyte is used to determine ovarian status. Either the average of the three largest oocytes (at least one ovariole per ovary) or the average of all of the eight terminal oocytes is equally effective in determining ovarian activation. Ready-to-lay eggs are approximately 3 mm long. Resorption of eggs appears as deformation in the shape of the terminal oocytes. Oc, oocyte (egg cell); tr, tropocyte (nursing cell).

of their spermatheca compared to virgin queens (Greeff and Schmid-Hempel, 2008), and multiply mated queens have reduced survival during diapause (Baer and Schmid-Hempel, 2004). Overall, mating was reported to have more negative than positive effects on queens, in contrast to the hypothesis that sexual conflict between males and females should be low in monandrous species (Strassmann, 2001).

3.2 Diapause in newly mated queens

Shortly after the queens mate, they enter a winter diapause that can last 6–9 months (Alford, 1969b). The factors regulating the onset and termination of diapause are still elusive. While diapause in most insects is generally triggered by temperature and/or day length (Denlinger, 2002; Denlinger et al., 2012; Sim and Denlinger, 2013), the onset of diapause in bumble bees is not likely to be induced by day length, while the role of temperature changes as prime cause for entering or exiting diapause remains to be examined (Alford, 1969b; Yeninar et al., 2000).

Environmental conditions, such as temperatures and diapause duration, interacting with queen nutritional stores, seem to be largely responsible for queen survival during diapause. *B. terrestris* queens exhibit lower survival rates at longer diapause (Beekman et al., 1998; Gosterit and Gurel, 2009; Yoon et al., 2010), whereas diapause temperature had either a negative effect on queen survival (Yoon et al., 2010) or no effect at all (Beekman et al., 1998). Although queens of *B. terrestris* below a threshold weight of 0.6 g did not survive diapause (different bumble bee species may have a different thresholds as function of the temperature of their habitats), above that threshold, no correlations were found between body mass and survival (Beekman et al., 1998) or between post-diapause queen weight and colony development (Gosterit and Gurel, 2007). Thus, queens need to reach a minimal threshold weight to survive diapause. This may also explain why workers cannot survive diapause: their small size does not allow them to store enough reserves to survive the long winter. Since the queen and worker castes only differ in body size, with queens being three times larger than the largest workers (Goulson, 2010; Michener, 1974), selection for body size, and thus the ability to survive diapause, was probably the first step in the divergence of castes in bumble bees and other primitively eusocial insects that go through winter diapause (Hunt et al., 2007). Indeed, size differences between the castes are simply the result of longer instar duration in prospective queen larvae (Cnaani and Hefetz, 2001). Thus, *B. terrestris* workers are basically dwarf queens and size differences may be the result of a queen manipulation during larval development. Indeed, close contact with the queen was found to influence larval development and worker body size (Shpigler et al., 2013) and bees that were cared for by 10 workers without a queen were significantly larger than bees cared for by 10 workers with a queen (Shpigler et al., 2013).

Queens use a variety of nutritional reserves to survive diapause. The accumulation of sufficient quantity of fat body reserves prior to diapause, mainly lipids and glycogen, is critical (Fliszkiewicz and Wilkaniec, 2007). Beside glycogen and lipids, queens also store nectar in their crop prior to entering diapause. Up to 80% of the reserves are consumed during diapause (Alford, 1969b). It was suggested that the metabolism, although suppressed during diapause, still responds to temperature, which is consistent with the observed higher consumption of reserves during the first half of diapause, when temperatures are also higher. However, different temperatures during diapause in *B. terrestris* did not have a significant effect on queen survival (Beekman et al., 1998; Vesterlund et al., 2014; but see Yoon et al., 2010).

When mated queens are treated with CO_2, a method that was developed to bypass diapause and induce oogenesis (Roseler, 1985), their survival is higher and the time to initiate egg laying is shorter compared to non-narcotized non-diapause queens (Tasei, 1994). It was suggested that CO_2 narcosis inhibits the formation of fat body reserves and increases the size of the corpora allata (CA), the source of the gonadotropin JH (Roseler, 1985; Roseler and Roseler, 1984). A recent study corroborated that CO_2 narcosis decreases lipid mass and increases JH levels, as well as affects the expression levels of selected candidate genes related to reproduction, hormone response, immunity, stress, and neural response (Amsalem, E., Galbraith, D., Cnaani, J., Teal, P., Grozinger, C.M., unpublished data). However, the pertinent mechanisms underlying the effects of CO_2 narcosis remain to be elucidated.

Different diapause lengths or the lack of diapause (in either 'natural' non-diapausing (Beekman et al., 1999) or CO_2-treated queens) affects not only queen survival and physiology but may also affect long-term characteristics of colonies headed by such queens, including colony size and the number of males and gynes produced. However, several studies provide contradicting results pertaining to these effects. Lack of diapause affected queen characteristics; non-diapause non-narcotized queens produced the lowest number of workers but the highest number of gynes (Beekman and van Stratum, 2000) and non-diapause narcotized queens produced significantly more workers, males, and gynes than the diapause queens (Gosterit and Gurel, 2009). While some studies did not find any differences between colonies headed by queens that were subjected to different diapause lengths (Beekman and van Stratum, 2000; Beekman et al., 1998; Gosterit and Gurel, 2009), another study found several differences: queens under longer diapause produced more brood during the first and the second generations, reached the switch point earlier, and produced colonies that reared more males and fewer gynes (Duchateau et al., 2004), consistent with the idea that colonies under low resource conditions are male biased (Beekman et al., 1998). However, in more recent studies, where the effect of diapause length on the number of reproductives produced was tested in controlled environments (under different colony size and photoperiod regimes), the effects seemed to be much more complicated: both colony size and diapause length affected the number of gynes (but not the number of males) and queen longevity in various ways and no interaction was found between colony size and diapause length (Amin et al., 2011). In a second study conducted by the same authors, gyne production was dependent on photoperiod but not on diapause duration, and both factors affected male production (Amin et al., 2007b). Different temperatures and photoperiodic regimes lead to different amount of weight loss in queens (Amin et al., 2007a; Yoon et al., 2010), and heavier queens produce larger brood in the first generation, but other colony parameters are not affected (Gosterit and Gurel, 2007). Several other factors such as pathogen and parasite loads may impact the queen post-diapause performances. For example, *Crithidia bombi*, a bumble bee parasite, had strong effect on survival and colony development: the pathogen was found to decrease the ability of queens to successfully establish a colony, colony size, number of males, and total fitness (Brown et al., 2003b). Infection with *Nosema bombi*, a gut parasite, did not affect colony structure but decreased the number of queens that successfully established a colony (Steen, 2008).

3.3 Genomic mechanisms underlying the queen solitary phase

Genomics studies of bumble bee queen behaviour and physiology have largely focused on the early stages of the queen's life cycle. A whole-body transcriptomics analysis demonstrated that several of the top 10 genes whose expression best differentiated virgin queens from males and workers are involved in nutrient storage (e.g. hexamerins) and fatty acid biosynthesis (Colgan et al., 2011); differential expression of these genes may prepare queens for diapause. Expressed sequence tag analysis of diapausing queen in *B. ignitus* revealed changes in expression levels of several candidate genes (Kim et al.,

2006). Another study focusing on three heat shock protein genes (*shsp*, *hsc70*, and *hsp90*) found dynamic changes in expression pattern that varied across the different queen tissues during diapause (Kim et al., 2008). We recently demonstrated that ~120 genes are consistently differentially expressed (either up- or down-regulated) in the fat bodies of diapausing queens relative to both mated and foundress queens, and these genes function in nutrient storage, metabolic processes, and stress resistance (Amsalem, E., Galbraith, D., Cnaani, J., Teal, P., Grozinger, C.M., unpublished data). Gene expression patterns in foundress queens differ substantially from virgin and mated queens in both the brain (Woodard et al., 2014) and fat body (Amsalem, E., Galbraith, D., Cnaani, J., Teal, P., Grozinger, C.M., unpublished data), respectively. However, when comparing laying queens in young colonies with and without workers, though the presence of workers substantially changed queen behaviour (queens laid more eggs and tended brood less), there was little effect of worker presence on queen brain gene expression (Woodard et al., 2013). Overall, these data indicate that there are substantial differences in gene expression profiles associated with the different physiological states that queens experience during their lives (virgin, mated, diapausing, egg laying), but other factors (perhaps changes in neuroendocrine signalling) underpin behavioural changes in laying queens in young colonies (e.g. the transition from both reproduction and maternal care to reproduction only). However, gene expression differences between pre-CP and CP queens have not been assessed, and this would be an exciting area for future research, particularly if queen-produced pheromones mediate caste differentiation in larvae. In honey bees, transcriptional profiles in pheromone-producing exocrine glands are strongly correlated with social environment (Malka et al., 2014).

4. BEHAVIOURAL, PHYSIOLOGICAL, AND GENETIC TRAITS ASSOCIATED WITH THE EUSOCIAL PHASE

As previously discussed, the *eusocial phase* in the bumble bee colony can be divided into two periods: the pre-CP and the CP. The pre-CP is characterized by social harmony and a complete reproductive division of labour between the queen and workers. In contrast, the CP is characterized by overt conflict over reproduction between the two castes. This provides an excellent opportunity to test evolutionary hypotheses stemming from kin selection as well as investigate the underlying proximate mechanisms.

There are two contrasting hypotheses regarding the ultimate cause for the sharp reproductive skew in social insects. The first was termed 'queen control' because it postulates that the queen coercively inhibits worker reproduction, either by behavioural or by pheromonal means. The second, termed 'queen signal' (or 'worker control'), postulates that workers have control over their own reproduction and behave, reproductively, in a manner that maximizes their own inclusive fitness, and the queen produces an honest signal of her reproductive abilities (Keller and Nonacs, 1993; Kocher and Grozinger, 2011). Since it is very difficult, if not impossible to discern between these two ultimate causes by proximate experiments, we will refrain from deciding which prevails but discuss the queen's effect on worker reproduction without referring to the ultimate cause.

4.1 Who wins the conflict over male production?

When discussing the outcome of the queen–worker conflict over male parentage, a distinction, at least for *B. terrestris*, between the pre-CP and the CP must be considered. All previous studies demonstrate that before the CP all eggs, irrespective of ploidy, are queen-produced. However, even in the pre-CP, a proportion of the workers possess activated ovaries (for *B. terrestris*, Alaux et al., 2004b; Amsalem et al., 2009; Duchateau and Velthuis, 1989; for *B. impatiens*, Cnaani et al., 2002; Jandt and Dornhaus, 2009) and thus apparently refrain from egg laying. Queen domination of male production also seems to prevail after the competition point. Behavioural observations followed by male genotyping showed that although over 38% of the workers were egg-layers, over 95% of the adult males were queen-born (Alaux et al., 2004b). This was also confirmed both by behavioural observation studies of laboratory-reared colonies (Duchateau and Velthuis, 1989) and by molecular studies of field-collected colonies (O'Connor et al., 2013). Two reasons for the queen domination of male production in *B. terrestris* were provided. First, the queen is far more fecund than workers and performed more active oophagy compared to workers (Alaux et al., 2004b; Bloch and Hefetz, 1999b). Second, during the CP, workers preferentially devour other worker's eggs, but not the queen's eggs (Zanette et al., 2012).

The outcome of queen–worker conflict over male parentage is more variable in the other species of bumble bees that have been examined. Queen domination of male production also seems to occur in *Bombus melnopygus* (Owen and Plowright, 1982) and *Bombus polaris* (Richards, 1977), but there was no information regarding the temporal dynamics of egg laying (e.g. before or during the CP). However, in the case of *B. hypnorum*, there is evidence that workers can successfully produce male eggs, though this may vary with queen mating number. In the first study (Paxton et al., 2001), the queen was multiply mated (effective mating frequency of 1.26 harmonic mean) and male genotyping revealed that they were all queen-produced, while worker-laid eggs were presumably destroyed by the queen (under this effective mating frequency, worker policing is not selectively favourable; Ratnieks, 1988). In the second study (Brown et al., 2003a), all queens were singly inseminated, and over 20% of the males produced in QR colonies were worker derived. Thus, in colonies headed by singly mated queens, *B. hypnorum* workers are better competitors than *B. terrestris* workers. Finally, in the case of neotropical bumble bees, workers dominate male production in both *B. wilmatte* (Huth-Schwarz et al., 2011a) and *B. attratus* (Zucci, 1973). Thus, comparative studies with additional bumble bee species may provide great insight into the mechanisms mediating queen–worker conflict.

4.2 Underlying hormonal mechanisms and queen effect on worker reproduction

Before addressing the relative roles of the queen and workers in the above conflict and its resolution, it is useful to describe briefly the hormonal mechanisms underlying ovary activation and reproduction in *B. terrestris*. Several independent studies have demonstrated that in *B. terrestris*, similar to many non-social insect species, JH acts as a gonadotropin. In gynes, JH titers are low, which largely affects the queen's metabolism, directing it into lipid and carbohydrate storage rather than ovary activation. Experimentally elevating JH blocks the synthesis of these storage products and enhances ovary activation (Roseler and Roseler, 1988). Since JH titers are constantly high in functional, egg-laying queens (Bloch, 1997; Bloch et al., 2000a), it is more informative to study the role of JH and its dynamics in workers before and during ovary activation. Early studies have shown that when workers are rendered queenless (QL), JH levels rise (Roseler, 1977; Roseler and Roseler, 1978), but these studies used the *Galleria* bioassay, which is only semi-quantitative. A more quantitative assessment of JH levels was subsequently performed using the highly sensitive radioimmunoassay and examined both JH biosynthesis and secretion rates by the CA and its titers in the hemolymph (Bloch et al., 1996, 2000a). The gonadotropin role of JH was also recently confirmed in experiments where the CA were either surgically removed (Shpigler et al., 2014) or impaired by application of precocene-I (Amsalem et al., 2014c). All the above studies demonstrated that when workers are rendered QL, JH biosynthesis and JH hemolymph levels increase which is followed by ovarian activation. Furthermore, activation of the CA and subsequent increase in JH titers induces the biosynthesis of the egg-yolk protein vitellogenin (Vg), which is then incorporated into the developing oocyte (Roseler and Roseler, 1988). However, the exact relationship between ovary activation, JH and Vg biosynthesis, and particularly the direct regulation of JH on Vg levels, are not fully resolved yet and will be discussed in Section 5.5.

Since in QR workers JH levels are low and ovaries inactive, it was postulated that the queen effect is manifested by inhibiting the activation of the CA and thereby blocking the subsequent cascade of events (reviewed in Roseler and Van Honk, 1990). To explain worker reproduction during the CP, it was postulated that the queen loses her inhibitory capacity, allowing the workers that have already developed ovaries, the so-called elite workers, to start laying eggs (Doorn van and Heringa, 1986). However, studies demonstrated that workers housed with mated pre-CP and CP queens had reduced JH titers and inactive ovaries relative to QL workers or workers houses with virgin queens (Bloch et al., 1996). Thus, in small groups, the inhibitory effect of the queen is intact. It is possible that as worker population increases near the end of the colony cycle, the queen's inhibitory capacity is diluted, which would be the case if the queen used aggression or pheromones to inhibit worker reproduction (see below, for further discussion). In a subsequent set of experiments, Bloch and Hefetz (1999b) measured worker reproduction, JH levels, and oocyte development in workers that were introduced into intact colonies at various stages of colony social development, before and after dequeening. In the pre-CP colonies before dequeening, both JH levels and oocyte activation in the introduced workers were low, indicative of reproductive inhibition. In contrast, after dequeening, workers introduced into pre-CP colonies had high JH and many of them had chorionated, ready-to-lay eggs. When introduced into colony at the verge of the CP, 90% of the introduced workers had activated ovaries, and a few had chorionated eggs even in the presence of the queen. Finally, nearly all workers introduced into colonies during the CP had low reproductive capacities, only about 4% had chorionated eggs. There was no difference in the distribution of ovary activation among the introduced workers whether the queen was present or absent. This indicates that during the CP, the workers that are already reproducing can inhibit reproduction of their sisters. The latter was further confirmed by the finding that old reproductive workers inhibit reproduction of introduced callow workers in small QL groups (Bloch and Hefetz, 1999b).

4.3 The chemistry underlying queen–worker conflict over reproduction—Is there a queen pheromone?

The queen mandibular gland secretion was the first candidate tested for its possible mediation of the queen inhibitory effect. Colonies headed by queens with extirpated mandibular glands started egg laying much sooner (i.e. the competition point was advanced) than the control colonies headed by sham-operated queens (van Honk et al., 1980). In a subsequent experiment, there was a significant effect of the mandibular gland extract or total body wash on the volume of the CA of workers as well as the rate of JH biosynthesis *in vitro* (Roseler et al., 1981). Since in these studies queen mandibular gland extracts or cuticular washes had similar inhibitory effects to those of live queens, it was suggested that the putative queen pheromone is produced in these glands and then spread over the queen body by self-grooming (Roseler and Van Honk, 1990). The mandibular gland's secretion is composed mostly of a series of 3-hydroxy acids (Hefetz et al., 1996) accompanied by lesser amounts of alcohols, 2-ketones, alkenes, alkanes, and fatty acyl esters. In an attempt to pinpoint the chemicals that may be involved in queen inhibition of worker reproduction, a re-evaluation of the queen's glands, body extracts, or washes were tested by assaying *in vitro* JH production by the CA (Bloch and Hefetz, 1999a). Using a Soxhlet washed queen (a method that thoroughly removes the lipids from the body surface) as a dummy onto which the secretion was applied, it was demonstrated that none of the glandular secretions was effective in inhibiting the activation of the CA in QL workers. This refuted the prevailing hypothesis that the mandibular glands are the sole source of such a pheromone. Notwithstanding, both body extract and body washes had an inhibitory effect on CA activation. A comparative study of cuticular washes of nest-searching queens (not reproductive yet) and egg-laying queens showed distinct cuticular profiles, suggesting that the latter possess a fertility signal (for review of the chemical ecology of bumble bees, see Ayasse and Jarau, 2014; Sramkova et al., 2008). It was further demonstrated that although egg-laying workers were distinct from the queens, their cuticular profile was more similar to egg-laying queens than non-laying workers, supporting the existence of fertility signals. While this comparative study provided correlative evidence for a queen pheromone, the direct causal experiment to test any of the compounds or the blends was not performed. A recent study tested the effect of several compounds of the above blends that were significantly more abundant in queens than in workers (Van Oystaeyen et al., 2014). Of these only pentacosane showed some activity, it acted to enhance oocyte resorption but not to inhibit ovary activation. Since multiple studies have shown that the queen does inhibit ovary activation in pre-CP, which pentacosane seemingly failed to mimic, there are likely additional pheromones involved. Moreover, as described above, during the CP, there is also high competition among workers, and ovary resorption could have resulted from worker–worker inhibition. Finally, there is no information in the study regarding whether all workers were dissected and what proportion had developed ovaries. This is important since, as described above, not all the workers in the colony develop ovaries even during the CP.

Other exocrine glands that were chemically analysed in queens under various social conditions are Dufour's and labial glands. There were no qualitative differences in Dufour's gland secretion of pre-CP compared to CP queens (Amsalem et al., 2009). In the labial gland, there were quantitative differences in the quantities of dodecyl esters between pre-CP and CP queens (Amsalem et al., 2014a). However, since this secretion had no effect on worker reproductive inhibition (Bloch and Hefetz, 1999a), these differences may reflect a different function (see discussion in Section 6.7).

Additional studies examined the nature of the queen signal, and whether *B. terrestris* uses a queen control versus queen signal system. To address this question, nests were split into two compartments: one contained the queen brood and workers (queenright compartment, QRC), while the second contained a group of five workers (queenless compartment, QLC). Groups of five workers housed in a separate nest box served as completely QL control (Alaux et al., 2004a). In the first experiment, the compartments were separated with a double screen so that all contact between the QRC and QLC was prevented, but emitted volatiles could spread into both compartments. Under these conditions, workers in the QLC possess fully activated ovaries 7 days post-group establishment (the time needed for oocyte maturation), which was equivalent to the QL control group. Workers in the QRC laid eggs only after about 3 weeks, as expected from the timing of the CP in non-manipulated colonies. Thus, if a queen pheromone exists it is clearly not volatile. In the second experiment, the compartments were separated by a queen excluder that allowed the workers, but not the queen, to travel between compartments. To entice workers to travel between compartments, the food supply was placed in the queen-excluded compartment (QEC). The hypothesis was that if the queen coercively inhibits worker reproduction, the QEC could provide a refuge for workers wishing to escape the queen's control. However, all workers moved back and forth between compartments indicating that none 'sought refuge' from the queen.

Moreover, ovary activation of the workers was inhibited relative to workers that were completely QL, and although egg laying occurred in the QEC earlier than in the QRC, it was still near the CP and much later than egg laying by workers that were completely QL. These results lend credence to the hypothesis that workers control their own reproduction and *B. terrestris* uses a 'queen signal'. In a subset of this experiment, five workers were confined to the QEC by glueing a bar to their thorax to prevent them from moving between compartments. These workers behaved like QL workers in that they established a dominance hierarchy among themselves and laid eggs about a week after group establishment. These results indicate that the information of the queens' presence has to be individually acquired by each worker and cannot be transmitted between workers. It further showed that even when egg laying by workers occurs in a maladaptive timing with respect to other workers in the colony, they do not police either the egg-laying workers or their laid eggs.

4.4 Genomic factors underlying queen–worker conflict

Studies examining the genes that regulate queen–worker conflict in bumble bees have largely focused on *Krüppel-homolog 1* (*Kr-h1*). *Kr-h1* is a zinc finger transcription factor whose brain expression levels are down-regulated by queen presence in *B. terrestris* workers (Shpigler et al., 2010) and queen pheromone in honey bees (Grozinger and Robinson, 2007; Grozinger et al., 2003). The effects of the queen on *Kr-h1* expression levels are likely mediated by JH. JH titers are lower in QR versus QL workers in both bumble bees (Bloch et al., 2000a) and honey bees (Hartfelder et al., 2015; Pankiw et al., 1998). Furthermore, *Kr-h1* RNA levels in the brains and fat bodies of bumble bee (Shpigler et al., 2010, 2014) and honey bee workers (Grozinger and Robinson, 2007) are up-regulated by JH treatment. In other insects, it has been demonstrated that transcription of *Kr-h1* is activated when JH binds to the JH receptor complex (consisting of the methoprene-tolerant transcription factor and its steroid receptor co-activator, with facilitation from Hsp83), and this complex binds to the E-box motif in the promoter region of *Kr-h1 in Locusta migratoria* (Song et al., 2014), *Bombyx mori* (Kayukawa et al., 2012), *Aedes aegypti* (Cui et al., 2014), and *Drosophila melanogaster* (He et al., 2014). Studies in honey bees suggest that *Kr-h1* expression may also be regulated by cGMP (Fussnecker and Grozinger, 2008). In other insects, *Kr-h1* mediates many of JH's effects on developmental processes (Minakuchi et al., 2009), vitellogenesis, and oogenesis (Song et al., 2014), and also can affect neuronal morphology (Shi et al., 2007). As a transcription factor, *Kr-h1*'s effects are due to its ability to modulate expression of other genes, including the transcription factor *E93* (Belles and Santos, 2014), though it does not appear to directly regulate expression of *vitellogenin* (*vg*) (Song et al., 2014). Recent studies in *B. impatiens* have demonstrated that *Kr-h1* expression is down-regulated in workers exposed to both uncaged queens and caged queens (which cannot approach and attack workers), suggesting that, at least in this species, *Kr-h1* expression responds to queen-produced social cues aside from aggression (Padilla, M., Amsalem, E., Hefetz, A., Grozinger, C.M., unpublished data).

5. WORKER–WORKER CONFLICT OVER REPRODUCTION

As previously described, workers compete over male production during the CP not only with the queen but also among themselves (Duchateau and Velthuis, 1988). Bumble bee workers are all morphologically capable of reproduction (Fig. 2). They have the same number of ovarioles as the queen and can produce relatively high number of eggs in a short period (a group of 10 callow QL workers that were kept for 15 days laid altogether 97 eggs, Amsalem et al., 2013a). Although workers still possess spermatheca, they do not mate and thus produce haploid males to whom they are more genetically related ($r=0.5$) than to their brothers ($r=0.25$) or nephews ($r=0.375$). However, there are several constraints to the time-window available for worker reproduction. Seasonal limitations, such as the availability of resources needed to rear brood, predict that only male eggs that are laid early enough in the season will successfully develop into adulthood as well as being synchronized with the mating period of the gynes. Inclusive fitness constraints predict that workers will not start reproducing before gynes are produced by the colony (Alaux et al., 2006). Furthermore, both the queen and other workers will destroy worker-laid eggs: while queen-laid eggs are only rarely eaten during the CP (Zanette et al., 2012). Consequently, only a small portion of the eggs laid by workers will successfully develop into adult males (Alaux et al., 2004b; Brown et al., 2003a; Owen and Plowright, 1982; Paxton et al., 2001). Thus, although all workers are capable of reproduction, there is a clear advantage for workers who reproduce as early as possible, predicting a high competition between workers over male production even before the competition actually starts.

The competition among workers is largely regulated by aggression, dominance, and hierarchy formation. QL conditions and specifically small, same age groups of workers provide an excellent opportunity to (1) test dyadic relationships in more detail, (2) understand how reproduction is regulated, (3) unravel the role of aggression in mediating hierarchy formation, (4) explore how dominance behaviour may act as a driving force for more advanced social features such as division of labour, and (5) the study of QL groups (in particular larger groups) can also be extrapolated to understand events occurring during the CP in QR colonies, and thus, provide a powerful tool to examine social behaviour in bumble bees.

5.1 Rates of worker reproduction

Worker reproduction is widespread in Hymenoptera and is particularly common in bumble bees (Bourke, 1988). In B. terrestris, in line with the abovementioned theoretical predictions, it is limited and only around 40% of workers eventually lay eggs either during the CP or under QL conditions. Despite different approaches for estimating worker reproduction in the literature, similar results were obtained for entire colonies or small QL groups. For example, Alaux et al. (2004b) found 64% of the workers with ovaries in stage 3–4 (terminal oocyte size >1.27 mm) and 38% egg-layers in colonies. Duchateau and Velthuis (1989) found 36% of workers with ovaries in stage 4–5 (terminal oocyte size >2.43 mm) in colonies. Amsalem et al. (2009) found 42% of workers with terminal oocyte size larger than 1 mm in either QL or CP colonies, and Amsalem and Hefetz (2011) found 45% with terminal oocyte >0.5 mm in 5-day-old workers from same-aged QL groups containing 10 workers.

5.2 Worker reproductive hierarchy

Hierarchy among workers can be defined in terms of either aggressive behaviour or reproductive output. The two hierarchies are highly correlated but not necessarily identical, suggesting that they are regulated by different pathways that interact with each other. In QL groups, the most aggressive worker is commonly the most reproductive, with a 94% match between aggression and reproduction in paired bees (Amsalem and Hefetz, 2010) and an 80% match in three-worker groups (Amsalem et al., 2014c). Supporting the disparate regulation of behavioural and reproductive dominances is the finding that ovariectomized workers, although unable to lay eggs and do not construct egg cells, perform the same ritualistic dominance behaviours as non-operated dominant workers (Doorn van, 1989).

Consistent with worker–worker competition, dominant workers are able to inhibit ovarian activation in nestmates (Amsalem et al., 2013a; Bloch and Hefetz, 1999b; Cnaani et al., 2007; van Honk and Hogeweg, 1981). However, ovarian suppression is not absolute, and in larger groups, it is less effective; so instead of one worker dominating reproduction, a few workers will share egg laying (Amsalem and Hefetz, 2011). *B. impatiens*, given enough time subordinate workers in small groups (even in paired bees), can activate their ovaries and lay eggs. Moreover, a dominant worker that had many sons had also many nephews, indicating that a very fertile worker stimulates her group-mates to lay more eggs (Sibbald and Plowright, 2013, 2014).

Interestingly, both ovary activation in dominant workers and suppression in subordinate workers depend on physical contact: workers kept in pairs but separated by a single or a double mesh show low ovarian activation, but when given the opportunity to interact by removing the mesh, aggression ensues resulting in a clear reproductive hierarchy formation (Amsalem and Hefetz, 2010). Furthermore, workers seem to require these aggressive social stimuli to initiate reproduction as ovarian activation and egg laying are greatly delayed in isolated workers (Amsalem and Hefetz, 2010; Amsalem et al., 2009; Duchateau and Velthuis, 1989). Larger groups enhance ovarian activation in group-mates compared to smaller groups (Duchateau and Velthuis, 1989), but the effect is rank dependent: while the α-worker rapidly activates her ovaries to the maximum irrespective to the group size, workers of lower ranks in large groups activate their ovaries faster than same-rank workers in smaller groups (Amsalem and Hefetz, 2011), emphasizing the importance of social interactions for reproduction.

5.3 Aggression and hierarchy formation

Although overt worker–worker aggression is observed primarily in the CP, it is likely that worker reproductive and behavioural hierarchies are formed earlier in the colony cycle. Even during the pre-CP, a distinct group comprising around 40% of the workers tends to interact more frequently among themselves (and with the queen) than with the other workers, forming an 'elite worker group' (Doorn van and Heringa, 1986; van Honk and Hogeweg, 1981). This designation describes workers that have activated ovaries (though they do not appear to lay eggs), reside near the queen, and later, during the CP, become egg-layers and display the most aggressive behaviours towards other elite workers as well as the queen (Doorn van and

Heringa, 1986). The queen may maintain her α-position for a limited period (Doorn van and Heringa, 1986), but during the CP, she often loses her superiority and can die or be killed by the workers (Bourke, 1994). Upon queen loss, aggression and the formation of dominance hierarchy among workers also occur in full-fledged colonies (Amsalem et al., 2013a; Doorn van, 1986, 1989; Doorn van and Heringa, 1986; van Honk and Hogeweg, 1981; van Honk et al., 1981).

Dyadic interactions among workers is the foremost method used to describe the formation and maintenance of the dominance hierarchy. These interactions were incorporated into mathematical models to describe the complex relationship between individuals in entire colonies or large groups (Amsalem et al., 2013a; Hogeweg and Hesper, 1983; van Honk and Hogeweg, 1981) or used to quantify aggression in small QL groups in one of two indices: the 'aggression index' and the 'dominance index'. The aggression index scores three types of aggressive behaviours. The two that are considered threatening behaviours are *Humming*, defined as short bursts of wing vibrations directed at another bee, and *Darting*, defined as a sudden and directed movement towards another bee without making actual contact. The third behaviour is an escalated *Attack*, defined as physical contact resulting in biting, pushing, dragging, struggling, or attempted stinging. The aggression index is the sum of these three behaviours. Workers that score high on the aggression index possess higher level of ovarian activation, higher JH levels, higher *vg* expression levels in both fat body and head, and produced less (or no) sterility signal in their Dufour's gland (Amsalem and Hefetz, 2010, 2011; Amsalem et al., 2014b,c). When aggression is scored solely based on the number of physical attacks, it is strongly correlated with the number of observed oviposition events per bee (Doorn van and Heringa, 1986) and with ecdysteroid titers in the ovaries and hemolymph (Geva et al., 2005). For the dominance index, aggression is scored based on the number of advances versus retreats each individual bee displays in each dyadic encounter (Doorn van and Heringa, 1986; van Honk and Hogeweg, 1981). The dominance index is defined as [1 − (retreats/total encounters)] (Bloch et al., 1996). This dominance index yielded correlations with *Kr-h1* expression levels, a gene that is regulated by JH (Shpigler et al., 2010).

Levels of aggression change dynamically during the course of dominance hierarchy establishment in QL worker groups. There is a peak of aggression during the first days post-group establishment that declines until there is little aggression apparent once the hierarchy is established. Similar results were obtained using either the aggression index or the dominance index

(Amsalem and Hefetz, 2010, 2011; Bloch et al., 1996). Aggression is also highly variable among both individuals and groups, with some groups showing high aggression levels while others show none. Since aggression starts very early, it seemingly precedes both ovarian activation and most of the hormonal changes in workers (Amsalem and Hefetz, 2010). In most cases, a single bee performs most of the antagonistic behaviours (Amsalem and Hefetz, 2011). A simulation model based on *B. terrestris* biology aiming to test the interplay between aggression and reproduction suggested that aggression and the outcome of previous dyadic encounters are the major determinants of ovarian activation and not vice versa (Amsalem et al., 2013a). Workers may gain a head start in the competition by fortuitously winning their first encounters, or due to variation in their innate aggressive tendencies, or because of any other physiological and morphological differences between them.

5.4 Chemical communication versus dominance behaviour

The decreased aggression observed after dominance hierarchy establishment in QL worker groups may indicate that subordinate workers have succumbed to their role, and/or social signals, such as pheromones, have replaced aggression to maintain the social structure while minimizing the cost of aggression. Indeed, there are several lines of evidence that indicate that dominant workers, subordinate workers, and brood produce a variety of pheromone signals (reviewed in Ayasse and Jarau, 2014). Dominant workers can be discriminated from sterile workers by several chemical markers that are associated with either their fecundity, aggressiveness, or both. Group-specific odour bouquets (cuticular and gland extracts as well as headspace volatiles) are found in dominant versus subordinate workers of *B. hypnorum* (Ayasse et al., 1995). In *B. terrestris*, cuticular hydrocarbons were found to steadily increase in workers with activated ovaries during the CP (Sramkova et al., 2008) and cuticular lipids were associated with reproductive status (Blacher et al., 2013a). The latter led to higher rejection and aggression towards fertile versus sterile workers by non-nestmate workers (Blacher et al., 2013a). Furthermore, the Dufour's gland secretion in *B. terrestris* shows a caste- and reproductive-state-specific composition. While the queen's secretion is composed of alkanes and alkenes, workers possess an additional series of octyl esters. These octyl esters are also negatively correlated with ovarian activation, declining to nearly undetectable levels in workers with activated ovaries (Amsalem et al., 2009). When placed in pairs, aggression towards subordinates was negatively correlated with the proportion of octyl esters they produced (Amsalem and Hefetz, 2010), suggesting these esters signal sterility to dominant workers and queens, rendering subordinate workers safe from attacks. A similar reproductive-state- and caste-related secretion composition also exists in the labial glands of *B. terrestris*. Infertile individuals (both workers and virgin queens) produced a series of dodecyl esters and several other hydrocarbons, the amount of which is negatively correlated with ovarian activation (Amsalem et al., 2014a).

It is fascinating to consider the evolutionary mechanisms underlying the phenomenon of signalling sterility. Although advertising sterility or fertility are two sides of the same coin, it is intriguing that *B. terrestris* workers actively invest in producing sterility-specific compounds. While the advantage of evolving a fertility signal is evident, how advertising sterility confers an advantage to the signalling bees is not intuitively obvious. We suggest that in small-size eusocial colonies that exhibit high aggression towards workers with developed ovaries and where workers may have an equivalent reproductive potential, as in the case of *B. terrestris*, signalling of sterility confers a high advantage to all partners in reducing unnecessary harassment of bees that are not in competition for reproduction and thus enhances social harmony in the seemingly chaotic CP. Indeed, the sterility signal along with a possible fertility signal in workers (Sramkova et al., 2008) provides a means for fine-tuning the division of labour in CP colonies, enabling the successful rearing of reproductives (males and gynes) by allowing continuous provision by unintimidated foragers. Unfortunately, the glandular source of the observed fertility signal in *B. terrestris* is still elusive, preventing any investigation into the mechanisms underlying the evolution of such dual pheromonal regulation. Similarly, it is interesting that both the Dufour's and labial gland secretions appear to serve the same function. We can attribute such redundancy for fine-tuning the delicate regulation of division of labour. However, it is also likely that this is the result of ongoing queen–worker and worker–worker social conflicts and represents an escalated means to resolve these conflicts (i.e. an arms race). As one party evolve means (e.g. pheromones) to outcompete its competitors, the latter evolve means to neutralize them (e.g. developing insensitivity to the means), leading to the evolution of new means (new pheromones) and so forth (Katzav-Gozansky, 2006).

Dominance, reproductive status, and aggression may be also modulated by chemicals produced by developing brood. When pairs of QL *B. impatiens* workers are exposed to developing brood, they exhibit reduced frequency and duration of aggressive behaviours (Sibbald and Plowright, 2014), suggesting that developing brood may trigger rearing behaviours and therefore disrupt formation of dominance hierarchies. Developing brood may signal to workers that the queen is still functioning, and it would be selectively advantageous for them to continue to rear sisters, or that the reproductive dominance hierarchy among QL workers has already been established.

5.5 Hormonal regulation of reproduction and dominance

Hormones play an important role in regulating reproduction and dominance behaviour in many insects (Hartfelder, 2000; Nijhout, 1994). As described above, JH in particular was extensively studied in bumble bees, and its role as a gonadotropin is well established using both correlative (Bloch et al., 1996; Roseler, 1977; Roseler and Roseler, 1978, 1988) and direct evidence. Both surgical removal of the CA (allatectomy) (Shpigler et al., 2014) and treatment with precocene-I (a JH inhibitor) (Amsalem et al., 2014c) reduce JH levels and consequently ovarian activation in QL workers, and the latter was rescued by a topical application of JH-III. However, it was long debated if JH also affects aggressive behaviour in bumble bees as it does in primitively eusocial wasps (Barth et al., 1975; Roseler et al., 1980, 1984). Studies found positive correlations between JH levels as well as the expression levels of *Kr-h1*, a gene that is regulated by JH and dominance behaviour (Amsalem et al., 2014c; Bloch et al., 1996; Shpigler et al., 2010). However, these correlations could be mediated by ovarian activation, since aggression precedes ovarian activation in QL workers and is highly correlated with it in advanced age (Amsalem and Hefetz, 2010). Direct manipulation in which a random worker was treated with JH in four-worker groups failed to show an increase in the treated worker's chances to gain dominance over her groupmates (Doorn van, 1989), but these experiments are still inconclusive since JH-I was used rather than JH-III, the natural hormone in *B. terrestris* (Bloch et al., 2000a). In a slightly different experimental design where callow workers were introduced to peer (two workers of the same age and size) or established groups (two older workers), the introduced workers did not show an increase in aggression as a function of the JH-III treatment, irrespective of the social set-up they were exposed to (Amsalem et al., 2014b). Furthermore, although precocene-I decreases aggression in workers, a single treatment with JH did not reverse this effect (Amsalem et al., 2014c). Thus, all the evidence so far supports a model in which a change in aggression leads to a change in JH levels, but JH does not directly

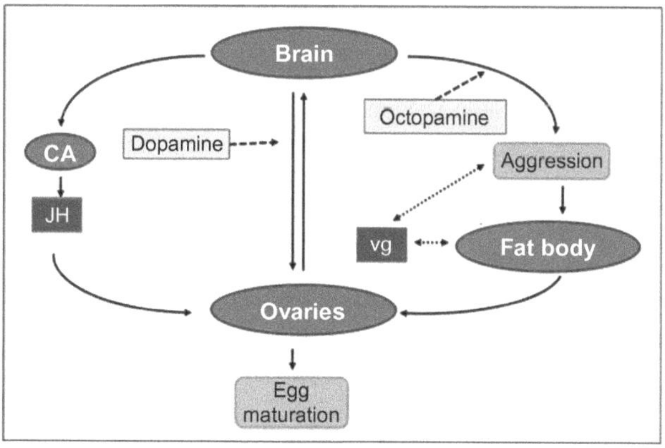

Figure 3 Simplified schematic representation of the hypothesized pathways regulating aggression and ovarian activation *in B. terrestris* workers. Signals from the brain stimulate the corpora allata (CA) to produce juvenile hormone (JH). High JH titers stimulate ovarian activation. Independently, a second set of signals from the brain (likely mediated by octopamine signalling pathways) triggers aggression. Aggression is correlated with high expression levels of vitellogenin (vg) in the fat bodies, though it is unknown whether high vg triggers aggression or vice versa. High levels of aggression and/or vg lead to ovarian activation and to a change in JH via a possible positive feedback between the ovaries and the brain. Egg maturation and egg laying are also possibly mediated by a positive feedback between the ovaries and the brain via dopamine signalling pathways.

regulate aggression (Fig. 3). Since sterility in *B. terrestris* is also tightly related to the division of labour among workers (Amsalem et al., 2013b; Roseler and Van Honk, 1990), JH was tested for its possible role in regulating task. However, unlike the honey bee, JH does not play a role in regulating division of labour in bumble bees, with foragers and nurses showing similar JH titers (Cameron and Robinson, 1990; Doorn van, 1986).

Both the behavioural and physiological evidence indicate high regulation by brain factor, most probably biogenic amines (BA). However, we have little knowledge on how these affect both queen and worker behaviour. Quantification of three major brain BA, dopamine, serotonin, and octopamine revealed that there were differences between virgin and mated, egg-laying queens (Bloch et al., 2000c). Workers on the other hand showed differences with respect to both reproductive and behavioural dominance. Reproductive workers, irrespective of their dominance status, showed higher dopamine levels in the last stages of oocyte development, while dominant workers, irrespective of their reproductive state, showed higher levels of octopamine. Serotonin levels were similar in all workers. While these findings indicate the differential roles of dopamine and octopamine in the reproductive competition among workers, it awaits more causative experimentation.

Vg in most insects is the major yolk protein, and accordingly, the expression levels of its gene (*vg*) are regulated by JH (Chen and Hillen, 1983; Engelmann, 1979; Hagedorn and Kunkel, 1979). In social insects (e.g. the honey bee), Vg has assumed additional regulatory roles (Amdam et al., 2004, 2006). This may be also the case in *B. terrestris*. The level of ovarian activation in *B. terrestris* workers showed only partial correlations with *vg* levels (Amsalem et al., 2014b), and *vg* expression levels were found to correlate with aggressiveness in QL workers that were kept in small controlled groups and were too young to show any significant differences in ovarian activation. Thus, *vg* expression levels in *B. terrestris* workers appear to primarily correlate with aggression and only secondary with reproduction (Fig. 3). Accordingly, the interrelation between JH and *vg* did not follow the common pattern in most insects, where JH directly regulates the levels of *vg*. Although there is an overall positive correlation between JH levels and vitellogenin (mRNA expression and protein level) in *B. terrestris* QL workers (Shpigler et al., 2014), a direct manipulation of JH in QL workers did not affect the *vg* expression levels (Amsalem et al., 2014b). While this suggests that JH and *vg* are uncoupled in *B. terrestris* and is consistent with the hypothesis that *Bombus* represents an intermediate stage in the evolution of eusociality, more genomic data are needed to fully explore this hypothesis.

Ecdysteroids also appear to regulate dominance behaviour in workers. In QL groups, high social status is associated with high ecdysteroid titers but is limited to a certain age and does not correlate with ovarian activation (Bloch et al., 2000b). In another study, ecdysteroid levels in both the ovaries and hemolymph positively correlated with different stages of ovarian activation, albeit only weakly with the terminal oocyte size. Likewise, ecdysteroid titers were correlated with overt aggression, but not with threatening behaviours (Geva et al., 2005). Brood presence and group size also influenced the level of ecdysteroids in QL workers (Geva et al., 2005). Thus, ecdysteroids titers are clearly sensitive to social interactions, but its role in mediating reproductive dominance remains to be deciphered.

5.6 Genomic mechanisms underlying worker–worker conflict

Genomic approaches have provided insights into the mechanisms mediating worker–worker conflict over reproduction in QL *B. terrestris* worker groups. Dominant workers have significantly higher expression levels of *Kr-h1* in their brains than subordinate workers, which mirrors the higher levels of *Kr-h1* found in the brains of QL versus QR workers (Shpigler et al., 2010). Allatectomy of young QL bumble bee workers results in decreased fat body expression levels of *Kr-h1* and *vg*, along with a decrease in JH titers and ovary activation in 7-day-old bees. Exogenous treatment of intact (non-operated) QL workers with a single JH dose increases *Kr-h1* expression (Shpigler et al., 2010) but does not increase *vg* levels (Amsalem et al., 2014b), which is consistent with *Kr-h1* functioning upstream of *vg*. These data, coupled with the studies of JH, aggression, ovary activation (see above), suggest that in bumble bees *vg* levels are either indirectly regulated by JH (likely through *Kr-h1*) or co-regulated by another pathway that is associated with aggression (see Fig. 3).

DNA methylation may also play a role in worker–worker conflict in *B. terrestris*. Using methylation-sensitive amplified fragment length polymorphism methodology, it was found that genome-wide DNA methylation patterns differ significantly between QL reproductive workers and both QL and QR non-reproductive workers (Amarasinghe et al., 2014); however, this study needs to be repeated using more accurate sequencing-based methods. Treatment of young (but not mature) QL worker bees with an inhibitor of DNA methylation resulted in increased aggression and ovary activation levels (Amarasinghe et al., 2014). However, the changes in methylation levels in the loci examined were complex—the majority of loci showed no change, and an equivalent number of loci (9 and 10, respectively) were hyper- or hypo-methylated in treated bees versus controls, and thus it is not clear if DNA methylation actually has a functional role in this process. The function of DNA methylation in insects is still under investigation, but it may primarily modulate splicing rather than transcription (Flores et al., 2012; Li-Byarlay et al., 2013; but see Wang et al., 2013). Interestingly, alternative splicing of the transcription factor *gemini* can lead to worker ovary activation in honey bees (Jarosch et al., 2011). Other studies have indicated that decreasing methylation levels biases developing honey bee larvae to queen rather than worker development (Kucharski et al., 2008), suggesting that methylation plays a general role in regulating reproductive potential in bees. For a comprehensive review of the effect of epigenetics on social behaviour, see Glastad et al. (2015).

5.7 Factors affecting dominance and reproduction in workers

Factors determining dominance rank and reproductive potential of workers largely depend on morphological and physiological characteristics such as age, body size, and task, but also on external factors including colony demography and spatial organization of workers in the nest. In the following paragraphs, we will briefly present recent findings relevant to each of these factors and their effects on worker reproduction.

Internal factors – Age has a strong effect on the likelihood of a worker to achieve dominance, though this depends on the social context. Under QL conditions, when workers of similar sizes but of different ages are housed together, reproduction is often dominated by the older worker (Amsalem et al., 2014b; Doorn van, 1989). Even age difference of as little as 5 h can increase the probability of a worker to become dominant (Doorn van, 1986). In QR colonies during the CP, the first egg-layers are often the oldest workers who emerged in the first or second brood. However, once the competition starts, worker age seems to no longer play an important role in the probability of workers to become egg-layers (Doorn van and Heringa, 1986; van Honk et al., 1981).

Bumble bees are often typified by size polymorphism among workers, and the fact that aggression is a major means by which workers may gain reproductive dominance suggests that size may have an impact on worker probability to become dominant (for discussion on the factors affecting worker size polymorphism in bumble bees, see Couvillon and Dornhaus, 2010; Couvillon et al., 2010; Duchateau and Velthuis, 1988; Owen, 1988, 1989; Shpigler et al., 2013; Sutcliffe and Plowright, 1988). Size seems to have a small effect mainly under QL conditions: in small groups of four workers of unequal body size, there was a positive correlation between body size and dominance rank, but the probability of the largest bee in each group to become dominant was not raised significantly (Doorn van, 1989). Under QR conditions, worker size was only important for the workers that emerged from the first batch of eggs (van Honk et al., 1981), but otherwise body size did not seem to play an important role in determining which worker will be an egg-layer (Duchateau and Velthuis, 1989).

Task allocation among workers in a QR colony may also affect the probability to become dominant. Although previous studies suggested that worker division of labour in bumble bees is more flexible than many other social insects (Doorn van and Heringa, 1986; Jandt and Dornhaus, 2009; O'Donnell et al., 2000; Yerushalmi et al., 2006), bumble bee workers do exhibit task specialization (Doorn van, 1986) that is loosely dependent on adult size (alloethism) rather than on worker age (polyethism) (Cameron, 1989; Cameron and Robinson, 1990; Goulson et al., 2002). In general, smaller workers tend to perform nurse-like tasks, while larger workers tend to be foragers but age at task varies widely between individuals (for detailed discussion on the effect of worker size and age on task allocation, see Amsalem et al., 2013b; Cameron, 1989; Cnaani and Hefetz, 1994; Doorn van, 1986; Goulson et al., 2002; O'Donnell et al., 2000; Spaethe and Chittka, 2003; Spaethe and Weidenmuller, 2002; Worden et al., 2005; Yerushalmi et al., 2006).

In *B. terrestris*, worker task strongly affects the likelihood that workers become reproductively dominant. A non-forager bee is twice as likely to become an egg-layer during the CP than a forager bee, and workers that have been observed to forage prior to the day they were killed had lower ovarian activation compared to workers that did not forage (Duchateau and Velthuis, 1989). Most likely, foragers simply have much less energy to invest into ovarian activation than nurse bees (Foster et al., 2004). *B. terrestris* foragers have also been found to produce larger amounts of octyl esters in their Dufour's gland (the previously discussed sterility signal) compared to nurses, and these quantities correlate with their number of foraging flights (Amsalem et al., 2013b).

As described previously, aggression is a decisive factor in establishing dominance hierarchy. When workers are kept in small controlled groups, they exhibit aggression relatively early in a predictable timetable. This aggression precedes most of the physiological changes in workers and is strikingly diverse even between workers who share the same physiological, morphological, and social conditions. It is therefore likely that aggression will be regulated by intrinsic factors, possibly genetic, that determine which bee will be more aggressive and thus will gain dominance and lay eggs.

External factors – The variability in worker aggression and reproduction can also be partially explained by external factors such as the population composition and in-nest organization, which may affect not only the chances of workers to be egg-layers but also determine their final size and role. *B. impatiens* larvae at the nest centre are better fed and grow larger, while those in the periphery of the nest are fed less and develop into smaller adults (Couvillon and Dornhaus, 2009). The smaller adult workers, on the other hand, tend to reside near the centre of the nest and perform nurse-like tasks such as larval feeding, while the larger workers tend to reside on the nest edges and are more likely to be foragers (Jandt and Dornhaus, 2009). Thus, it is likely that in *B. impatiens*, larvae reared in the centre of the nest are more likely to become non-reproductively active workers.

Examination of the effects of brood presence on worker reproduction showed ambiguous results: ovarian activation was not significantly different when workers were exposed to larvae (Duchateau and Velthuis, 1989) or kept in pairs with or without brood (Sibbald and Plowright, 2013). However, in the presence of brood, ovarian mass was higher (Sibbald and Plowright, 2014). The presence of gynes is highly correlated with worker reproduction in CP colonies (Cnaani et al., 2000a; Duchateau and Velthuis, 1988; Lopez-Vaamonde et al., 2007).

Another important social context affecting the probabilities of workers to reproduce is group size, via its effect on the cost of aggression versus gain in reproduction. The frequency of aggressive behaviour in the α-workers increased proportionally with group size (Amsalem and Hefetz, 2011), suggesting that the α-worker adjusts her level of behavioural dominance to control groups of different size. However, the ability of the α-worker to control her group-mate's reproduction is not unlimited. In larger groups, although the α-worker seems to attain her maximal aggression level, it is still insufficient because other members of the group possess activated ovaries. Therefore, there seems to be a trade-off between the cost of being aggressive and the gain of being the only reproductive, which imposes limits on aggressiveness. Nonetheless, the findings that both aggressive and aggressed workers show higher levels of *vg* expression and ovarian activation compared to passive workers (Amsalem et al., 2014b) raise the possibility that increased social interactions enhance ovarian activation.

5.8 Drifting bees: Fortuitous error or intra-specific parasitism?

Drifting of workers from their natal colony to a host colony may be the consequence of navigation errors coupled with mis-identification of their own nest location. However, molecular studies revealed that such drifter might lay eggs and contribute to male production of the host colony. Thus, drifting can be considered as an alternative reproductive strategy and social parasitism. Drifting and subsequent egg-laying by non-natal workers has been recorded in several bumble bee species. In wild and laboratory-reared, free-foraging *B. terrestris* colonies, drifting was estimated at 2–3%, and drifting bees produced twice as many eggs as resident workers (Lopez-Vaamonde et al., 2004; O'Connor et al., 2013; Zanette et al., 2014). In a study of *B. deuteronymus*, 3 out of 11 colonies were invaded by non-natal workers that produced 19% of the males, compared to 17% of male production by the resident workers (Takahashi et al., 2010). In managed bumble bee colonies, drifting was reported to be more common: in *B. impatiens* and *B. occidentalis*, 28% of the bees drifted (Birmingham and Winston, 2004; Birmingham et al., 2004) and the numbers were higher for managed *B. terrestris* colonies, reaching up to 50% when the colonies were vertically stacked (Lefebvre and Pierre, 2007). Thus, drifting may be greatly influenced by the geographical distance between colonies, which in managed bees tend to be much smaller compared to natural conditions. Interestingly, 60% of a cohort of marked reproductive or subordinate *B. terrestris* workers visited other nests, with the fertile workers drifting at higher rate compared to the unfertile (Blacher et al., 2013b).

The question of whether drifting is a result of fortuitous error or social parasitism still remains to be determined. Increased drifting as a function of nest proximity lends credence to the hypothesis that drifting is a result of navigation errors. Negating this hypothesis is the fact that in nature, nests are more dispersed and there is low probability of navigation error, and yet drifting is still observed in wild colonies. Furthermore, reproductive workers are more likely to drift, though reproductive workers are also more likely to refrain from foraging, suggesting that when they exit their natal nest it is specifically to search for another bumble bee colony. The question of why drifter eggs are not destroyed by the resident bees also remains unanswered. During the CP, oophagy by the resident bees is mostly directed to other workers' eggs but not those of the queen (Zanette et al., 2012), raising the possibility that drifters somehow mask their eggs to become indistinguishable from queen eggs.

5.9 Extrapolating from QL groups to whole QR colonies

The social structure and division of labour in *B. terrestris* QR colonies is maintained not only during the eusocial phase but also under the chaotic conditions of the CP. Moreover, under QL conditions, lethal aggression is rare and larvae are still fed by subordinate helpers, and many tasks are performed continuously (although some of them decreased as the colony ages) (Foster et al., 2004; O'Donnell et al., 2000). We propose that the study of QL groups can be extrapolated to understand events occurring during the eusocial phase of bumble bee colonies.

While small controlled groups of three workers provide insight into the mechanisms regulating hierarchy and reproduction among nestmates, they can hardly simulate whole colonies. For example, patterns of reproduction and pheromone production in 3- and 5-worker groups were clearly different from those of 10-worker group or whole colonies (Amsalem and Hefetz, 2011). However, as previously mentioned, increase in group size is correlated with elevated aggression and lower reproductive suppression (Amsalem and Hefetz, 2011), providing an insight into the mechanisms by which the queen may inhibit worker reproduction during the pre-CP when worker reproduction starts due to the queen's inability to physically inhibit worker oviposition. These findings emphasize the role of aggression in shaping and maintaining the social structure.

Task allocation correlates in workers may also resemble the behavioural transition of queens from the solitary to the eusocial phases. Foraging, being risky, is, in most social insects, the last task that a worker performs in her life. In contrast, in bumble bees, foraging is usually the first task workers perform. Workers start foraging at the age of 2–3 days and when older they may switch to in-nest tasks. Foragers are typically sterile and non-aggressive, contrary to the typically old and aggressive, in-nest house bees. Similarly, post-diapause queens forage during the solitary phase, but switch into in-nest tasks with the establishment of the new colony and egg laying, suggesting that there might be similar physiological processes involved in these transitions.

6. LARVAL DEVELOPMENT AND MECHANISMS UNDERLYING CASTE DETERMINATION

One of the features that characterize insect eusociality is clear bimodal size distribution (queens are much larger than workers with almost no overlap) and/or distinct morphology between castes. Depending on the species, size differences between bumble bee queens and workers may be continuous (as exemplified in the North American species *Bombus nevadensis*, *Bombus borealis*, and *Bombus perplexus* and European species *B. hypnorum*) or bimodal (in the North American species *Bombus terricola*, *B. ternarius*, and *Bombus rufocinctus* and European species *B. terrestris*) (Plowright and Jay, 1968) (figure 3.1 in Goulson, 2010; Roseler, 1989). The factors affecting worker and queen sizes may be numerous and differ according to the species, including quantitatively differential feeding, worker/queen-derived factors, or the consequences of colony demographic development (Goulson, 2010 and references therein). Here, we will focus the discussion mostly on *B. terrestris* for which there is voluminous information as to the factors affecting caste determination, that is, the bifurcation in diploid female larvae developmental pathways to either worker or queen.

6.1 Larval development

Larval developmental time in *B. terrestris* from hatching to end of the feeding period is about 18 days for queens and 15 days for workers. There is also a difference in duration from the pre-pupal stage to adulthood, again longer for queens than for workers (~13 and ~10 for queens and workers, respectively Cnaani et al., 2000b). Similar differences between queens and worker developmental duration were also reported for *B. impatiens* (Cnaani et al., 2002). The larvae, irrespective of caste, undergo four larval instars as determined by the distribution of the width of the larval head capsule (the head capsule is the only sclerotized part of the larvae which does not change in size during the instar and therefore represents larval size between instars, irrespective of the increase in weight and other body parts). There is a complete size overlap between worker- and queen-destined larval size distributions during the first two instars, but the distributions begin to separate in the third instar until finally by the fourth instar there is no overlap in sizes between the castes (Cnaani, 1998; Roseler, 1989). Accordingly, the final larval weight for queens (1.3 g) is much greater than workers (0.3 g). There are, however, no differences in growth rate between castes, indicating that the greater size and weight of queen larvae is due to prolonged developmental time rather than speed of growth. The duration of each of the last three instars in queens is longer than those in workers (Cnaani et al., 1997).

6.2 The critical period for caste determination

The critical period for caste determination is during the second larval instar, when the larvae are approximately 5 days old. To elucidate the critical age for queen determination, larvae at different ages were transferred from young colonies (worker-development conducive conditions) to QL groups (queen development conducive conditions). Almost all (80–100%) of the 1- to 4-day-old larvae developed into queens, whereas almost all (80–100%) of the 6- to 8-day-old larvae developed into workers. Among the 5-day-old larvae transferred, about half developed into queens and half into workers (Cnaani et al., 2000b). To reveal whether and when the queen presence can reverse larval determination, larvae reared under queen-conducive conditions (housed with 12 old workers that have reared queens before) were transferred to worker-conducive conditions (young colonies that rear almost exclusively workers). Among the 5-day-old queen-destined larvae, only 30% developed into queens, compared to 60% for such 6-day-old larvae (Cnaani, 1998).

Earlier studies with *B. terrestris* gave different result (Roseler, 1970, 1976); the critical period for queen determination was found to be earlier (3.5 days as opposed to 5 days) and an attempt to reverse queen development to worker development under QR conditions resulted in a queen size bee, although they were physiologically and behaviourally worker-like. We can attribute this difference to differences in the method for determining the critical period as well as possible between-population differences: Central Europe versus the much more southern population in Israel.

6.3 Hormonal and genomic regulation of caste determination

In addition to their roles in regulating behaviour and reproduction in adult bumble bees, JH and ecdysteroids also regulate larval development and caste differentiation in bumble bees; indeed, these are the major hormones regulating larval development in most insects (Hartfelder et al., 2015; Nijhout and Wheeler, 1982; Wheeler, 1986). In social insects, the endocrine system is presumably responsive to the nutritional state of the larvae as well as the social conditions under which they grow. A detailed study of JH biosynthesis during larval development in *B. terrestris* was performed by Cnaani et al. (1997). To neutralize the differences in instar duration and larval weight between queen and workers, JH biosynthesis rates were compared between queen and worker-destined larvae of the same physiological state (for definition of physiological state, see Cnaani et al., 1997). In worker-destined larvae, JH biosynthesis rates were consistently low throughout the entire larval development, but those of queens showed two distinct peaks. The first peak started in the middle of the first instar and declined to worker levels by the end of the instar. The second peak started from the middle of the second instar and declined to worker levels by the end of the third instar and through the fourth instar. A subsequent study also measured the actual JH titers in the blood during larval development revealing high correlation between the two methods of JH measurements (Cnaani et al., 2000a). Previous studies that measured JH levels were limited to the pre-pupal stage and gave similar differences between queens and workers (Strambi et al., 1984).

The first JH peak roughly coincides with the critical period for caste determination, and can be considered as queen development commitment peak, although experimental evidence is lacking. The second larger JH peak occurs after the larva has committed to queen development. It is assumed that these high JH levels delay moulting to the next instar, resulting in the prolonged instars observed in queen-destined larvae, and consequently greater growth. In Lepidoptera, JH is known to inhibit the prothoracicotropic hormone and with that the events cascade leading to moulting (Riddiford, 1981).

The role of increasing JH titers in triggering the onset of queen-destined development was confirmed by examining the effects of external application of JH (Bortolotti et al., 2001). JH applied to workers reared under conditions that are conducive to worker development in the first or second instar (attempting to mimic the two JH peaks in queens as determined by Bortolotti et al., 2001; Cnaani et al., 1997) resulted mostly in queen development. Interestingly, larvae of the second brood were more responsive to the JH treatment than those of the third brood, and among the former, treatments of second instar were more effective than treatments of the first instar larvae in turning worker-destined into queen-destined larvae.

Ecdysteroid is another hormone that shows differential titers between queen- and worker-destined larvae. For *B. terrestris*, two separate studies provided an almost complete picture of ecdysteroid titers during larval development (Hartfelder et al., 2000; Strambi et al., 1984). There are two clear peaks roughly in the middle of the second and third instars. Though generally queens demonstrated higher titers, this difference is significant only during the second instar. These peaks seem to be equivalent to the general pre-moult increase in ecdysteroid in holometablous insects. Likewise, during the fourth instar, there is a small peak that could be analogous to the prepupal commitment peak that was described for the hawkmoth *Manduca sexta* (Riddiford, 1981). This ecdysteroid peak is delayed in queens until they reached a greater weight compared to workers, consistent with the prolonged instar duration, suggesting that its function in committing the larvae to metamorphose. Finally, there is a large ecdysteroid peak in the pre-pupal stage, probably inducing pupation. This peak in queens is delayed until they reach the appropriate weight for pupation, again consistent with the size differences between queens and workers.

Comparing the differences between queen- and worker-destined larvae in JH and ecdysteroid titers revealed that these differences are synchronized during larval development. Thus, although JH is traditionally considered as the regulator of caste determination, whereas ecdysteroids regulate moulting, their synchronization indicates that both may be involved in caste determination and further suggest that they interact during this process (Hartfelder et al., 2000).

Thus far, only one study has sought to identify genes involved in caste determination in *B. terrestris*. Pereboom et al. (2005) used subtractive hybridization to identify 12 genes differentially expressed between early and late instars of queen- and worker-destined larvae. Interestingly, it seems that temporal pattern of gene expression, rather than expression *per se*, is important for driving caste differentiation: for example, genes expressed at high levels in early-instar queens versus workers were subsequently expressed at high levels in late-instar workers versus queens. Since only a few genes were identified, it is difficult to determine what general processes underlie caste determination, but several identified genes are involved in nutrient storage and metabolism. Four of the genes (*hexamerin, ATP-synthase beta subunit, Cytochrome Oxidase I*, and *Larval Cuticle Protein*) were previously found to be differentially expressed in worker- versus queen-destined honey bee larvae (Corona et al., 1999; Evans and Wheeler, 1999), but the expression patterns differed from bumble bees. Larger scale studies using genome-wide gene expression will be necessary to comprehensively examine the processes underlying caste differentiation and how conserved these are across Hymenoptera.

6.4 Physiological and social factors affecting caste determination

As noted above, there is a conflict between the queens and workers over the timing of gyne production (Bourke and Ratnieks, 1999). Theoretically, the queen, the workers, and the developing female larvae can play a role in determining the fate of the larvae. The queen may modulate caste determination by adjusting the type of egg she lays, e.g., changing oocyte composition both with respect to nutritional reserves and molecular factors that may affect gene expression. The queen may also indirectly affect larval development by regulating worker brood care behaviour (although this can be counteracted by worker behaviour if it negates their self-interest). The workers, being the prime caretakers of the brood, may affect larval destination by regulating its nutrition, in particular in species where there is a clear size difference between the castes. Overfeeding may lead to queen determination, while underfeeding may lead to worker determination. Workers can also potentially mix endogenous factors with the food which induce/inhibit particular developmental pathways. Theoretically, the best interest of the totipotent diploid larva is to develop into queen, since as queen it will be more related to its own female (would be queens) offspring

($r = 0.5$) than to female offspring of any of its sister gynes $r < 0.5$) (Bourke and Ratnieks, 1999). In pocket maker bumble bee species, larvae can compete over food and achieve differential growth rate so that some develop into large workers, i.e., queens. In pollen storing species, however, larvae are fed progressively and have little or no control over their growth rates. Nonetheless, they can potentially still eavesdrop on colony environment and direct their own fate accordingly. Below, we discuss some of the factors that may affect larval developmental pathway to either worker or queen.

6.5 Nutrition as a factor affecting caste determination

In most species, the queen is larger in size than workers, suggesting that feeding rates and feeding amounts may affect caste determination. Support for this hypothesis comes from several studies, mostly in species in which caste determination occur only in their last stages of larval development, e.g., the European *B. pratorum* (Free, 1955), *B. hypnorum* (Free, 1955; Roseler, 1970), and *B. pascuorum* (Reuter and Schwammberger, 1999) and the North American *B. rufocinctus* and *B. ternarius* (Plowright and Jay, 1968). Moreover, manipulation of larval feeding rate in *B. rufocinctus* demonstrated that well-fed larvae developed into queens, whereas underfed larvae developed into workers (Plowright and Jay, 1977). Likewise, in *B. hypnorum*, if worker-destined larvae are well fed, they develop into queens (Roseler, 1989).

In contrast, caste determination in *B. terrestris* occurs early in larval development, and so far no effect of diet was found. Feeding duration of female larvae increases with colony development, with workers of the third brood receiving longer duration feedings compared to the first and second brood. However, there were no differences in feeding duration between queen- and worker-destined larvae (Ribeiro, 1999). Feeding frequencies were rather variable across individuals over time but overall queen larvae in the last phase of development were fed more frequently than worker larvae (Ribeiro et al., 1999). This is probably the consequence of caste determination, since in *B. terrestris* caste determination occurs at earlier phase of development. One possible explanation is that workers respond to larval 'hunger signal' and feed larvae. Queen larvae, due to their larger size, may metabolize the food quicker and therefore signal for food more frequently (Pereboom et al., 2003).

Food type may constitute another factor that directs caste differentiation, as was shown for the honey bee (Kamakura, 2011). Analysis of the three major components in larval food, sucrose, pollen, and proteins, revealed that although workers add proteins (presumably gland derived) to the regurgitated food, there were no differences in food composition between worker- and queen-destined larvae (Pereboom, 2000). These analyses however were too general to conclusively state that workers do not add some 'determination factors' into the larval food.

6.6 The effect of queen–worker conflict on caste determination

In most species of bumble bees, gyne production occurs towards the end of colony life cycle, and as seen in *B. terrestris* is tightly linked with the eruption of queen–worker conflict over male production (Cnaani et al., 2000a; Duchateau and Velthuis, 1988). As discussed above, queens and workers have different optima regarding the timing of gyne production, providing the basis for another social conflict, and raising the question of who controls the onset of gyne production.

Support for the 'social effect on caste determination' came from experiments where egg cells from very young colonies (thus normally destined to develop into workers) were transplanted into host colonies at various stages of development. Eggs transplanted into host colonies that were up to 5 days before the onset of the CP developed into workers, whereas those transplanted into colonies closer to the onset of the CP developed into queens (Cnaani et al., 2000a). Evidence for queen control over gyne production came from experiments where uncommitted larvae were reared with queens either before or after the CP in combination with either naïve or experienced (in rearing gynes) workers. Irrespective of worker type, larvae in the presence of a pre-CP queen mostly developed into workers compared to larvae in the presence of CP queens that developed mostly into queens (Cnaani et al., 2000a). Interestingly, in this experiment, experienced workers were better at rearing queens in the presence of CP queens. Thus, it is tempting to speculate that the queen effect is nevertheless mediated through worker behaviour and that experienced workers may supplement the larvae with some nutritional factors that are absent in naive workers.

Earlier studies of caste determination in *B. terrestris* hypothesized that the queen controls the developmental fate of the larvae through a pheromone, and that as the colony grows, the effect of this pheromone declines and the developing larvae consequently switch to queen development (Roseler, 1970, 1976, 1989). Supporting the idea of queen influence on larval fate is the fact that uncommitted larvae reared under QL conditions, thence in the absence of the putative queen pheromone, develop mostly into queens (Cnaani et al., 1997; Roseler, 1970). Uncommitted larvae that were reared with workers but isolated from the queen developed into gynes, whereas if they were separated from the queen by a queen excluder (allowing workers to travel between compartments), they developed into workers. However, if some of the workers were transferred every 24 h between the QR and QL compartments (completely separated to prevent any contact between compartments), the larvae in the QL compartment still developed into queens. The author concluded that the queen produces a pheromone that affects worker behaviour directing them to rear either queens or workers, and that this pheromone is not transferable to the workers that were not in direct contact with the queen (Roseler, 1970). This study further revealed that queens taken from colonies at the CP were not able to inhibit gyne production, suggesting a decline in pheromone capacity. The correlation between the onset of gyne production and the onset of the CP leads to a further conclusion; the same pheromone inhibits both queen determination and worker reproduction, and that the queen stops producing the pheromone as a consequence of her loss of dominance (Duchateau and Velthuis, 1988; Roseler, 1989). This conclusion was later challenged by Bourke and Ratnieks (2001) on theoretical grounds.

Queen versus worker effects on caste determination were further investigated by creating young colonies headed by old queens and, vice versa, old colonies headed by young queens (Alaux et al., 2005). When colony composition was completely standardized (i.e. each colony composed of only 20 callow workers), queen age was a decisive factor in caste determination. Larvae housed with a young queen developed into workers, while larvae housed with an old queen developed into gynes. When the host colony social composition remained unchanged (i.e. the number of workers and their age and the larvae/worker ratios), the results were mixed. In young colonies that were headed by old queens, gyne production was advanced, but the inverse was not true. Old colonies headed by a young queen still produced queens. It is possible in this case that workers from old colonies underwent an irreversible change that rendered them insensitive to the change in the queen's pheromonal output, or that the unusually high number of workers signalled to the queen that the colony is at an advanced stage of development and induced her to change her pheromone output. This latter explanation is consistent with earlier findings that doubling the worker population advanced the onset of gyne production (Bloch, 1999). It was further revealed in this study that the queen calculates her age from the first worker emergence, that is, from the onset of the eusocial phase (Alaux et al., 2005).

6.7 Is there a queen pheromone that regulates caste determination?

The evidence presented above suggests that the regulation of caste determination in *B. terrestris* is pheromonally mediated, but neither the nature nor the glandular source of this pheromone is known. Below we list, on the basis of the present knowledge, the characteristics of such a putative pheromone(s) and suggest its possible nature and glandular source.

The mesh and queen excluder experiments indicate that the pheromone is not volatile and is not transferable between workers (Roseler, 1970), suggesting that the queen, while travelling in the nest, may add some of the secretion to the developing larvae or at least deposit the pheromone in the cell wax. The queen exchange experiments further indicate that the queen's pheromonal output changes both with age and social conditions. Young queens are able to direct larval development into workers, and if we postulate that queen development is the default developmental pathway (Cnaani and Hefetz, 2001) this is presumably achieved via inhibition of queen development. Since in the presence of old queens this presumed inhibition is alleviated, it can be concluded that such queens either stop producing the pheromone or change its composition. Recent analysis of labial gland secretion of queens and workers under various social conditions points to its possible role in regulating caste determination. The secretion is endowed with hydrocarbons and a plethora of esters of fatty acids (Amsalem et al., 2014a). Of particular interest is a series of dodecyl esters ranging from dodecyl hexanoate to dodecyl octadecenoate, which exhibit complex patterns. In queens, they are in high amounts in virgin and in young pre-CP fertile queens, but decline considerably (by 80%) in queens at the CP. In parallel, there is a change in the corresponding acids, which are high in old CP queens and low in young pre-CP queens. These changes are uncorrelated with ovarian activation since ovarian activation was low in virgin queens and comparably high in the two queen types (pre-CP and CP). This suggests that as queens' age and/or the changes in colony social status, the esters are hydrolyzed and consequently the quantities of the corresponding acids increase. It is tempting to postulate that these changes are perceived by either the workers or the larvae or both, thus influencing, directly or indirectly, the developmental fate of the larvae.

Unlike in queens, the changes in the labial gland dodecyl esters of workers are correlated with ovary activation. Sterile workers possess relatively high amounts of dodecyl esters (but still much lower than in queens), irrespective of whether taken from pre-CP or CP colonies, but they decline by more than half in egg-laying workers. We do not fully understand the meaning of these differences, but it may point to a dual function of these esters and the possibility that they work on concert with other pheromones (e.g. Dufour's gland secretion).

Whether the decline in dodecyl esters and corresponding increase in acids precedes the CP and queen production, or not, requires a more precise time-dependent study. Nonetheless, the findings that workers that are exposed to either queen larvae or queens taken from gyne-producing colonies advance the competition point (Alaux et al., 2006), coupled with the findings that the larvae switch to the queen developmental pathway a few days before the onset of the CP indicate that the switch in queen pheromonal composition precedes and triggers all of these events. The question of whether the queen exerts her regulation directly on the larval development or through affecting worker brood-care behaviour, as well as the question of whether the same pheromone is responsible for regulating worker reproduction and gyne commitment, remain open.

7. CONCLUSIONS AND FUTURE DIRECTIONS

In addition to their importance in pollination and agriculture, bumble bees are also an excellent system to examine and test the proximate and ultimate mechanisms underlying the organization and evolution of social behaviour. While sophisticated theories have been developed regarding the ultimate mechanisms driving social behaviour during the different colony phases, our understanding of the associated communication, physiological, and genomic mechanisms remain largely in their infancy. However, detailed knowledge of these mechanisms can greatly inform our understanding of the evolutionary processes; for example, until empirical studies revealed the presence of a 'sterility signal' in bumble bee workers, such a communication signal was not predicted. Below, we highlight outstanding questions in the field of bumble bee biology.

(1) What are the factors and mechanisms underlying the myriad of life history strategies used by different bumble bee species? How do ecological, social, and physiological factors interact to shape these different strategies, and to what extent are these life history strategies evolutionarily constrained or plastic?

(2) What is the nature of the queen fertility signal? Despite years of research in this area, we are still lacking a complete understanding of the nature of the social signal underlying queen reproductive dominance during the pre-CP phase, and how this dominance is lost during the CP phase.

(3) How did the worker sterility signal evolve? Is this signal present in other species that display reproductive dominance hierarchies among workers or queens?

(4) How do social cues interact with physiological and genetic mechanisms to establish and maintain dominance hierarchies, ovarian activation, and egg laying?

(5) How have ancient pathways (genetic toolkits) been modified and reorganized during the evolution of social behaviour? For example, vg has evolved novel functions in highly eusocial honey bees compared to solitary species, and the regulation of *vg* by JH has changed concurrently. In bumble bees, the function of JH and vg and their regulatory interactions appear to represent an intermediate state (Amsalem et al., 2014b). How have other pathways been modified to generate complex social behaviours?

(6) Are there epigenetic mechanisms that underpin social behaviour? Recent studies suggest that DNA methylation regulates worker reproduction (Amarasinghe et al., 2014), but the underlying mechanisms (which genes? How does methylation impact gene function?) remain to be determined. Is there intra-genomic conflict between maternally derived and paternally derived alleles (Queller, 2003) and if so, how is this regulated?

(7) What social factors trigger caste differentiation? What is the signal produced by the queen that alters larval developmental fate? Is there also a signal produced by the queen or larvae that alters worker feeding-behaviour? What molecular mechanisms trigger caste differentiation? Are these conserved across Hymenopteran species? Are these derived from diapause-related pathways (Hunt et al., 2007)?

Addressing these questions will require an integrative multi-disciplinary approach, spanning behaviour, neurobiology, physiology, chemical ecology, and genomics. Recent advances in genomic resources (Sadd et al., 2015), genomic technologies (including high-throughput sequencing of transcriptomes, genomes, and methylomes), and genetic technologies allowing the modulation of gene function (such as RNAi, TALENs, and CRISPRs) have greatly facilitated our ability to probe the molecular mechanisms underlying these behaviours and examine how these have evolved across species. Finally, we should take advantage of the breadth and diversity of natural histories and life history strategies displayed by different bumble bee species, to better understand the mechanisms underlying both social behaviour and adaptation to different ecological niches.

REFERENCES

Abbot, P., Abe, J., Alcock, J., et al., 2011. Inclusive fitness theory and eusociality. Nature 471, E1–E4.

Alaux, C., Jaisson, P., Hefetz, A., 2004a. Queen influence on worker reproduction in bumblebees (*Bombus terrestris*) colonies. Insect. Soc. 51, 287–293.

Alaux, C., Savarit, F., Jaisson, P., Hefetz, A., 2004b. Does the queen win it all? Queen-worker conflict over male production in the bumblebee, *Bombus terrestris*. Naturwissenschaften 91, 400–403.

Alaux, C., Jaisson, P., Hefetz, A., 2005. Reproductive decision-making in semelparous colonies of the bumblebee *Bombus terrestris*. Behav. Ecol. Sociobiol. 59, 270–277.

Alaux, C., Jaisson, P., Hefetz, A., 2006. Regulation of worker reproduction in bumblebees (*Bombus terrestris*): workers eavesdrop on a queen signal. Behav. Ecol. Sociobiol. 60, 439–446.

Alford, D.V., 1969a. Studies on the fat-body of adult bumblebees. J. Apic. Res. 8, 37–38.

Alford, D.V., 1969b. A study of the hibernation of bumblebees (Hymenoptera: Bombidae) in Southern England. J. Anim. Ecol. 38, 149–170.

Alford, D.V., 1978. The Life of the Bumblebee. Davis-Poynter, London.

Allen, B., Nowak, M.A., Wilson, E.O., 2013. Limitations of inclusive fitness. Proc. Natl. Acad. Sci. U.S.A. 110, 20135–20139.

Amarasinghe, H.E., Clayton, C.I., Mallon, E.B., 2014. Methylation and worker reproduction in the bumble-bee (*Bombus terrestris*). Proc. R. Soc. B Biol. Sci. 281, 20132502.

Amdam, G.V., Norberg, K., Fondrk, M.K., Page, R.E., 2004. Reproductive ground plan may mediate colony-level selection effects on individual foraging behavior in honey bees. Proc. Natl. Acad. Sci. U. S. A. 101, 11350–11355.

Amdam, G.V., Csondes, A., Fondrk, M.K., Page, R.E., 2006. Complex social behaviour derived from maternal reproductive traits. Nature 439, 76–78.

Amin, M.R., Kwon, Y.J., Suh, S.J., 2007a. Photoperiodic influence on the body mass of bumblebee, *Bombus terrestris* and its copulation duration. J. Appl. Entomol. 131, 537–541.

Amin, M.R., Suh, S.J., Kwon, Y.J., 2007b. Impact of artificial photoperiodism on the colony development of the bumblebee *Bombus terrestris*. Entomol. Sci. 10, 315–321.

Amin, M.R., Kwon, Y.J., Thet, Z.M., 2011. Effect of worker number and diapause duration on colony parameters of the bumblebee, *Bombus terrestris* (Hymenoptera: Apidae). J. Asia Pacif. Entomol. 14, 155–458.

Amsalem, E., Hefetz, A., 2010. The appeasement effect of sterility signaling in dominance contests among *Bombus terrestris* workers. Behav. Ecol. Sociobiol. 64, 1685–1694.

Amsalem, E., Hefetz, A., 2011. The effect of group size on the interplay between dominance and reproduction in *Bombus terrestris*. PLoS One 6, e18238.

Amsalem, E., Twele, R., Francke, W., Hefetz, A., 2009. Reproductive competition in the bumble-bee *Bombus terrestris*: do workers advertise sterility? Proc. R. Soc. B Biol. Sci. 276, 1295–1304.

Amsalem, E., Shamia, D., Hefetz, A., 2013a. Aggression or ovarian development as determinants of reproductive dominance in *Bombus terrestris*: interpretation using a simulation model. Insect. Soc. 60, 213–222.

Amsalem, E., Shpigler, H., Bloch, G., Hefetz, A., 2013b. Dufour's gland secretion, sterility and foraging behavior: correlated behavior traits in bumblebee workers. J. Insect Physiol. 59, 1250–1255.

Amsalem, E., Kiefer, J., Schulz, S., Hefetz, A., 2014a. The effect of caste and reproductive state on the chemistry of the cephalic labial glands secretion of *Bombus terrestris*. J. Chem. Ecol. 40, 900–912.

Amsalem, E., Malka, O., Grozinger, C., Hefetz, A., 2014b. Exploring the role of juvenile hormone and vitellogenin in reproduction and social behavior in bumble bees. BMC Evol. Biol. 14, 45.

Amsalem, E., Teal, P., Grozinger, C.M., Hefetz, A., 2014c. Precocene-I inhibits juvenile hormone biosynthesis, ovarian activation, aggression and alters sterility signal production in bumble bee (*Bombus terrestris*) workers. J. Exp. Biol. 217, 3178–3185.

Arnqvist, G., Nilsson, T., 2000. The evolution of polyandry: multiple mating and female fitness in insects. Anim. Behav. 60, 145–164.

Ayasse, M., Jarau, S., 2014. Chemical ecology of bumble bees. Annu. Rev. Entomol. 59, 299–319.

Ayasse, M., Marlovits, T., Tengö, J., Taghisadeh, T., Francke, W., 1995. Are there pheromonal dominance signals in the bumblebee *Bombus hypnorum* L. (Hymenoptera, Apidae)? Apidologie 26, 163–180.

Baer, B., Schmid-Hempel, P., 2004. Sperm influences female hibernation success, survival and fitness in the bumble-bee *Bombus terrestris*. Proc. R. Soc. B Biol. Sci. 272, 319–323.

Barth, R.H., Lester, L.J., Sroka, P., Kessler, T., Hearn, H., 1975. Juvenile hormone promotes dominance behavior and ovarian development in social wasps (*Polistes annularis*). Experientia 31, 691–692.

Beekman, M., van Stratum, P., 1998. Bumblebee sex ratios: why do bumblebees produce so many males? Proc. R. Soc. B Biol. Sci. 265, 1535–1543.

Beekman, M., van Stratum, P., 2000. Does the diapause experience of bumblebee queens *Bombus terrestris* affect colony characteristics? Ecol. Entomol. 25, 1–6.

Beekman, M., van Stratum, P., Lingeman, R., 1998. Diapause survival and post-diapause performance in bumblebee queens (*Bombus terrestris*). Entomol. Exp. Appl. 89, 207–214.

Beekman, M., van Stratum, P., Veerman, A., 1999. Selection for non-diapause in the bumblebee *Bombus terrestris*, with notes on the effect of inbreeding. Entomol. Exp. Appl. 93, 69–75.

Belles, X., Santos, C.G., 2014. The MEKRE93 (methoprene tolerant-Krüppel homolog 1-E93) pathway in the regulation of insect metamorphosis, and the homology of the pupal stage. Insect Biochem. Mol. Biol. 52, 60–68.

Bergman, P.A., Bergström, G., 1997. Scent marking, scent origin, and species specificity in male premating behavior of two Scandinavian bumblebees. J. Chem. Ecol. 23, 1235–1251.

Birmingham, A.L., Winston, M.L., 2004. Orientation and drifting behaviour of bumblebees (Hymenoptera: Apidae) in commercial tomato greenhouses. Can. J. Zool. 82, 52–59.

Birmingham, A.L., Hoover, S.E., Winston, M.L., Ydenberg, R.C., 2004. Drifting bumble bee (Hymenoptera: Apidae) workers in commercial greenhouses may be social parasites. Can. J. Zool. 82, 1843–1853.

Blacher, P., Boreggio, L., Leroy, C., Devienne, P., Chaline, N., Chameron, S., 2013a. Specific recognition of reproductive parasite workers by nest-entrance guards in the bumble bee *Bombus terrestris*. Front. Zool. 10, 74.

Blacher, P., Yagound, B., Lecoutey, E., Devienne, P., Chameron, S., Chaline, N., 2013b. Drifting behaviour as an alternative reproductive strategy for social insect workers. Proc. R. Soc. B Biol. Sci. 280, 1888.

Bloch, G., 1997. Regulation of Reproduction and Colony Development in the Bumblebee *Bombus terrestris*. (Ph.D. thesis). Tel Aviv University, Tel Aviv, Israel, 131 p.

Bloch, G., 1999. Regulation of queen-worker conflict in bumble-bee (*Bombus terrestris*) colonies. Proc. R. Soc. B Biol. Sci. 266, 2465–2469.

Bloch, G., Hefetz, A., 1999a. Reevaluation of the role of mandibular glands in regulation of reproduction in bumblebee colonies. J. Chem. Ecol. 25, 881–896.
Bloch, G., Hefetz, A., 1999b. Regulation of reproduction by dominant workers in bumblebee (*Bombus terrestris*) queenright colonies. Behav. Ecol. Sociobiol. 45, 125–135.
Bloch, G., Borst, D.W., Huang, Z.Y., Robinson, G.E., Hefetz, A., 1996. Effects of social conditions on juvenile hormone mediated reproductive development in *Bombus terrestris* workers. Physiol. Entomol. 21, 257–267.
Bloch, G., Borst, D.W., Huang, Z.Y., Robinson, G.E., Cnaani, J., Hefetz, A., 2000a. Juvenile hormone titers, juvenile hormone biosynthesis, ovarian development and social environment in *Bombus terrestris*. J. Insect Physiol. 46, 47–57.
Bloch, G., Hefetz, A., Hartfelder, K., 2000b. Ecdysteroid titer, ovary status, and dominance in adult worker and queen bumble bees (*Bombus terrestris*). J. Insect Physiol. 46, 1033–1040.
Bloch, G., Simon, T., Robinson, G.E., Hefetz, A., 2000c. Brain biogenic amines and reproductive dominance in bumble bees (*Bombus terrestris*). J. Comp. Physiol. A. 186, 261–268.
Bortolotti, L., Duchateau, M.J., Sbrenna, G., 2001. Effect of juvenile hormone on caste determination and colony processes in the bumblebee *Bombus terrestris*. Entomol. Exp. Appl. 101, 143–158.
Bourke, A.F.G., 1988. Worker reproduction in the higher eusocial hymenoptera. Q. Rev. Biol. 63, 291–311.
Bourke, A.F.G., 1994. Worker matricide in social bee and wasps. J. Theor. Biol. 167, 283–292.
Bourke, A.F.G., 1997. Sex ratios in bumble bees. Philos. Trans. R. Soc. B Biol. Sci. 352, 1921–1932.
Bourke, A.F.G., 2014. Hamilton's rule and the causes of social evolution. Philos. Trans. R. Soc. B Biol. Sci. 369, 20130362.
Bourke, A.F.G., Ratnieks, F.L.W., 1999. Kin conflict over caste determination in social Hymenoptera. Behav. Ecol. Sociobiol. 46, 287–297.
Bourke, A.F., Ratnieks, F.L., 2001. Kin-selected conflict in the bumble-bee *Bombus terrestris* (Hymenoptera: Apidae). Proc. R. Soc. B Biol. Sci. 268, 347–355.
Brown, M.J.F., Schmid-Hempel, R., Schmid Hempel, P., 2002. Dynamics of multiple-mating in the bumble bee *Bombus hypnorum*. Insect. Soc. 49, 315–319.
Brown, M.J.F., Schmid-Hempel, R., Schmid-Hempel, P., 2003a. Queen-controlled sex ratios and worker reproduction in the bumble bee *Bombus hypnorum*, as revealed by microsatellites. Mol. Ecol. 12, 1599–1605.
Brown, M.J.F., Schmid-Hempel, R., Schmid-Hempel, P., 2003b. Strong context-dependent virulence in a host–parasite system: reconciling genetic evidence with theory. J. Anim. Ecol. 72, 994–1002.
Bulmer, M.G., 1981. Worker-queen conflict in annual social hymenoptera. J. Theor. Biol. 93, 239–251.
Buttermore, R.E., 1997. Observations of successful *Bombus terrestris* (L.) (Hymenoptera: Apidae) colonies in Southern Tasmania. Aust. J. Entomol. 36, 251–254.
Cameron, S.A., 1989. Temporal Patterns of division of labor among workers in the primitively eusocial bumble bee, *Bombus griseocollis* (Hymenoptera: Apidae). Ethology 80, 137–151.
Cameron, S.A., Jost, M.C., 1998. Mediators of dominance and reproductive success among queens in the cyclically polygynous neotropical bumble bee *Bombus atratus* Franklin. Insect. Soc. 45, 135–149.
Cameron, S.A., Robinson, G.E., 1990. Juvenile hormone does not affect division of labor in bumble bee colonies. Ann. Entomol. Soc. Am. 83, 626–631.

Cameron, S.A., Hines, H.M., Williams, P.H., 2007. A comprehensive phylogeny of the bumble bees (*Bombus*). Biol. J. Linn. Soc. 91, 161–188.

Chapman, T., Liddle, L.F., Kalb, J.M., Wolfner, M.F., Partridge, L., 1995. Cost of mating in *Drosophila melanogaster* females is mediated by male accessory gland products. Nature 373, 241–244.

Chen, T.T., Hillen, L.J., 1983. Expression of the vitellogenin genes in insects. Gamete Res. 7, 179–196.

Cnaani, J., 1998. Larval development and caste determination in the bumble bee *Bombus terrestris*. (Ph.D. thesis). Tel Aviv University, Tel Aviv, Israel, 77 pp.

Cnaani, J., Hefetz, A., 1994. The effect of workers size frequency distribution on colony development in *Bombus terrestris*. Insect. Soc. 41, 301–307.

Cnaani, J., Hefetz, A., 2001. Are queen *Bombus terrestris* giant workers or are workers dwarf queens? Solving the 'chicken and egg' problem in a bumblebee species. Naturwissenschaften 88, 85–87.

Cnaani, J., Borst, D.W., Huang, Z.Y., Robinson, G.E., Hefetz, A., 1997. Caste determination in *Bombus terrestris*: differences in development and rates of JH biosynthesis between queen and worker larvae. J. Insect Physiol. 43, 373–381.

Cnaani, J., Robinson, G.E., Bloch, G., Borst, D., Hefetz, A., 2000a. The effect of queen-worker conflict on caste determination in the bumblebee *Bombus terrestris*. Behav. Ecol. Sociobiol. 47, 346–352.

Cnaani, J., Robinson, G.E., Hefetz, A., 2000b. The critical period for caste determination in *Bombus terrestris* and its juvenile hormone correlates. J. Comp. Physiol. A. 186, 1089–1094.

Cnaani, J., Schmid-Hempel, R., Schmidt, J.O., 2002. Colony development, larval development and worker reproduction in *Bombus impatiens* Cresson. Insect. Soc. 49, 164–170.

Cnaani, J., Wong, A., Thomson, J.D., 2007. Effect of group size on ovarian development in bumblebee workers (Hymenoptera: Apidae: *Bombus*). Entomol. Gen. 29, 305–314.

Colgan, T.J., Carolan, J.C., Bridgett, S.J., Sumner, S., Blaxter, M.L., Brown, M.J., 2011. Polyphenism in social insects: insights from a transcriptome-wide analysis of gene expression in the life stages of the key pollinator, *Bombus terrestris*. BMC Genomics 12, 623.

Corona, M., Estrada, E., Zurita, M., 1999. Differential expression of mitochondrial genes between queens and workers during caste determination in the honeybee *Apis mellifera*. J. Exp. Biol. 202, 929–938.

Couvillon, M.J., Dornhaus, A., 2009. Location, location, location: larvae position inside the nest is correlated with adult body size in worker bumble-bees (*Bombus impatiens*). Proc. R. Soc. B Biol. Sci. 276, 2411–2418.

Couvillon, M.J., Dornhaus, A., 2010. Small worker bumble bees (*Bombus impatiens*) are hardier against starvation than their larger sisters. Insect. Soc. 57, 193–197.

Couvillon, M.J., Jandt, J.M., Duong, N., Dornhaus, A., 2010. Ontogeny of worker body size distribution in bumble bee (*Bombus impatiens*) colonies. Ecol. Entomol. 35, 424–435.

Cui, Y., Sui, Y., Xu, J., Zhu, F., Palli, S.R., 2014. Juvenile hormone regulates *Aedes aegypti* Krüppel homolog 1 through a conserved E box motif. Insect Biochem. Mol. Biol. 52, 23–32.

Denlinger, D.L., 2002. Regulation of diapause. Annu. Rev. Entomol. 47, 93–122.

Denlinger, D.L., Yocum, G.D., Rinehart, J.P., 2012. Hormonal control of diapause. In: Gilbert, L.I. (Ed.), Insect Endocrinology. Elsevier, Amsterdam, pp. 430–463.

Doorn van, A., 1986. Investigations into the regulation of dominance behaviour and of the division of labour in bumblebee colonies (*Bombus terrestris*). Neth. J. Zool. 37, 255–276.

Doorn van, A., 1989. Factors influencing dominance behavior in queenless bumblebee workers *Bombus terrestris*. Physiol. Entomol. 14, 211–222.

Doorn van, A., Heringa, J., 1986. The ontogeny of a dominance hierarchy in colonies of the bumblebee *Bombus terrestris* (Hymenoptera, Apidae). Insect. Soc. 33, 3–25.

Duchateau, M.J., Velthuis, H.H.W., 1988. Development and reproductive strategies in *Bombus terrestris* colonies. Behaviour 107, 186–207.
Duchateau, M.J., Velthuis, H.H.W., 1989. Ovarian development and egg laying in workers of *Bombus terrestris*. Entomol. Exp. Appl. 51, 199 213.
Duchateau, M.J., Velthuis, H.H.W., Boomsma, J.J., 2004. Sex ratio variation in the bumblebee *Bombus terrestris*. Behav. Ecol. 15, 71–82.
Engelmann, F., 1979. Insect vitellogenin: identification, biosynthesis, and role in vitellogenesis. In: Treherne, J.E., Berridge, M.J., Wigglesworth, V.B. (Eds.), Advances in Insect Physiology. Academic Press, London, pp. 49–108.
Estoup, A., Scholl, A., Pouvreau, A., Solignac, M., 1995. Monoandry and polyandry in bumble bees (Hymenoptera: Bombinae) as evidenced by highly variable microsatellites. Mol. Ecol. 4, 89–93.
Evans, J.D., Wheeler, D.E., 1999. Differential gene expression between developing queens and workers in the honey bee, *Apis Mellifera*. Proc. Natl. Acad. Sci. U.S.A. 96, 5575–5580.
Fliszkiewicz, M., Wilkaniec, Z., 2007. Fatty acids and amino acids in the fat body of bumblebee *Bombus terrestris* (L.) in diapausing and non-diapausing queens. J. Apic. Sci. 51, 55–63.
Flores, K., Wolschin, F., Corneveaux, J.J., Allen, A.N., Huentelman, M.J., Amdam, G.V., 2012. Genome-wide association between DNA methylation and alternative splicing in an invertebrate. BMC Genomics 15, 480.
Foster, R.L., Brunskill, A., Verdirame, D., O'Donnell, S., 2004. Reproductive physiology, dominance interactions, and division of labour among bumble bee workers. Physiol. Entomol. 29, 327–334.
Free, J.B., 1955. Queen production in colonies of bumblebees. Proc. R. Entomol. Soc. A 30, 19–25.
Fussnecker, B., Grozinger, C.M., 2008. Dissecting the role of Kr-h1 brain gene expression in foraging behavior in honey bees (*Apis mellifera*). Insect Mol. Biol. 17, 515–522.
Garofalo, C.A., Zucchi, R., Muccillo, G., 1986. Reproductive studies of a neotropical bumblebee, *Bombus atratus* (Hymenoptera, Apidae). Rev. Brasil. Genet. IX 2, 231–243.
Geva, S., Hartfelder, K., Bloch, G., 2005. Reproductive division of labor, dominance, and ecdysteroid levels in hemolymph and ovary of the bumble bee *Bombus terrestris*. J. Insect Physiol. 51, 811–823.
Gillott, C., 2003. Male accessory gland secretions: modulators of female reproductive physiology and behavior. Annu. Rev. Entomol. 48, 163–184.
Glastad, K.M., Chau, L., Goodisman, M.A.D., 2015. Epigenetics in social insects. Adv. Insect Physiol. 48, 227–269.
Gosterit, A., Gurel, F., 2007. Effects of weight of queens after diapause on colony development in the bumblebee, *Bombus terrestris* L. Akdeniz Üniversitesi Ziraat Fakültesi Dergisi 20, 67–70.
Gosterit, A., Gurel, F., 2009. Effect of different diapause regimes on survival and colony development in the bumble bee, *Bombus terrestris*. J. Apic. Res. 48, 279–283.
Goulson, D., 2003. Bumblebees Behavior and Ecology. Oxford University Press, Oxford, England.
Goulson, D., 2010. Bumblebees: Behaviour, Ecology, and Conservation. Oxford University Press, Oxford, England.
Goulson, D., Peat, J., Stout, J.C., Tucker, J., Darvill, B., Derwent, L.C., Hughes, W.O.H., 2002. Can alloethism in workers of the bumblebee, *bombus terrestris*, be explained in terms of foraging efficiency? Anim. Behav. 64, 123–130.
Greeff, M., Schmid-Hempel, P., 2008. Sperm reduces female longevity and increases melanization of the spermatheca in the bumblebee *Bombus terrestris* L. Insect. Soc. 55, 313–319.

Grozinger, C.M., Robinson, G.E., 2007. Endocrine modulation of a pheromone-responsive gene in the honey bee brain. J. Comp. Physiol. A. 193, 461–470.
Grozinger, C.M., Sharabash, N.M., Whitfield, C.W., Robinson, G.E., 2003. Pheromone-mediated gene expression in the honey bee brain. Proc. Natl. Acad. Sci. U.S.A. 100, 14519–14525.
Hagedorn, H.H., Kunkel, J.G., 1979. Vitellogenin and vitellin in insects. Annu. Rev. Entomol. 24, 475–505.
Hamilton, W.D., 1964. The genetical evolution of social behaviour. I, II. J. Theor. Biol. 7, 1–52.
Hamilton, W.D., 1972. Altruism and related phenomena mainly in the social insects. Annu. Rev. Ecol. Syst. 2, 193–232.
Hammond, R.L., Keller, L., 2004. Conflict over male parentage in social insects. PLoS Biol. 2, 1472–1482.
Hartfelder, K., 2000. Insect juvenile hormone: from "Status Quo" to high society. Braz. J. Med. Biol. Res. 33, 157–177.
Hartfelder, K., Cnaani, J., Hefetz, A., 2000. Caste-specific differences in ecdysteroid titers in early larval stages of the bumblebee *Bombus terrestris*. J. Insect Physiol. 46, 1433–1439.
Hartfelder, K., Lazzarini-Guidugli, K.R., Cervoni, M.S., Santos, D.E., Humann, F.C., 2015. Old threads make new tapestry—rewiring of signaling pathways underlies caste phenotypic plasticity in the honey bee, Apis mellifera L. Adv. Insect Physiol. 48, 1–36.
He, Q., Wen, D., Jia, Q., Cui, C., Wang, J., Palli, S.R., Li, S., 2014. Heat shock protein 83 (Hsp83) facilitates methoprene-tolerant (Met) nuclear import to modulate juvenile hormone signaling. J. Biol. Chem. 289, 27874–27885.
Hefetz, A., Taghizadeh, T., Francke, W., 1996. The exocrinology of the queen bumble bee *Bombus terrestris* (Hymenoptera: Apidae, Bombini). Z. Naturforsch. 51, 409–422.
Hogeweg, P., Hesper, B., 1983. The ontogeny of the interaction structure in bumble bee colonies: a MIRROR model. Behav. Ecol. Sociobiol. 12, 271–283.
Hunt, J.H., Kensinger, B.J., Kossuth, J.A., Henshaw, M.T., Norberg, K., Wolschin, F., Amdam, G.V., 2007. A diapause pathway underlies the gyne phenotype in *Polistes* wasps, revealing an evolutionary route to caste-containing insect societies. Proc. Natl. Acad. Sci. U.S.A. 104, 14020–14025.
Huth-Schwarz, A., Leon, A., Vandame, R., Moritz, R.F.A., Kraus, F.B., 2011a. Workers dominate male production in the neotropical bumblebee *Bombus wilmattae* (Hymenoptera: Apidae). Front. Zool. 8, 13.
Huth-Schwarz, A., León, A., Vandame, R., Moritz, R.F.A., Kraus, F.B., 2011b. Mating frequency and genetic colony structure of the neotropical bumblebee *Bombus wilmattae* (Hymenoptera: Apidae). Apidologie 42, 519–525.
Jandt, J.M., Dornhaus, A., 2009. Spatial organization and division of labour in the bumblebee *Bombus impatiens*. Anim. Behav. 77, 641–651.
Jarosch, A., Stolle, E., Crewe, R.M., Moritz, R.F., 2011. Alternative splicing of a single transcription factor drives selfish reproductive behavior in honeybee workers (*Apis mellifera*). Proc. Natl. Acad. Sci. U.S.A. 108, 15282–15287.
Kamakura, M., 2011. Royalactin induces queen differentiation in honeybees. Nature 473, 478–483.
Katzav-Gozansky, T., 2006. The evolution of honeybee multiple queen-pheromones—a consequence of a queen-worker arms race? Braz. J. Morphol. Sci. 23, 129–136.
Kayukawa, T., Minakuchi, C., Namiki, T., et al., 2012. Transcriptional regulation of juvenile hormone-mediated induction of Krüppel homolog 1, a repressor of insect metamorphosis. Proc. Natl. Acad. Sci. U.S.A. 109, 11729–11734.
Keller, L., Nonacs, P., 1993. The role of queen pheromones in social insects: queen control or queen signal? Anim. Behav. 45, 787–794.

Kim, Y.-J., Hwang, J.-S., Yoon, H.-J., et al., 2006. Expressed sequence tag analysis of the diapausing queen of the bumblebee *Bombus ignitus*. Entomol. Res. 36, 191–195.
Kim, B.-G., Shim, J.-K., Kim, D.-W., Kwon, Y.J., Lee, K.-Y., 2008. Tissue-specific variation of heat shock protein gene expression in relation to diapause in the bumblebee *Bombus terrestris*. Entomol. Res. 38, 10–16.
Kocher, S.D., Grozinger, C.M., 2011. Cooperation, conflict, and the evolution of queen pheromones. J. Chem. Ecol. 37, 1263–1275.
Kocher, S.D., Richard, F.J., Tarpy, D.R., Grozinger, C.M., 2008. Genomic analysis of post-mating changes in the honey bee queen (*Apis mellifera*). BMC Genomics 9, 232.
Krieger, G., Duchateau, M.-J., Van Doorn, A., Ibarra, F., Francke, W., Ayasse, M., 2006. Identification of queen sex pheromone components of the bumblebee *Bombus terrestris*. J. Chem. Ecol. 32, 453–471.
Kucharski, R., Maleszka, J., Foret, S., Maleszka, R., 2008. Nutritional control of reproductive status in honeybees via DNA methylation. Science 319, 1827–1830.
Kullenberg, B., Bergström, G., Stalpberg-Stenhagen, S., 1970. Volatile components of the cephalic marking secretion of male bumble bees. Acta Chem. Scand. 24, 1481–1483.
Larrere, M., Lavenseau, L., Tasei, J.N., Couillaud, F., 1993. Juvenile-hormone biosynthesis and diapause termination in *Bombus terrestris*. Invertebr. Reprod. Dev. 23, 7–14.
Lecocq, T., Vereecken, N.J., Michez, D., Dellicour, S., Lhomme, P., Valterová, I., Rasplus, J.Y., Rasmont, P., 2013. Patterns of genetic and reproductive traits differentiation in mainland vs. corsican populations of bumblebees. PLoS One 8, e65642.
Lefebvre, D., Pierre, J., 2007. Demographic consequences of drift in contiguous hives of *Bombus terrestris*. J. Econ. Entomol. 100, 1756–1763.
Li-Byarlay, H., Li, Y., Stroud, H., et al., 2013. RNA interference knockdown of DNA methyl-transferase 3 affects gene alternative splicing in the honey bee. Proc. Natl. Acad. Sci. U.S.A. 110, 12750–12755.
Lopez-Vaamonde, C., Koning, J.W., Brown, R.M., Jordan, W.C., Bourke, A.F.G., 2004. Social parasitism by male-producing reproductive workers in a eusocial insect. Nature 430, 557–560.
Lopez-Vaamonde, C., Brown, R.M., Lucas, E.R., Pereboom, J.J.M., Jordan, W.C., Bourke, A.F.G., 2007. Effect of the queen on worker reproduction and new queen production in the bumble bee Bombus terrestris. Apidologie 38, 171–180.
Macevicz, S., Oster, G., 1976. Modeling social insect populations II: optimal reproductive strategies in annual eusocial insect colonies. Behav. Ecol. Sociobiol. 1, 265–282.
Malka, O., Nino, E.L., Grozinger, C.M., Hefetz, A., 2014. Genomic analysis of the interactions between social environment and social communication systems in honey bees (*Apis mellifera*). Insect Biochem. Mol. Biol. 47C, 36–45.
Michener, C.D., 1974. The Social Behavior of the Bees. A Comparative Study. Harvard University Press, Cambridge, MA.
Minakuchi, C., Namiki, T., Shinoda, T., 2009. Kruppel homolog 1, an early juvenile hormone-response gene downstream of methoprene-tolerant, mediates its anti-metamorphic action in the red flour beetle *Tribolium castaneum*. Dev. Biol. 325, 341–350.
Muller, C.B., Shykoff, J.A., Sutcliffe, G.H., 1992. Life history patterns and opportunities for queen-worker conflict in bumblebees (Hymenoptera, Apidae). Oikos 65, 242–248.
Nijhout, H.F., 1994. Insect Hormones. Princeton University Press, Princeton, NJ.
Nijhout, H.F., Wheeler, D.E., 1982. Juvenile-hormone and the physiological-basis of insect polymorphisms. Q. Rev. Biol. 57, 109–133.
Nowak, M.A., Tarnita, C.E., Wilson, E.O., 2010. The evolution of eusociality. Nature 466, 1057–1062.
O'Donnell, S., Reichardt, M., Foster, R., 2000. Individual and colony factors in bumble bee division of labor (*Bombus bifarius* nearcticus Handl; Hymenoptera, Apidae). Insect. Soc. 47, 164–170.

O'Connor, S., Park, K.J., Goulson, D., 2013. Worker drift and egg dumping by queens in wild *Bombus terrestris* colonies. Behav. Ecol. Sociobiol. 67, 621–627.
Oster, G.F., Wilson, E.O., 1978. Caste and ecology in the social insects. Princeton University Press, Princeton, NJ.
Owen, R.E., 1988. Body size variation and optimal body size of bumble bee queens (Hymenoptera: Apidae). Can. Entomol. 120, 19–27.
Owen, R.E., 1989. Differential size variation of male and female bumblebees. J. Hered. 80, 39–43.
Owen, R.E., Plowright, R.C., 1982. Worker-queen conflict and male parentage in bumble bees. Behav. Ecol. Sociobiol. 11, 91–99.
Owen, R.E., Whidden, T.L., 2013. Monandry and polyandry in three species of North American bumble bees (*Bombus*) determined using microsatellite DNA markers. Can. J. Zool. 91, 523–528.
Owen, R., Rodd, F.H., Plowright, R.C., 1980. Sex ratios in bumble bee colonies: complications due to orphaning? Behav. Ecol. Sociobiol. 7, 287–291.
Pankiw, T., Huang, Z.Y., Winston, M.L., Robinson, G.E., 1998. Queen mandibular gland pheromone influences worker honey bee (*Apis mellifera* L.) foraging ontogeny and juvenile hormone titers. J. Insect Physiol. 44, 685–692.
Paxton, R.J., Thoren, P.A., Estoup, A., Tengo, J., 2001. Queen-worker conflict over male production and the sex ratio in a facultatively polyandrous bumblebee, *Bombus hypnorum*: the consequences of nest usurpation. Mol. Ecol. 10, 2489–2498.
Payne, C.M., Laverty, T.M., Lachance, M.A., 2003. The frequency of multiple paternity in bumble bee (*Bombus*) colonies based on microsatellite DNA at the B10 locus. Insect. Soc. 50, 375–378.
Pereboom, J.J.M., 2000. The composition of larval food and the significance of exocrine secretions in the bumblebee *Bombus terrestris*. Insect. Soc. 47, 11–20.
Pereboom, J.J.M., Velthuis, H.H.W., Duchateau, M.J., 2003. The organisation of larval feeding in bumblebees (Hymenoptera, Apidae) and its significance to caste differentiation. Insect. Soc. 50, 127–133.
Pereboom, J.J.M., Jordan, W.C., Sumner, S., Hammond, R.L., Bourke, A.F.G., 2005. Differential gene expression in queen-worker caste determination in bumble-bees. Proc. R. Soc. B Biol. Sci. 272, 1145–1152.
Plowright, R.C., Jay, S.C., 1968. Caste differentiation in bumblebees (*Bombus* Latr.: Hym.). I. The determination of female size. Insect. Soc. 15, 171–192.
Plowright, R.C., Jay, S.C., 1977. On the size determination of bumble bee castes (Hymenoptera: Apidae). Can. J. Zool. 55, 1133–1138.
Plowright, R.C., Lavert, Y.T.M., 1984. The ecology and sociobiology of bumble bees. Annu. Rev. Entomol. 29, 175–199.
Queller, D.C., 2003. Theory of genomic imprinting conflict in social insects. BMC Evol. Biol. 3, 15.
Ratnieks, F.L.W., 1988. Reproductive harmony via mutual policing by workers in eusocial Hymenoptera. Am. Nat. 132, 217–236.
Reuter, K., Schwammberger, K.H., 1999. Food-dependent caste determination in the bumblebee *Bombus pascuorum* (Hymenoptera: Apidae). Entomol. Gen. 24, 293–307.
Ribeiro, M.F., 1999. Long-duration feedings and caste differentiation in *Bombus terrestris* larvae. Insect. Soc. 46, 315–322.
Ribeiro, M.F., Velthuis, H.H.W., Duchateau, M.J., Van Der Tweel, I., 1999. Feeding frequency and caste differentiation in *Bombus terrestris* larvae. Insect. Soc. 46, 306–314.
Richards, K.W., 1977. Ovarian development of queen and worker bumble bees (Hmenoptera: Apidae) in southern Alberta. Can. Entomol. 109, 109–116.
Riddiford, L.M., 1981. Hormonal control of epidermal cell development. Integr. Comp. Biol. 21, 751–762.

Roseler, P.F., 1970. Differences in the caste determination between the bumblebee species *Bombus hypnorum* and *Bombus terrestris*. Z. Naturforsch. 25, 543–548.
Roseler, P.F., 1976. Juvenile hormone and queen rearing in bumblebees. In: Lüscher, M. (Ed.), Phase and Caste Determination in Insects: Endocrine Aspects. Pergamon Press, New York, NY, pp. 55–61.
Roseler, P.F., 1977. Juvenile hormone control of oogenesis in bumblebee workers, *Bombus terrestris*. J. Insect Physiol. 23, 985–992.
Roseler, P.F., 1985. A technique for year-round rearing of *Bombus terrestris* (Apidae, Bombini) colonies in captivity. Apidologie 16, 165–170.
Roseler, P.F., 1989. Roles of morphogenetic hormones in caste polymorphism in bumble bees. In: Gupta, A.P. (Ed.), Morphogentic of Arthropods: Roles in Histogenesis, Organogenesis, and Morphogenesis. Rutgers University Press, New Brunswick, NJ, pp. 385–399.
Roseler, I., Roseler, P.F., 1978. Studies on the regulation of juvenile hormone titer in bumblebee workers, *Bombus terrestris*. J. Insect Physiol. 24, 707–713.
Roseler, P.F., Roseler, I., 1984. Effects of carbon dioxide and brain cauterization on corpora allata activity and oogenesis in bumblebees (*Bombus hypnorum* and *Bombus terrestris*). Zool. Jb. Physiol. 88, 237–246.
Roseler, P.F., Roseler, I., 1988. Influence of juvenile hormone on fat-body metabolism in ovariolectomized queens of the bumblebee, *Bombus terrestris*. Insect Biochem. 18, 557–563.
Roseler, P.F., Van Honk, C.G.J., 1990. Cast and reproduction in bumblebees. In: Engels, W. (Ed.), Social Insects: An Evolutionary Approach to Castes and Reproduction. Springer Verlag, Berlin, pp. 147–166.
Roseler, P.F., Roseler, I., Strambi, A., 1980. The activity of corpora allata in dominant and subordinated females of the wasp *Polistes gallicus*. Insect. Soc. 27, 97–107.
Roseler, P.F., Roseler, I., van Honk, C.G.J., 1981. Evidence for inhibition of corpora allata in workers of *Bombus terrestris* by a pheromone from the queen's mandibular gland. Experientia 37, 348–351.
Roseler, P.F., Roseler, I., Strambi, A., Augier, R., 1984. Influence of insect hormones on the establishment of dominance hierarchies among foundresses of the paper wasp, *Polistes gallicus*. Behav. Ecol. Sociobiol. 15, 133–142.
Sadd, B.M., Barribeau, S.M., et al., 2015. The genomes of two key bumblebee species with primitive eusocial organisation. Genome Biol. (under review).
Sakagami, S.F., 1976. Specific differences in the bionomic characters of bumblebees. A comparative review. J. Fac. Sci. Hokkaido Univ. 20, 390–447.
Schmid-Hempel, R., Schmid-Hempel, P., 2000. Female mating frequencies in *Bombus* spp. from Central Europe. Insect. Soc. 47, 36–41.
Shi, L., Lin, S., Grinberg, Y., Beck, Y., Grozinger, C.M., Robinson, G.E., Lee, T., 2007. Roles of *Drosophila* Kruppel-homolog 1 in neuronal morphogenesis. Dev. Neurobiol. 67, 1614–1626.
Shpigler, H., Patch, H.M., Cohen, M., Fan, Y., Grozinger, C.M., Bloch, G., 2010. The transcription factor Kruppel homolog 1 is linked to hormone mediated social organization in bees. BMC Evol. Biol. 10, 120.
Shpigler, H., Tamarkin, M., Gruber, Y., Poleg, M., Siegel, A.J., Bloch, G., 2013. Social influences on body size and developmental time in the bumblebee *Bombus terrestris*. Behav. Ecol. Sociobiol. 67, 1601–1612.
Shpigler, H., Amsalem, E., Huang, Z.Y., Cohen, M., Siegel, A.J., Hefetz, A., Bloch, G., 2014. Gonadotropic and physiological functions of juvenile hormone in bumblebee (*Bombus terrestris*) workers. PLoS One 9, e100650.
Sibbald, E.D., Plowright, C.M.S., 2013. On the relationship between aggression and reproduction in pairs of orphaned worker bumblebees (*Bombus impatiens*). Insect. Soc. 60, 23–30.

Sibbald, E.D., Plowright, C.M., 2014. Social interactions and their connection to aggression and ovarian development in orphaned worker bumblebees (*Bombus impatiens*). Behav. Processes 103, 150–155.

Sim, C., Denlinger, D.L., 2013. Insulin signaling and the regulation of insect diapause. Front. Physiol. 4, 189.

Skyrm, K., Rao, S., Stephen, W., 2012. A scientific note on a trend towards bivoltinism in Western North American bumblebees. Apidologie 43, 82–84.

Song, J., Wu, Z., Wang, Z., Deng, S., Zhou, S., 2014. Krüppel-homolog 1 mediates juvenile hormone action to promote vitellogenesis and oocyte maturation in the migratory locust. Insect Biochem. Mol. Biol. 52, 94–101.

Spaethe, J., Chittka, L., 2003. Interindividual variation of eye optics and single object resolution in bumblebees. J. Exp. Biol. 206, 3447–3453.

Spaethe, J., Weidenmuller, A., 2002. Size variation and foraging rate in bumblebees (*Bombus terrestris*). Insect. Soc. 49, 142–146.

Sramkova, A., Schulz, C., Twele, R., Francke, W., Ayasse, M., 2008. Fertility signals in the bumblebee *Bombus terrestris* (Hymenoptera: Apidae). Naturwissenschaften 95, 515–522.

Steen, J.J.M.V.D., 2008. Infection and transmission of *Nosema bombi* in *Bombus terrestris* colonies and its effect on hibernation, mating and colony founding. Apidologie 39, 273–282.

Stelzer, R.J., Chittka, L., Carlton, M., Ings, T.C., 2010. Winter active bumblebees (*Bombus terrestris*) achieve high foraging rates in urban Britain. PLoS One 5, e9559.

Strambi, A., Strambi, C., Roseler, P.F., Roseler, I., 1984. Simultaneous determination of juvenile hormone and ecdysteroid titers in the Hemolymph of bumblebee prepupae (*Bombus hypnorum* and *B. terrestris*). Gen. Comp. Endocrinol. 55, 83–88.

Strassmann, J., 2001. The rarity of multiple mating by females in the social Hymenoptera. Insect. Soc. 48, 1–13.

Sutcliffe, G.H., Plowright, R.C., 1988. The effects of food supply on adult size in the bumblebee *Bombus terricola*. Can. Entomol. 120, 1051–1058.

Takahashi, J.I., Itoh, M., Shimizu, I., Ono, M., 2008. Male parentage and queen mating frequency in the bumblebee *Bombus ignitus* (Hymenoptera: Bombinae). Ecol. Res. 23, 937–942.

Takahashi, J.-I., Martin, S., Ono, M., Shimizu, I., 2010. Male production by non-natal workers in the bumblebee, *Bombus deuteronymus* (Hymenoptera: Apidae). J. Ethol. 28, 61–66.

Tasei, J.N., 1994. Effect of different narcosis procedures on initiating oviposition of prediapausing *Bombus terrestris* queens. Entomol. Exp. Appl. 72, 273–279.

Trivers, R.L., Hare, H., 1976. Haplodiploidy and the evolution of the social insects. Science 191, 249–263.

van Honk, C., Hogeweg, P., 1981. The ontogeny of the social structure in a captive *Bombus terrestris* colony. Behav. Ecol. Sociobiol. 9, 111–119.

van Honk, C.G.J., Velthuis, H.H.W., Röseler, P.F., 1978. A sex pheromone from the mandibular glands in bumblebee queens. Experientia 34, 838–839.

van Honk, C.G.J., Velthuis, H.H.W., Roseler, P.F., Malotaux, M.E., 1980. The mandibular glands of *Bombus terrestris* queens as a source of queen pheromone. Entomol. Exp. Appl. 28, 191–198.

van Honk, C.G.J., Röseler, P.F., Velthuis, H.H.W., Hoogeveen, J.C., 1981. Factors influencing the egg laying of workers in a captive *Bombus terrestris* colony. Behav. Ecol. Sociobiol. 9, 9–14.

Van Oystaeyen, A., Oliveira, R.C., Holman, L., et al., 2014. Conserved class of queen pheromones stops social insect workers from reproducing. Science 343, 287–290.

Vesterlund, S.R., Lilley, T.M., van Ooik, T., Sorvari, J., 2014. The effect of overwintering temperature on the body energy reserves and phenoloxidase activity of bumblebee *Bombus lucorum* queens. Insect. Soc. 61, 265–272.
Wang, X., Wheeler, D., Avery, A., Rago, A., Choi, J.H., Colbourne, J.K., Clark, A.G., Werren, J.H., 2013. Function and evolution of DNA methylation in *Nasonia vitripennis*. PLoS Genet. 9, e1003872.
Wheeler, D.E., 1986. Developmental and physiological determinants of caste in social Hymenoptera: evolutionary implications. Am. Nat. 128, 13–34.
Williams, P.H., Cameron, S.A., Hines, H.M., Cederberg, B., Rasmont, P., 2008. A simplified subgeneric classification of the bumblebees (genus *Bombus*). Apidologie 39, 46–74.
Wilson, E.O., 1971. The Insect Societies. Belknap Press of Harvard University press, Cambridge, MA.
Wilson, E.O., Nowak, M.A., 2014. Natural selection drives the evolution of ant life cycles. Proc. Natl. Acad. Sci. U.S.A. 111, 12585–12590.
Woodard, S.H., Bloch, G., Band, M.R., Robinson, G.E., 2013. Social regulation of maternal traits in nest-founding bumble bee (*Bombus terrestris*) queens. J. Exp. Biol. 216, 3474–3482.
Woodard, S.H., Bloch, G.M., Band, M.R., Robinson, G.E., 2014. Molecular heterochrony and the evolution of sociality in bumblebees (*Bombus terrestris*). Proc. R. Soc. B Biol. Sci. 281, 20132419.
Worden, B.D., Skemp, A.K., Papaj, D.R., 2005. Learning in two contexts: the effects of interference and body size in bumblebees. J. Exp. Biol. 208, 2045–2053.
Yeninar, H., Duchateau, M.J., Kaftanoglu, O., Velthuis, H., 2000. Colony developmental patterns in different local populations of the Turkish bumble bee, *Bombus terrestris dalmatinus*. J. Apic. Res. 39, 107–116.
Yerushalmi, S., Bodenhaimer, S., Bloch, G., 2006. Developmentally determined attenuation in circadian rhythms links chronobiology to social organization in bees. J. Exp. Biol. 209, 1044–1051.
Yoon, H.J., Lee, K.Y., Hwang, J.S., Park, I.G., 2010. Chilling temperature and humidity to break diapause of the bumblebee queen *Bombus terrestris*. Int. J. Indust. Entomol. 20, 93–98.
Zanette, L.R., Miller, S.D., Faria, C.M., Almond, E.J., Huggins, T.J., Jordan, W.C., Bourke, A.F., 2012. Reproductive conflict in bumblebees and the evolution of worker policing. Evolution 66, 3765–3777.
Zanette, L.R.S., Miller, S.D.L., Faria, C.M.A., Lopez-Vaamonde, C., Bourke, A.F.G., 2014. Bumble bee workers drift to conspecific nests at field scales. Ecol. Entomol. 39, 347–354.
Zucci, R., 1973. Aspecto bionomicos de *Exomalopsis aureopilosa* e *Bombus atratus* incluindo consideracoes sobre a evolucao do comportamento social (Hymenoptera, Apoidea). PhD Thesis, Faculdad de Filosofia, Cienca e Letras de Ribeirao Preto, SP, Brasil, 172 pp.

CHAPTER THREE

Physiological and Genomic Mechanisms of Social Organization in Wasps (Family: Vespidae)

Jennifer M. Jandt*, Amy L. Toth*,†,1
*Department of Ecology, Evolution, and Organismal Biology, Iowa State University, Ames, Iowa, USA
†Department of Entomology, Iowa State University, Ames, Iowa, USA
[1]Corresponding author: e-mail address: amytoth@iastate.edu

Contents

Advances in Insect Physiology, Volume 48
ISSN 0065-2806
http://dx.doi.org/10.1016/bs.aiip.2015.01.003

Abstract

The family Vespidae provides an excellent group for studying transitional steps associated with the evolution of social behaviour. The family contains species with a wide range of social behaviour, from solitary to highly eusocial, with variation both across and within social subfamilies related to the extent of differentiation of reproductive queen and non-reproductive worker castes. Here, we explore physiological and genomic mechanisms that influence the development of reproductive castes—in both larval and adult stages—and mechanisms that influence non-reproductive division of labour among workers. We synthesize what is known to begin to understand wasp social evolution from a mechanistic perspective. However, since most studies on mechanisms of social organization in wasps have focused on the model genus *Polistes*, we point out a need for more studies on solitary and advanced eusocial species. We emphasize that by filling these gaps, future comparative studies of these mechanisms will provide key insights into hypotheses underlying the evolution of sociality—such as solitary ground plan, novel genes, and genetic toolkit hypotheses.

1. INTRODUCTION

1.1 Why study physiology and genomics of Vespidae?

Mechanisms that influence social behaviour have been studied most extensively in other hymenopteran families, especially the bees (in the family Apidae) and ants (family Formicidae). However, if the goal is to glean evolutionary insights, the family Vespidae is an excellent study system due to the presence of solitary through highly social species (Fig. 1). By understanding the mechanisms, such as nutritional and reproductive physiology, hormones, and genes, that regulate social behaviour across species with varying degrees of sociality, we can gain valuable insights into *how* eusociality may have evolved. Mechanistic hypotheses developed in wasps (West-Eberhard, 1996) have been fruitfully used to generate new, empirically testable hypotheses about social evolution (Johnson and Linksvayer, 2010; Linksvayer and Wade, 2005). Studies of wasps, paired with those of other independently evolved social lineages such as bees and ants, provide an informative 'natural experiment' to examine whether convergently evolved phenotypes are associated with the same or different physiological and genetic mechanisms.

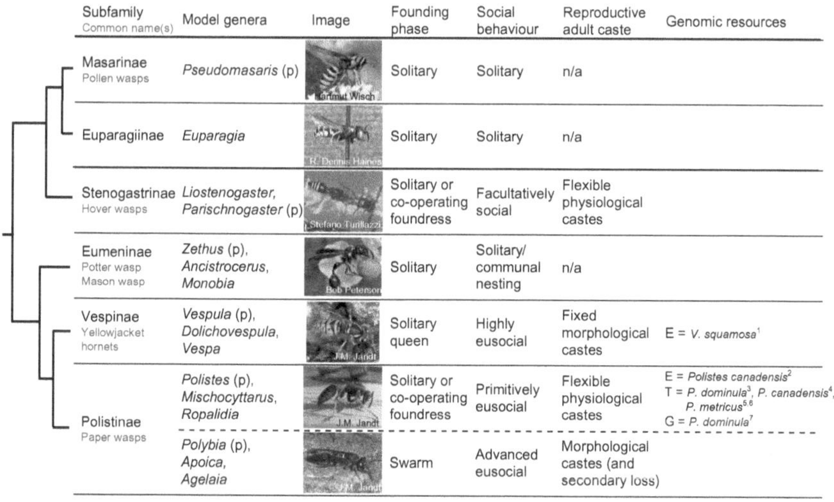

Subfamily Common name(s)	Model genera	Image	Founding phase	Social behaviour	Reproductive adult caste	Genomic resources
Masarinae Pollen wasps	*Pseudomasaris* (p)		Solitary	Solitary	n/a	
Euparagiinae	*Euparagia*		Solitary	Solitary	n/a	
Stenogastrinae Hover wasps	*Liostenogaster*, *Parischnogaster* (p)		Solitary or co-operating foundress	Facultatively social	Flexible physiological castes	
Eumeninae Potter wasp Mason wasp	*Zethus* (p), *Ancistrocerus*, *Monobia*		Solitary	Solitary/ communal nesting	n/a	
Vespinae Yellowjacket hornets	*Vespula* (p), *Dolichovespula*, *Vespa*		Solitary queen	Highly eusocial	Fixed morphological castes	E = *V. squamosa*[1]
Polistinae Paper wasps	*Polistes* (p), *Mischocyttarus*, *Ropalidia*		Solitary or co-operating foundress	Primitively eusocial	Flexible physiological castes	E = *Polistes canadensis*[2] T = *P. dominula*[3], *P. canadensis*[4], *P. metricus*[5,6] G = *P. dominula*[7]
	Polybia (p), *Apoica*, *Agelaia*		Swarm	Advanced eusocial	Morphological castes (and secondary loss)	

Figure 1 Social characteristics and available genomic information for the six subfamilies (and most commonly studied genera) of vespid wasps. (p), genus in photo image; n/a, not applicable (because solitary); E, some EST sequences; T, transcriptome; G, genome. References for genomic resources: [1](Hoffman and Goodisman, 2007), [2](Sumner et al., 2007), [3](A.L. Toth, unpublished data), [4](Ferreira et al., 2013), [5](Toth et al., 2007), [6](Berens et al., 2015), [7](A.L Toth, unpublished data).

Within the Vespidae, a complete spectrum of solitary to primitively and advanced eusocial species is represented within a single monophyletic lineage, making it one of the best systems for studying the evolution of sociality (see Table 1 for descriptions of levels of sociality). Some phylogenies of the Vespidae, based on morphological and behavioural traits, indicate a single origin of sociality in the family (reviewed by Carpenter, 1991). A different study, based solely on molecular data, suggests two independent origins (Hines et al., 2007), although this interpretation has been challenged (Pickett and Carpenter, 2010). More data are needed to resolve these relationships, but if the two-origin hypothesis is correct, then the transition to eusociality can be studied in two lineages within the family. In addition, there is evidence that morphological castes have been gained and lost multiple times within the vespids (Noll and Wenzel, 2008; Noll et al., 2004). Thus, the vespids provide unparalleled opportunities to engage in comparative, phylogenetically controlled analyses of the mechanisms underlying transitions in social evolution.

Vespid wasps have been important models for studying the evolution of cooperation and eusociality. There has been much emphasis on studies of the model genus *Polistes*, which exhibits an intermediate form of social

Table 1 A summary and brief description of the types of social behaviour found in Vespidae and the wasp taxa in which each type of behaviour are found

Form of social behaviour	Description	Example
Subsocial	Maternal behaviour with prolonged mother–offspring interactions, no reproductive castes	Eumeninae
'Nest sharing'	Multiple reproductive females on the same nest, no evidence of sterile workers	Eumeninae
Facultatively eusocial	Reproductive castes sometimes present (flexible within a species)	Stenogastrinae
Primitively eusocial	Reproductive castes always present but caste switching can occur, no morphological differences between castes	Solitary-founding Polistinae
Highly eusocial	Reproductive castes always present, morphological differences between castes, nests founded by single female	Vespinae
Advanced eusocial	Reproductive castes always present, morphological differences between castes,[a] nests founded by swarms	Swarm-founding Polistinae

[a]Secondary loss in some species.

behaviour referred to as primitive eusociality (Table 1) that involves a unique blend of both cooperation and conflict. This has led to a wealth of studies elucidating both the importance of inclusive fitness (Field et al., 1998; Klahn, 1979; Reeve, 1991; Ross and Gamboa, 1981; Seppä et al., 2002; Strassmann, 1981), as well as direct benefits to the evolution of cooperation (Queller et al., 2000; Reeve, 1991; Zanette and Field, 2008). However, beyond *Polistes*, there is much more to be learned about the evolution of sociality by studying other vespid wasps, and in particular via a deeper understanding of the physiological and genomic mechanisms of sociality in this group of insects.

Although there have been numerous studies on the physiological and genomic basis of social organization in primitively eusocial species in the genus *Polistes* (reviewed in Jandt et al., 2014), relatively few of these studies have utilized mechanistic data for an integrative understanding of the evolution of sociality. Fewer still have considered the roles of physiological and genomic mechanisms over the course of social evolution by comparing taxa at different levels of sociality (Table 1).

1.2 The purpose of this review

Here, we summarize what is known about the physiological and genomic basis of sociality throughout the family Vespidae. Specifically, we explore the mechanisms that influence reproductive division of labour—in both the larval and adult stages—as well as non-reproductive division of labour among workers. For each, we summarize what is known for each of the major subfamilies. Upon reviewing what is known on these factors, we highlight the imbalance in the amount of information available for certain species relative to others. Finally, we synthesize this information in light of some of the major mechanistic hypotheses for the evolution of sociality and point out gaps in our knowledge, along with exciting opportunities for future research in which we suggest that comparative studies within Vespidae can provide new insights.

1.3 The major groups of vespids and their social behaviour

There are six recognized subfamilies of vespid wasps: Eumeninae, Euparagiinae, Masarinae, Stenogastrinae, Polistinae, and Vespinae (Fig. 1). Because of a nearly complete lack of information on mechanisms of behaviour in two of these (Euparagiinae and Masarinae), our review will focus on the remaining four subfamilies. Throughout the review, we will cover each of these four subfamilies, roughly in order of increasing social complexity (Eumeninae–Stenogastrinae–Vespinae–Polistinae, Table 1). For the purpose of this review, we provide a general overview of the biology of each of these subfamilies. Because this review is not intended to be a complete description of the social biology of wasps, we have necessarily omitted some details about the biology of each family, including references to species for which there are valuable behavioural data. Instead, we narrow our focus to species and genera for which there is available physiological and genetic data.

Among the solitary subfamilies, the majority of studies have been conducted on species in the subfamily Eumeninae (i.e. potter wasps and mason wasps). Among the eumenines, there are some species with well-developed maternal care that includes prolonged brood feeding by mother wasps (also called progressive provisioning)—an important precursor to the evolution of sociality (e.g. *Zethus*, *Eudynerus*, and *Ancistrocerus*, West-Eberhard,

1987a). There are also examples of eumenines with even more developed social behaviour, involving nest sharing by multiple reproductive females (e.g. *Zethus* and *Xenorhynchium*, West-Eberhard, 1987a,b) and even overlapping generations (e.g. *Montezumia*, West-Eberhard, 2005). Moreover, although they are solitary, *A. antelope* show a high degree of relatedness among nests within a population (indicative of high levels of inbreeding) (Chapman and Stewart, 1996). These data suggest that high relatedness could have been a pre-condition for the evolution of sociality in vespids. Here, we review the relatively scant literature on physiological factors in these genera that correlate with either reproductive or maternal care behaviour and focus on these factors because of their importance to understanding early stages of social evolution.

The distribution of Stenogastrinae (i.e. hover wasps) is limited to rainforests of the Indo-Pacific region (Turillazzi, 1991). Stenogastrines are considered facultatively eusocial in that female offspring may remain on the nest and care for sisters, and reproductive females may choose to cooperate and found nests together (Turillazzi, 1991). However, all females from a given nest, at any time of the year, can leave the nest, mate, and start a new colony. Colony sizes are very small, on the order of 1–5 individuals in *Parischnogaster nigricans serrei*, for example (Turillazzi, 1991). Among hover wasps, there are species that build nests out of mud or clay (similar to Eumeninae, but see Hermes et al., 2013) and other species that build nests out of plant fibres (similar to Polistinae and Vespinae). In fact, the genus *Liostenogaster* comprises species that do both. However, unlike characteristic polistine and vespine nests, stenogastrines do not construct a petiole from which their cells are built. Instead, the cells are attached directly to the substrate, occasionally scattered or arranged in rows (Turillazzi, 2013). Well-studied genera include *Liostenogaster* and *Parischnogaster* (Turillazzi, 2013).

The subfamily Vespinae (i.e. yellowjackets and hornets) is comprised primarily of social species, most of which are highly eusocial with moderate to large colony sizes (e.g. colonies can produce 50–2000 workers in *V. germanica*, Malham et al., 1991). Colonies generally live in temperate regions with an annual life cycle and are founded by a solitary queen (Matsuura and Yamane, 1990). Notably, there are some exceptions of perennial colonies, multi-queen species, or species with a secondary loss of sociality in which females lay eggs in the nests of other social species (called 'social parasites', Greene, 1991; Matsuura and Yamane, 1990). Genetic relatedness within colonies is generally low due to multiple mating by queens; this genetic diversity appears to provide fitness benefits to the colony (Goodisman et al., 2007). Queens and workers have distinct morphologies (called 'morphological castes'), and there is an age-related worker division of labour. Well-studied genera include: *Vespula*, *Dolichovespula*, and *Vespa* (Greene, 1991; Matsuura and Yamane, 1990).

Finally, the social subfamily Polistinae (i.e. paper wasps) can be further subdivided into two major groups: the independent-founding species (primitively eusocial, including some social parasites) and the swarm-founding species (in the tribe Epiponini, advanced eusocial, see Table 1). Within the primitively eusocial species, queen and worker castes are flexible (i.e. instead of morphological differences, queens and workers are behaviourally/physiologically different), and the primary reproductive is established through a dominance hierarchy (reviewed in Jandt et al., 2014). Colony sizes are small to intermediate (e.g. from 10 to 150 individuals in *Polistes dominula*, Hunt, 2007). Well-studied genera include *Polistes*, *Mischocyttarus*, and *Ropalidia*. In contrast, in the advanced eusocial Epiponini, colonies may be very large (up to 1 million individuals in *Agelaia vicina*, Hunt, 2007; Zucchi et al., 1995), workers specialize on different tasks, and there is a division of nest-building behaviour among workers. Morphological differences between castes (i.e. size dimorphic queens and workers) have been gained and lost within this tribe. Well-studied genera include *Polybia*, *Apoica*, and *Agelaia* (Jeanne, 1991).

2. LARVAL DEVELOPMENT AND ADULT REPRODUCTIVE STATE

Insect larvae undergo large genomic and physiological changes throughout development. Which genes are expressed, and the timing of expression throughout larval development can have significant repercussions on the reproductive physiology of the emerging adult (Pereboom et al., 2005). Here, we explore how physiological and genomic changes throughout larval development, many of which are dependent upon social interactions with adults or differential nutrition, influence reproductive state of adults, with particular attention to queen and worker reproductive caste differences in social species.

2.1 Eumeninae

Although there are no reproductive castes in solitary eumenine wasps, it can be informative to examine the influences of nutrition and hormones during larval development on adult phenotypes, especially in females. This is because comparing the ways in which solitary and social wasps respond to the nutritional and hormonal environment can provide insights into the ancestral physiological state of female wasps during development. In *Euodynerus foraminatus* and *Ancistrocerus adiabatus*, there appear to be nutrition-dependent effects on fertility in females. Fertilized (female) eggs are provisioned with more food than unfertilized male eggs (Cowan, 1981), which leads to disproportionately larger females (Cowan, 1983), and in *E. foraminatus*, the largest females tend to develop larger ovaries (Tibbetts et al., 2013). In another eumenine species (*Pachodynerus nasidens*), female eggs are larger than male eggs (Cowan, 1983).

Nearly all of the work done on hormones in vespids has focused on juvenile hormone (JH). JH is a terpenoid hormone with many roles in insects, one of the most important being that it is a main gonadotropic hormone in females. Typically, higher JH in larvae and/or adults results in more ovarioles and higher ovary activation (the chapter 'The Physiology and Genomics of Social Transitions in Aphids' by Abbot, this volume; the chapter 'Bumble Bee Sociobiology: The Physiological and Genomic Bases of Bumble Bee Social Behaviour' by Amsalem et al., this volume; the chapter 'Old Threads Make New Tapestry—Rewiring of Signalling Pathways Underlies Caste Phenotypic Plasticity in the Honey Bee, *Apis mellifera* L.' by Hartfelder et al., this volume; the chapter 'Juvenile Hormone: A Central Regulator of Termites' Caste Polyphenism' by Korb, this volume). In eumenine wasps, as is well known in many solitary insects, levels of JH are also correlated with ovary development. When treated with methoprene (a JH analog), *E. foraminatus* females develop larger ovaries (Tibbetts et al., 2013).

On the level of protein expression, in *Monobia quadridens*, storage proteins such as hexamerin increase in expression through female development, with a jump in expression as larvae transition from their final instar to prepupal larvae to pupae (Hunt et al., 2003). Moreover, hexamerin levels vary among newly emerged adults—small females show little to no expression whereas large females show much higher expression. Thus, overall in eumenine females, there is a typical solitary insect relationship between higher nutrition, higher levels of JH, increased storage protein expression, and higher fertility.

2.2 Stenogastrinae

Although there have been no direct studies of larval physiology in stenogastrines, the scant evidence available suggests there is no effect of larval nutrition on adult reproductive state and behaviour. Upon eclosion, *Liostenogaster flavolineata* adult females are totipotent regarding their reproductive potential—all females have the potential to leave the nest, mate, and found a new colony (Field and Foster, 1999). Further, there is no evidence that body size predicts whether females will leave to start their own nest or stay and help at the natal nest (i.e. larger females are not more likely to engage in nest-founding behaviour, Field et al., 1999).

2.3 Vespinae

In *Vespula* yellowjackets, there is a clear relationship between the size of female larvae and their adult morphological caste. Larvae reared in large nest cells, produced at the end of the colony cycle, are fed disproportionately more than larvae reared in 'worker-cells' (Schmidt et al., 2012), resulting in larger adult size and individuals that typically become 'gynes' or future reproductive queens. Larger larval size may be related to JH as well. Gyne larvae show delayed late instar moults, and it has been hypothesized that an increase in JH at a critical developmental period may be responsible (Greene, 1991; Nijhout and Wheeler, 1982), but we currently lack empirical tests of this hypothesis.

With respect to gene expression, an expressed sequence tag (EST)-based study in *Vespula squamosa* examined expression patterns of several hundred genes throughout larval development (Hoffman and Goodisman, 2007). The results indicate that gene expression patterns are quite similar in young queen and worker larvae, but diverge more as larvae develop. For example, later in development, queen larvae show higher expression of genes related to metabolic processes. Specific hexameric storage proteins show different patterns of expression depending on caste: e.g., *VSQ019* and *VSQ233* are up-regulated in queen-destined larvae, whereas *VSQ232* and *VSQ292* (both *hexamerin 70b*-like ESTs) are up-regulated in worker-destined larvae (Hoffman and Goodisman, 2007). Overall, the available data from Vespinae suggest that, as in Eumeninae, higher larval nourishment is associated with larger size, increased expression of hexameric storage proteins, and a hypothesized connection to higher JH.

2.4 Polistinae

In primitively eusocial polistines, there is also a connection between higher female larval nourishment and increased adult reproduction. In *Ropalidia marginata*, better fed larvae often become reproductively mature earlier (Gadagkar, 2009). In *Polistes* larvae, there is a well-documented connection between larval feeding rate (which is highly correlated with the stage of colony development, being lower in early development), and the adult caste fate of these larvae. Gyne-destined larvae experience a higher adult:larva ratio, higher feeding rates (Hunt, 2007), and have significantly higher lipid stores than worker-destined larvae (Fig. 2; Jandt et al., in review). Experimental nourishment deprivation of *P. metricus* larvae led to some worker-like changes in ovarian physiology and development time (Judd et al., 2015). There is less work on the connection between larval nutrition and caste in the advanced eusocial epiponines. Observations of feeding rate suggest that morphological differences between castes in *Polybia sericea* are also the result of differential feeding (Desuó et al., 2011). Although more studies are needed on epiponines, including species with more pronounced morphological caste differences than seen in *Polybia*, there is no reason to expect differential larval nutrition is not also, at least partially, responsible for caste differences.

With respect to gene expression, we again see connections between nourishment level and expression of hexameric storage proteins in primitively eusocial polistines. In *P. metricus*, gyne-destined larvae exhibit higher levels of storage proteins, including hexameric storage proteins compared to worker-destined larvae (Hunt et al., 2007, 2010). Like Vespinae, storage

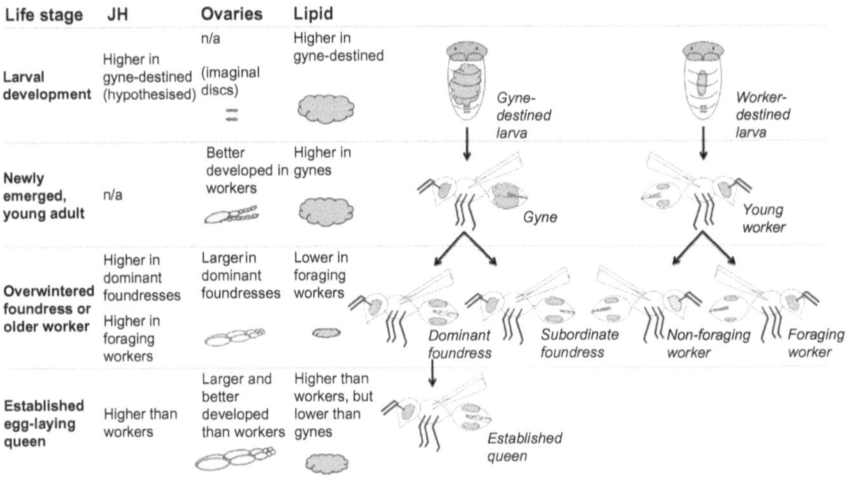

Figure 2 A summary of some of the physiological differences (juvenile hormone (JH), ovary, and lipid) in the development of queen and worker caste in vespid wasps (most data based off of studies with *Polistes*). Larvae referred to as 'gyne-destined' represent those that are likely to develop into future reproductive queens. n/a refers to instances in which data are not available. JH shows a dual function among adults, in a condition-dependent manner: it affects reproductive dominance among better-nourished foundresses and foraging division of labour among more poorly nourished workers.

proteins are synthesized towards the end of larval growth, and gradually diminish throughout pupation (Hunt et al., 2003). However, RNA-interference-based knockdown of expression levels of *hexamerin 2* had no influence on the development of worker/gyne phenotype (Hunt et al., 2011). Instead, *hexamerin 2* may influence larval development time and ovary development (Hunt et al., 2007, 2011). In *P. fuscatus*, gyne-destined larvae show higher expression of *heat-shock protein 90α*, *cytochrome P450*, *inositol oxygenase* (an oxidoreductase), and *long-wave opsin* (Jandt et al., in review). Large-scale genomic analyses of gene expression using RNA-sequencing have shown that many more genes beyond *hexamerins* are differentially expressed in gyne- and worker-destined larvae, and in response to differential nourishment. Berens et al. (2015) identified nearly 800 transcripts that differed in expression between gyne-destined and worker-destined *P. metricus* larvae, many of which related to oxidation–reduction processes, and lipid and carbohydrate metabolism pathways. In addition, experimental manipulations of nourishment level resulted in shifts in the expression of some caste-related genes (Berens et al., in review). These data implicate nutritional physiology and nutrition-related pathways as important for larval caste bias in *Polistes*. However, these data also showed that many caste-related genes are not responsive to nutritional state, suggesting other non-nutritional factors, including social environmental inputs such as vibrational communication (Suryanarayanan et al., 2011) affect caste differentiation in *Polistes* larvae.

3. ADULT REPRODUCTIVE STATE AND CASTE

As adults, female vespids vary in reproductive success and behaviour. This variation can result from both environmental inputs during larval development, but also from the abiotic and social environment surrounding them as adults. Below, we review what is known about the mechanisms underlying variation in reproductive success (in solitary species and among co-foundresses) or queen–worker caste differences (in eusocial species only). Note that in primitively eusocial vespids, there are also large gradients in female reproductive success related to dominance status within a social hierarchy, both within the queen and worker castes. This subject has received considerable attention in *Polistes*, and will not be covered in detail in this review; for more information, we refer the reader to a recent review that describes physiological, exocrine, and genomic mechanisms of reproductive dominance in this model genus (Jandt et al., 2014).

3.1 Eumeninae

As mentioned above, wasps in the subfamily Eumeninae do not produce queen and worker castes, but there are differences in reproductive potential among adults. Regardless of body size, adult females of *E. foraminatus* that consume prey produce eggs that are more than twice the size of those produced by females where prey is withheld (Chilcutt and Cowan, 1992). There are also reproduction-related differences among females in terms of their aggressive behaviour. *Symmorphus cristatus* females exhibit aggressive territorial behaviour over nesting sites (Sears et al., 2001). However, there are no studies that explore the physiological or genetic mechanisms of this aggressive behaviour in solitary vespids.

Despite lacking castes, there are observations from eumeninae that suggest cycles of reproductive and non-reproductive activity. *Zethus miniatus* females, after laying an egg and while the next egg matures, engage in behaviours more typical of 'workers'—e.g., nest cell building, nest defence (West-Eberhard, 1987a). In the group nesting *Montezumia cortesioides*, females alternate between brood care and aggressive competition for egg-laying opportunities (West-Eberhard, 2005). The possible existence of an ovarian cycle, while not well-documented in eumenines, has formed the basis of some important ideas about the evolution of sociality from maternal behaviour (i.e. the Ovarian Ground Plan, West-Eberhard, 1996, discussed below). Such transitions between a queen-like and a worker-like phase are not unlike those observed in the clonal ant *Cerapachys biroi*. Genomic analyses of *C. biroi* show that shifts between reproductive and foraging phases are accompanied by changes in the expression of genes known to be related to reproduction and foraging in honey bees (e.g. *vitellogenin*, *foraging*, and *malvolio*, Oxley et al., 2014). For eumenines, it will be valuable to focus future studies on possible phasic transitions between queen-like to worker-like behaviour. Specifically, we suggest evolutionary insights can be gained by investigating (a) whether phasic shifts are indeed common in eumenines and whether they are accompanied by physiological changes and (b) whether such phasic shifts are accompanied by changes in the expression of genes related to caste differentiation in social wasps.

3.2 Stenogastrinae

Stenogastrines can divide reproductive behaviour in two ways: between co-foundresses and between queens and daughters that remain at the nest. In *L. flavolineata*, there is no evidence that non-reproductive individuals are physiologically constrained from developing viable ovaries (Field and Foster, 1999). Among those individuals that remain at the nest, the reproductive dominance hierarchy is linear and age-based (Bridge and Field, 2007). In *Parischnogaster nigricans*, ovaries develop in females as they age, though some develop faster than others (Turillazzi, 1991). Moreover, in both *P. nigricans* and *L. vechti*, mated females tend to have better developed ovaries than unmated females, although it is unknown whether ovarian development increases the odds of mating, or if mating induces ovarian development (Turillazzi, 1985, 1991). Nothing is known about genetic correlates of adult caste; all that is currently known related to stenogastrine genetics is that in *Parischnogaster alternata*, colonies are comprised of related individuals (Bolton et al., 2006).

3.3 Vespinae

Vespine queens are allometrically morphologically different from workers. Before founding a new colony, queens undergo a period of hibernation, during which time the high fat stores accumulated pre-hibernation can be depleted by 30–40% (Matsuura and Yamane, 1990). The single reproductive queen in vespine colonies is the only individual in the colony that can mate and lay diploid (female) eggs. Workers may compete for reproductive opportunities, develop ovaries towards the end of the season, and lay unfertilized (male) eggs, but these are often eaten or 'policed' by other workers (Bonckaert et al., 2012).

In *Vespula maculifrons*, there is no evidence that morphological caste differences between queens and workers are based on hereditary genetic differences (Kovacs et al., 2010). Regarding differential gene expression, there has been a single study that has investigated caste-related expression in adults of three vespine species using an EST-based approach (Hunt and Goodisman, 2010). Across *V. maculifrons*, *V. squamosa*, and *Dolichovespula maculata*, caste-related gene expression levels between species vary widely. Out of seven genes that had caste-related gene expression in honeybees, two had conserved caste-specific patterns across the three vespines (Hunt and Goodisman, 2010). In general, this study found that caste-biased gene expression profiles in vespine wasps seem to have undergone greater rates of evolutionary change than sex-biased profiles. Thus, these data suggest that only a small subset of caste-related genes show conservation across species, providing limited support for the idea that a shared 'genetic toolkit' underlies caste differences in social insects (Hunt and Goodisman, 2010). The small amount of overlap across vespine species is not surprising; recent studies comparing wasps, ants, and bees have also found limited overlap in the expression of specific caste-related genes across species. Instead, there is similarity in which pathways and gene functions are related to caste differences across species (Berens et al., 2015; Toth et al., 2014).

3.4 Polistinae

In the tropical, primitively eusocial genera *Ropalidia* and *Mischocyttarus*, females that eclose at any time of the year have a chance of becoming queen. In tropical wasps, where habitats are less seasonal, wasps can initiate colonies at any time of the year, and reproductive physiology can be flexible, changing in response to social conditions. Still, in *R. marginata*, older females tend to have more well-developed ovaries, and wasps fed well are more likely to become egg layers than those that are not (Gadagkar, 2009). JH likely plays a role, as topical application of JH on *R. marginata* adults accelerates ovarian development (Agrahari and Gadagkar, 2003). In *M. mastigophorus*, there is no evidence that body size correlates with reproductive potential, but those individuals with higher fat stores tend to have better developed ovaries (Markiewicz and O'Donnell, 2001). Moreover, there are differences in brain neuroanatomy between queens and workers: the mushroom bodies (the brain region associated with sensory integration and learning) are larger in queens compared to young workers, but there is no difference between the brains of queens and older workers (which also have high reproductive potential, O'Donnell, 2006).

Temperate and tropical species of *Polistes* also possess more subtle, behavioural caste differences, and lack morphologically different queen and worker castes. The ultimate caste fate of females remains flexible throughout much of adult life and is sensitive to changes in the social environment (e.g. dominance hierarchy) as well as individual physiological state. In the temperate-zone *P. metricus*, the queen (mother) has the greatest ovarian development of any colony member. On average, queens have twice as much abdominal lipid compared to their daughter workers, but significantly lower lipids than their daughter gynes (females that will mate, over-winter, and found colonies the following spring). Furthermore, although they are unmated and rarely lay eggs, adult workers actually tend to have better developed ovaries than gynes (which do not activate their ovaries until nest founding the following spring). Instead, gynes have huge lipid stores compared to all other females (Fig. 2; Toth et al., 2009). Among queens and workers, there is a positive correlation between lipid stores and ovary development—those with greater lipid stores also tend to develop larger and better developed ovaries (Toth et al., 2009).

JH is directly linked to fertility in adult *Polistes* (Fig. 2; Röseler, 1991; Röseler et al., 1984; Tibbetts et al., 2011a). Moreover, experimentally increasing JH among foundresses leads to an increase in fertility as well (Tibbetts and Izzo, 2009; Tibbetts and Sheehan, 2012). Biogenic amines, such as dopamine and serotonin, have also been shown to be positively correlated with fertility in *P. chinensis* (Sasaki et al., 2009). Recently, Toth et al. (2014) found a link in the expression of genes with functions related to JH and biogenic amine synthesis and reproductive dominance status in queen and worker castes in *P. metricus*.

A number of studies have investigated differential gene expression in queen and worker *Polistes* wasps. In *Polistes canadensis*, several genes associated with metabolism exhibited twofold higher expression in whole bodies of queens compared to workers, whereas only two genes (those associated with heat shock and imaginal disk development) were up-regulated in workers (Sumner et al., 2006). Several genes that are well known to be associated with reproduction and caste differences in other species (reviewed in Smith et al., 2008) including *vitellogenin*, *major royal jelly protein*, and *hexamerin 2* also had higher expression in *P. canadensis* queens compared to workers (Sumner et al., 2006).

In *P. metricus* adult female brains and surrounding fat body, increased expression of *vitellogenin* and decreased expression of *insulin-like peptide 2* are associated with higher reproductive potential (Toth et al., 2009). In fact, queens have twofold higher expression in *vitellogenin* than workers, and also show higher expression in genes related to oxidation–reduction (Toth et al., 2014). Queens also exhibit higher ovary gene expression for genes associated with for protein folding, mitotic spindle organization, proteolysis, and metabolism (Toth et al., 2014). These data highlight the inter-relationship between genes and pathways related to nutrition, metabolism, and reproduction and their likely roles in caste differences across species (Smith et al., 2008).

Unlike independent-founding *Polistes* spp., swarm founding *Polybia* colonies can have multiple queens. A surprising and unique feature of some epiponine species is that the presence of multiple queens is associated with diminished morphological and physiological differences between queen and worker castes. In *Polybia ignobilis*, although queens are larger than workers, morphological differences become less apparent as the colony matures or when there are more queens (Desuó et al., 2011). However, this pattern is the opposite in other species. For example, in *Protonectarina sylveirae*, queens are smaller in the early colony stages before new workers emerge compared to colonies in later, male-producing stages (Shima et al., 2003). In *Po. micans* colonies, queens have higher JH titer only when the colony is in the founding phase or if it only has one queen. Otherwise, in established, multi-queen colonies, queens have higher ovarian development and higher ecdysteroid content in the ovaries, but not necessarily differences in JH or ecdysteroids in the haemolymph relative to workers (Kelstrup et al., 2014). Finally, in the temperate-zone *Parapolybia indica*, young females (though not the youngest) are more likely to replace the lost queens (Suzuki, 2003).

Across 12 genera of polistine wasps, including both independent and swarm-founding species, workers tend to invest more than queens in mushroom body brain tissue responsible for visual processing relative to antennal processing (O'Donnell et al., 2014). Queens, however, tend to have significantly larger mushroom body lip and collar regions than workers (the areas associated with learning and cognition), and this phenomenon is more pronounced among independent-founding species (O'Donnell et al., 2014).

4. ADULT WORKER DIVISION OF LABOUR

In addition to dividing reproductive tasks between queens and workers, in eusocial colonies, workers also divide colony maintenance tasks. One of the major divisions of labour is between foragers and non-foragers, which may spend more time on the nest also engaged in brood-tending or nest-building tasks. In many species, worker division of labour is age-related, called 'temporal polyethism'. In contrast, as they are solitary, by definition eumenine wasps do not have workers. Therefore, we focus here on the physiological and genomic mechanisms of worker behaviour only in subfamilies that divide non-reproductive tasks.

4.1 Stenogastrinae

Among stenogastrines, the probability that an individual forages is age-related and correlates with rank in the dominance hierarchy. For example, in *L. flavolineata*, the amount of time individuals spend foraging directly correlates with hierarchical rank and ovarian development (Turillazzi, 1991 and references therein). In *Parischnogaster gracilipes* and *P. alternata*, middle-aged individuals spend the most time foraging, whereas the oldest are egg layers, and the youngest are less capable of foraging (Turillazzi, 1991).

4.2 Vespinae

Vespula germanica workers exhibit an age-based division of labour such that younger individuals are more likely to engage in in-nest tasks, whereas older individuals are more likely to engage in foraging tasks (Hurd et al., 2007). On average, each individual tends to specialize on one type of task each day, but the specific mechanism that influences this specialization or regulates the transitions between tasks is yet unknown (Hurd et al., 2007). In a related species, *V. koreensis*, pulp foraging (a less costly task) is performed by younger foragers, whereas nectar and prey foraging (a task that requires more time and effort for heavier loads) is conducted by older foragers (Kim et al., 2012).

In *Vespula vulgaris*, reduced expression of a cGMP-dependent protein kinase (*Vvforaging*) is correlated with an increase in foraging behaviour among workers (Tobback et al., 2008). This is similar to what is observed in *Pogonomyrmex barbatus* (Ingram et al., 2005), but is the opposite direction of expression observed in *Apis mellifera*, in which expression is higher in foragers (Ben-Shahar et al., 2002). These results suggest a conservation of *foraging* gene function in division of labour across species, but a difference in the specific patterns of gene activity and brain function that regulate division of labour.

4.3 Polistinae

In *Ropalidia marginata*, there is also evidence of age-based polyethism. Workers transition through on-nest tasks (feeding larvae) to nest building tasks to foraging tasks (collecting nesting material or food, Gadagkar, 2009). However, although JH influences ovarian development, it does not influence the rate at which individuals transition through these tasks—the age that individuals begin to tend brood, build, or forage remains constant regardless of JH application (Agrahari and Gadagkar, 2003).

Mischocyttarus mastigophorus workers divide foraging tasks in accordance with dominance rank—socially dominant individuals spend more time foraging for wood pulp, subordinate individuals spent more time foraging for nectar and insect prey (O'Donnell, 1998). Lipid content is also related to foraging rate—those individuals with smaller fat bodies spend more time off the nest engaged in foraging (Markiewicz and O'Donnell, 2001). In addition to nutritional differences, older individuals (typically foragers) have larger mushroom body calyces, as do individuals that are more aggressive and socially dominant (Molina and O'Donnell, 2008).

In *Polistes*, there is also a loose worker temporal polyethism (Shorter and Tibbetts, 2009). In addition, dominance rank and usually also ovarian development are associated with worker task performance. Subordinates may divide non-reproductive tasks based on their relative rank in the hierarchy, the most dominant individuals engaging in the less risky behaviours (Jandt et al., 2014). *P. dominula* alpha subordinates, for example, are more likely to construct the nest, whereas beta subordinates are more likely to forage (Pratte, 1989). *P. metricus* and *P. ferreri* co-foundresses even divide foraging preferences, and these preferences correlate with dominance rank: dominant wasps are more likely to collect vegetable fibre, whereas subordinates will collect nectar or prey (De Souza et al., 2008, Gamboa et al., 1978). Similarly, in *Polistes instabilis*, ovary development and worker reproductive competition correlate negatively with foraging behaviour (Molina and O'Donnell, 2009).

Among co-foundresses, JH is directly linked to reproductive dominance rank (Tibbetts et al., 2011a). Among workers, however, the role of JH may be condition dependent—in larger, fatter workers, JH increases the fertility and dominance of workers, but in smaller workers, JH increases the probability of foraging (Fig. 2; Giray et al., 2005; Shorter and Tibbetts, 2009; Tibbetts and Izzo, 2009; Tibbetts et al., 2011b).

In *P. metricus*, foraging workers tend to have lower fat stores than non-foraging workers (Fig. 2; Toth et al., 2009). Moreover, individuals that are starved as adults not only have lower abdominal lipid, but exhibit an increase in foraging rate (Daugherty et al., 2011). This difference in foraging is also associated with numerous differences in brain gene expression, including several nutrition-related genes in the insulin pathway, and genes coding for oxidoreductases, and fatty acid-binding proteins (Daugherty et al., 2011).

A number of studies have explored the basis of worker division of labour in *Polybia* wasps. Workers can engage in one of three task groups: in-nest work (nest construction, brood care, etc.), on-nest work (receive materials from returning foragers), and foraging (Jeanne et al., 1988). Like other polistines, workers transition from in-nest to foraging tasks as they age, and this transition is correlated with a number of physiological traits: lipid content gradually decreases (O'Donnell and Jeanne, 1995b) and ovarian status decreases (O'Donnell, 2001). In *P. occidentalis*, smaller individuals transition from in-nest to out-of nest tasks more slowly, and are more likely to achieve dominant social status (O'Donnell and Jeanne, 1995a). Topical application of a JH analog (methoprene) on day-old *P. occidentalis* workers leads to accelerated temporal polyethism, suggesting that JH facilitates this transition (O'Donnell and Jeanne, 1993). In *Polybia* wasps, foragers also exhibit a division of labour for a preference of foraged materials: water, pulp, honey, and prey (O'Donnell and Jeanne, 1990). There is evidence that the probability to forage on a specific material may be genetically linked. In *Polybia aequatorialis*, forager specialization on pulp, water, nectar, or insect prey is associated with specific segments of DNA (based on an analysis of anonymous markers, O'Donnell, 1996).

5. TOWARDS A SYNTHESIS: UNDERSTANDING WASP SOCIAL EVOLUTION FROM A MECHANISTIC PERSPECTIVE

In the data reviewed above, we have simultaneously presented information on physiology and genomics. This was intentional—physiology is shaped by its underlying genomics, and in turn gene expression is responsive to the internal physiological environment. In addition, both physiology and genes can influence group dynamics, and at the same time the nest environment itself can potentially have important influences on organismal physiology and behaviour (Fig. 3). We wish to highlight the importance of considering physiology and genomics, rather than considering these as separate systems. In fact, this type of integrative perspective has a strong tradition of leading to novel ideas and important insights into social evolution. Below, we review some of these ideas, emphasizing the role that Vespidae have played, and as we project, will continue to play, in developing heuristic hypotheses about social evolution.

Paper wasps, especially primitively eusocial species in the genus *Polistes*, have provided excellent fodder for the development and empirical testing of hypotheses about the proximate and ultimate factors contributing to the evolution of eusociality (Table 2). Here, we highlight several emerging ideas relating to the evolution of eusociality that have been addressed using physiological and genomic data from vespid wasps (Table 2). The first, on the level of hormones, relates to the changing role of JH from gonadotropin to regulator of division of labour. The next three represent mechanistic hypotheses for how wasp castes evolved from solitary physiological and behavioural 'ground plans' (reviewed in Toth and Robinson, 2007). The final two ideas deal with the relative contributions of deeply conserved genes or recently evolved (novel) genes.

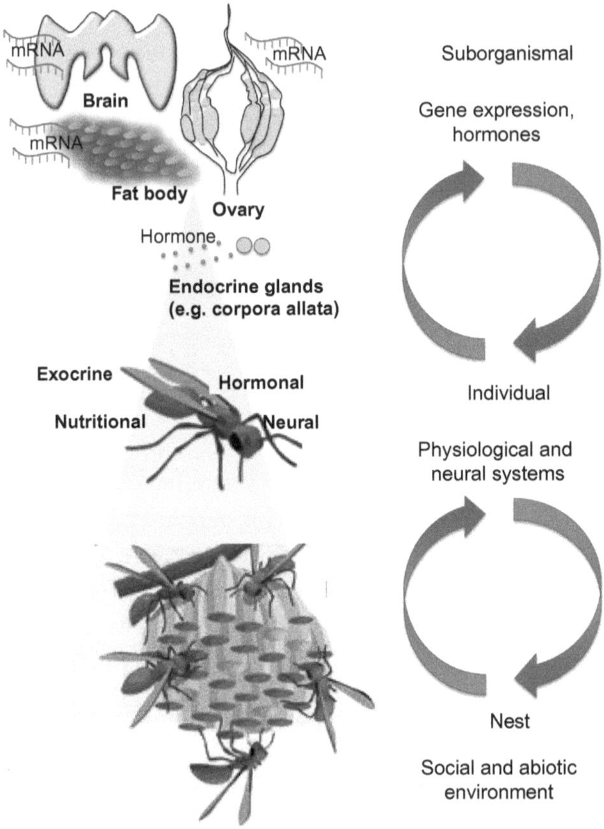

Figure 3 The dynamic interplay between multiple levels of biological organization, from genes to hormones to physiological systems, can affect social organization in paper wasps. Wasp drawings adapted from Hunt et al. (2011).

5.1 Split-function hypothesis

The role of JH is known to have changed radically during insect social evolution. In a wide variety of non-eusocial insects, JH is well known to be a powerful regulator of female reproductive state, and high JH during both pre-adult development and during adulthood are associated with increased ovary size, ovary activation, and egg-laying behaviour (Chapman, 2012). In striking contrast, JH is the major regulator of worker division of labour in adult honey bees and has no apparent association with reproduction (Hartfelder, 2000). Instead, in honey bees, high JH is associated with foraging (the chapter 'Old Threads Make New Tapestry—Rewiring of Signalling Pathways Underlies Caste Phenotypic Plasticity in the Honey

Table 2 Some mechanistic hypotheses for the evolution of sociality, associated predictions, and supporting studies from Vespidae

Hypothesis	Description	Prediction(s)	Supporting studies
Split-function hypothesis	The ancestral reproductive function of JH became split into a dual function, regulating reproduction in queens and behavioural division of labour among workers	1: JH regulates reproduction in solitary species 2: JH regulates both reproduction and worker division of labour in eusocial species	Giray et al. (2005), Shorter and Tibbetts (2009), and Tibbetts et al. (2013)
Ovarian ground plan hypothesis	The ancestral phasic shift between egg-laying/egg-development and brood care/foraging evolved into reproductive and non-reproductive castes, respectively	1: Gene expression patterns related to worker behaviour will be similar to those of maternal individuals 2: Maternal care related gene expression patterns from solitary species will resemble patterns of workers in eusocial species	Toth et al. (2007)
Diapause ground plan hypothesis	The ancestral diapausing phenotype evolved into the gyne phenotype	1: Genes related to diapause in solitary species will be related to gyne phenotype in eusocial species	Hunt et al. (2007) and Hunt et al. (2010)
Genetic toolkit hypothesis	Deeply conserved genes related to solitary behaviour (including maternal behaviour, reproduction, food-searching behaviour, aggression) are involved in the evolution of eusocial behaviour	1: Similar sets of genes will be differentially expressed in queens and workers from species with different origins of eusociality 2: Genes associated with behaviour such as maternal behaviour, feeding, aggression in solitary species will be related to worker behaviour, social foraging, and social aggression in eusocial species	Berens et al. (2015), Toth et al. (2014), but see: Hunt and Goodisman (2010), Toth et al. (2010)
Novel genes hypothesis	Novel social phenotypes (e.g. worker caste in social insects) are the result of new genes, most likely the result of rapid protein evolution	1: Genes that are differentially expressed between castes are likely to be taxonomically restricted genes, or genes with no homology to sequences from other taxa 2: Genes with caste-biased expression are more likely to show rapid rates of protein evolution	Ferreira et al. (2013)

Bee, *Apis mellifera* L.' by Hartfelder et al., this volume; Robinson, 1987) and experimental manipulations of JH levels can cause worker bees to accelerate the age-related shift to foraging behaviour (Sullivan et al., 2000). These observations led to the suggestion that during social evolution, the function of JH shifted from a gonadotropin to a regulator of worker foraging, and that we should be able to find intermediate species in which JH regulates both reproduction and worker behaviour in a condition-dependent manner. Studies on *Polistes* wasps have provided tests of this hypothesis. JH is well known to affect ovary activation and dominance status in *Polistes* (Röseler, 1991; Röseler et al., 1984; Tibbetts et al., 2011a), and subsequent work demonstrated that JH can also affect worker division of labour such that JH-treated workers are more aggressive (Giray et al., 2005) and more likely to initiate foraging at a young age (Fig. 2; Shorter and Tibbetts, 2009). In addition, the effects of JH appear to be condition-dependent; JH can apparently activate ovaries in wasps that have high nutritional state, but it is more likely to lead to changes in worker foraging behaviour in wasps that have low nutritional state (Shorter and Tibbetts, 2009). Overall, studies of *Polistes* support the hypothesis that JH, once freed from constraints on regulating reproduction in the physiological context of non-reproducing workers, was selected to take on new functions in regulating colony-level traits such as foraging division of labour. Subsequently, in some advanced eusocial species such as honey bees, the reproductive function of JH can be completely lost; whether the same is true in vespid wasps remains to be seen, but such evidence would represent a compelling example of convergent evolution of hormonal mechanisms across social insect lineages (Toth and Robinson, 2007; the chapter 'Old Threads Make New Tapestry—Rewiring of Signalling Pathways Underlies Caste Phenotypic Plasticity in the Honey Bee, *Apis mellifera* L.' by Hartfelder et al., this volume).

5.2 Ovarian ground plan hypothesis

Originally proposed by Evans and West-Eberhard (1973), the ovarian ground plan hypothesis (OGPH) posits that different elements or modules of solitary behaviour were uncoupled during evolution to produce distinct castes (West-Eberhard, 1996). For paper wasps, this idea was based on a proposed cycle of ovarian activation (egg laying) alternating with a period of ovarian inhibition, accompanied by foraging for larval provisions and maternal care of developing larvae. Specifically, the ovarian ground plan idea proposes that these two modules of solitary behaviour, specifically foraging/provisioning and egg laying/egg-development, were separated during evolution and manifest as two distinct groups of females—egg-laying 'queens' and non-egg-laying, foraging 'workers'. Along with these different behavioural modules, the OGPH suggests selection on correlated suites of behavioural, physiological, and gene expression traits produced two distinct female groups, and in this way the first rudimentary castes were born. This idea was supported by observations on solitary wasps such as *Zethus* (West-Eberhard, 1987a), but the objection has been raised that there may not be a clearly defined ovarian/foraging cycle that can be separated into distinct modules (Hunt, 2007). Nonetheless, even without a strict ovarian cycle, the general idea of the OGPH, specifically, that maternal/offspring care and egg-laying elements of solitary behaviour can be separated into distinct castes, has been one of the most influential ideas about how castes evolved in the Hymenoptera and has led to several related hypotheses.

A related idea, termed the maternal heterochrony hypothesis (MHH) focuses on the idea that worker behaviour arose due to a change in the timing of the expression of genes related to maternal care, to occur in females that had not yet mated nor reproduced (Linksvayer and Wade, 2005). Empirical evidence supporting the OGPH and MHH has come from studies on primitively eusocial *P. metricus*, which showed similar patterns of brain gene expression in maternal females (foundresses) and workers, compared to queens and gynes(Toth et al., 2007). A related hypothesis termed the reproductive ground plan hypothesis (also called the 'forager reproductive ground plan hypothesis', Oldroyd and Beekman, 2008) posits that regulation of foraging division of labour among workers arose from a further split of solitary gene networks related to reproduction (egg laying and maternal behaviour, Amdam et al., 2004, 2006). Evidence for this idea comes from data on the egg yolk protein *vitellogenin* from honey bees (Amdam et al., 2006), as well as data on the insect gonadotropic hormone JH from both honey bees and ants (Amdam et al., 2004; Dolezal et al., 2012). To date, there have been no direct tests of the forager reproductive ground plan hypothesis in paper wasps.

5.3 Diapause ground plan hypothesis

The diapause ground plan hypothesis (DGPH), proposed by Hunt and Amdam (2005), posits that the reproductive caste in paper wasps is related to the expression of diapause-related behaviour, physiology, and gene expression. In temperate paper wasps, female offspring reared later in the colony cycle show many characteristic traits of insect diapause: they are typically larger, with extremely high fat stores, and inactivated ovaries (Hunt and Amdam, 2005). These represent the gynes (future reproductive queens) that will over-winter in hibernaculae (sheltered areas with clusters of both related and unrelated wasps) and found colonies the following spring. Just as food availability during larval development determines whether individuals of some solitary insect species will develop into adults or enter diapause and develop into adults the following season, a similar process might regulate whether a social wasp larva develops into a non-reproductive or a reproductive adult. For example, *P. metricus* gynes are more well-nourished than workers, and they do enter a quiescent period prior to founding colonies and egg-laying the following season (Toth et al., 2009). According to this idea, the molecular machinery underlying diapause in solitary insects is involved in caste determination in social wasps (Hunt and Amdam, 2005; Hunt et al., 2007). Molecular level support for this notion comes from measurements of storage proteins known to mediate diapause, including hexameric storage proteins (Hunt et al., 2007, 2010). Gyne-destined larvae have higher levels of *hexamerin* mRNAs and protein expression than worker-destined larvae. In addition, development time in gyne-destined larvae is extended as predicted for a diapause phenotype (Hunt et al., 2007). Additional studies are needed, especially on tropical *Polistes* and solitary vespids, to better understand the relevance of the diapause phenotype to social evolution in a phylogenetic context.

Are the OGPH and DGPH mutually exclusive hypotheses? Not necessarily. There can potentially be multiple ground plans selected in concert, simultaneously acting to pull caste phenotypes apart by influencing multiple gene networks and physio-behavioural systems. New data on a genomic scale are beginning to provide some first hints at the complex and multifactorial nature of the genomic regulation of caste differences in wasps.

5.4 Genetic toolkit hypothesis

Solitary ground plan hypotheses generally focus on the principle of evolutionary co-option—specifically, that 'old' or deeply conserved genes involved in solitary phenotypes are retooled or simply shifted in their expression to regulate social phenotypes. A more general version of this idea has been put forward as the genetic toolkit hypothesis, based on principles from evolutionary developmental biology or 'evo-devo' (Toth and Robinson,

2007). This idea suggests that a small set of deeply conserved genes involved in solitary phenotypes were co-opted for the evolution of social traits such as caste-specific physiology, behavioural specialization, and social organization. These genes may have functions including reproduction and diapause, but are not limited to them; other fundamental solitary phenotypes include aggression and food-searching behaviour. This hypothesis predicts that similar genes and/or pathways will be associated with social behaviour across multiple, independently evolved social insect lineages. A series of studies that used *Polistes* wasps as a focal point compared gene expression patterns associated with maternal behaviour (Toth et al., 2007), foraging (Daugherty et al., 2011), reproduction (Toth et al., 2010), aggression (Toth et al., 2014), and caste differences (Berens et al., 2015) across social species. Overall, these studies found evidence for a relatively small, but statistically significant overlap in gene expression patterns associated with similar types of social behaviour across wasps, bees, and ants. In some cases, the overlap across species on the gene level was very small and not significant (Berens et al., 2015; Ferreira et al., 2013; Toth et al., 2014). However, an analysis on the pathway level showed that indeed there was significant overlap across lineages on the level of conserved enzymatic pathways (e.g. glycolysis) and functional types of genes (e.g. genes related to oxidation–reduction activity) associated with caste development across bees, ants, and wasps (Berens et al., 2015). These data provide some supporting evidence for the genetic toolkit hypothesis, but suggest that the toolkit itself is rather loose. Unlike the classic evo-devo cases of *Hox* genes being repeatedly involved in the generation of morphological novelty in development (Carroll, 2008), there are no specific 'major player genes' that are always associated with caste evolution in multiple lineages. Instead, data suggest that largely different, lineage-specific genes affect caste development, but that there are certain pathways and gene networks that are repeatedly recruited for caste differences across multiple origins of sociality.

5.5 Novel genes hypothesis

As an alternative to the focus on 'old genes' with the genetic toolkit hypothesis, an emerging idea focuses instead on 'new genes', that is, novel or taxonomically restricted genes that show no homology to known sequences from other species. Instead of novel phenotypes being the result of recycling old genes, the novel genes hypothesis suggests that novel social phenotypes are the result of new genes, most likely the result of rapid protein evolution.

Recent studies on honey bees have suggested that novel, rapidly evolving genes are more likely to be over-expressed in workers in advanced eusocial honey bees (Harpur et al., 2014; Johnson and Tsutsui, 2011). One study on *P. canadensis* also found novel genes to be more likely to be differentially expressed between adult queens and workers (Ferreira et al., 2013). However, a subsequent study on a different species of *Polistes* (*metricus*), did not find any association between novel genes and caste differences in larvae (Berens et al., 2015). A comparative study on ants (Simola et al., 2013) suggests that novel genes abound in every single examined lineage of ants. The ant comparisons suggest that, rather than being related to fundamental caste differences, these novel genes may be more likely to be involved in highly derived, lineage-specific social traits. At this time, it is still not completely clear whether conserved genes or novel genes contribute more to social evolution. Additional studies are needed, especially utilizing solitary and social species in direct comparisons within a monophyletic group, rather than relying on cross-family comparisons.

6. GAPS IN OUR UNDERSTANDING AND FUTURE DIRECTIONS

The Vespidae are an excellent, arguably one of the best, monophyletic groups for studying evolutionary transitions associated with eusocial evolution. This family contains representatives that are solitary, subsocial, facultatively social, primitively social, and advanced eusocial. Morphological castes have been gained and lost multiple times (Noll and Wenzel, 2008; Noll et al., 2004). Many vespid species are highly ecologically successful, and even invasive, and can be reared in semi-natural or laboratory settings (Jandt et al., in review) making them attractive model organisms (Jandt et al., 2014). Considerable progress has been made in recent years on understanding the physiological and genomic mechanisms of sociality in vespids. However, we suggest that the biggest and most informative insights are still awaiting us. This is because nearly all research to date has focused on a relatively small number of species, most of which are primitively eusocial, especially *Polistes*. While this genus is extremely informative for understanding some aspects of social evolution, we are in dire need of more studies on additional genera, especially solitary and swarm-founding advanced eusocial species (Table 3). While these species are often more difficult to locate (solitary), work with, and study (including some highly aggressive advanced eusocial species), a deeper knowledge of the mechanisms regulating pre-social and

Table 3 Available data relating to the physiological and genetic mechanisms underlying development, reproduction, and behaviour, in four subfamilies of vespid wasps

		Larval development and adult reproductive state		Adult reproductive state and caste		Adult division of worker behaviour	
	Genus	Physiology	Genetic	Physiology	Genetic	Physiology	Genetic
Eumeninae	*Euodynerus*	✓		✓		n/a	
	Monobia		✓			n/a	
Stenogastrinae	*Parischnogaster*			✓		✓	
	Liostenogaster			✓			
Vespinae	*Vespula*	✓	✓	✓	✓	✓	✓
	Dolichovespula				✓		
Polistinae	*Polistes*	✓	✓	✓	✓	✓	✓
	Ropalidia			✓		✓	
	Mischocyttarus			✓		✓	
	Polybia			✓		✓	✓

n/a, not applicable (solitary).

social behaviour in these groups will be very illuminating. Below, we detail several specific areas that we believe to represent some of the most promising future directions for vespid wasp research.

6.1 Ovarian ground plan hypothesis

The ovarian ground plan and maternal heterochrony hypotheses are based on underlying assumptions about ancestral/conserved roles of genes and hormones related to maternal care and reproduction in solitary species. However, to date, there are no studies that actually address the roles of important reproductive hormones (such as JH) or genes (such as genes with possible roles in maternal behaviour) in a comparative context across both solitary and social species. We suggest that studies on solitary vespids, especially species such as *Zethus* with well-defined maternal care in the form of progressive provisioning, will be critical to a more definitive test of the OGPH and MHH. Under these hypotheses, we would predict that genes with roles in maternal provisioning behaviour in solitary species will have roles in sibling care behaviour in social species.

6.2 Genetic toolkit hypothesis

Thus far, the idea of deeply conserved genes affecting the evolution of novel social traits has been tested using cross-lineage comparisons of, for example, one bee species, one ant species, and one wasp species (Berens et al., 2015). However, a more comprehensive test of the toolkit idea will also encompass multiple species within a lineage. For example, by comparing gene expression patterns associated with maternal behaviour, worker behaviour in primitively social species, and worker behaviour in advanced eusocial species, we can ask to what extent are ancestral mechanisms regulating maternal/sibling care behaviour retained throughout social evolution? Are there more similarities between solitary and primitively eusocial species, with more departure in advanced eusocial species?

6.3 Novel genes hypothesis

Thus far, it has been difficult to assess the importance of novel genes in social evolution, especially in wasps, because of a lack of genomic information. This situation is changing rapidly with the advent of next-generation genomic technologies, and many new insect genomes and transcriptomes, including those of vespids (A.L. Toth, unpublished data), are on the horizon. We suggest that a comprehensive comparison of novel genes, based on full transcriptome and/or genome sequencing, could be especially informative when including representatives of the major transitional states in social evolution: solitary, incipiently social, primitively eusocial, and advanced eusocial. Using such an approach, it will be possible to identify vespid-specific genes, as well as genus and/or species-specific genes, and assess whether these genes have associations with fundamental eusocial traits (such as caste differences), or whether they are related more to lineage-specific and derived social traits (such as swarm founding).

7. CONCLUSIONS

In summary, a deeper understanding of vespid physiology and genomics will provide not only a template for testing hypotheses on the origins of eusociality, but also on its subsequent elaboration into advanced eusocial traits, including morphological castes, swarm founding, and multiple queens. There is a rich ethological tradition with numerous species of social wasps not discussed in detail in this review, including multiple origins of complex traits such as swarm founding, morphological castes, and the loss of eusociality in social parasites (Jeanne and Hunt, 1992; Matsuura and Yamane, 1990; Noll and Wenzel, 2008; Noll et al., 2004; Smith et al., 2002). Studies of such species could be highly informative for understanding the mechanisms underlying the gain and loss of social traits. Recent technological advances have now made it possible to study non-model species on a genomic scale, and we hope that more attention will be paid on using these techniques to explore new vespid genomes and associated physiological mechanisms. In addition, although we have stressed the importance of nutritional physiology and nutrition-related genes in their influence on both reproductive caste and worker caste differences, it is important to think beyond nutrition—there is still a great deal to be learned about the environmental determinants of individual differences in social insects. We suggest that it will be fruitful in the future to more fully investigate feedback between the social environment (such as chemical and vibrational communication), individual development and physiology, epigenetics, and gene expression (Bengston and Jandt, 2014; Jeanne and Suryanarayanan, 2011; Weiner and Toth, 2012). Finally, although there are numerous mechanistic hypotheses that have been put forward to explain the evolution of sociality in vespids and other social insects, there is a need for a more synthetic approach (Johnson and Linksvayer, 2010) to tie these ideas together across multiple levels of biological organization. We suggest that phylogenetically controlled comparative studies within the Vespidae can provide the needed empirical and conceptual framework to move towards such a synthesis in the future.

ACKNOWLEDGEMENTS

We wish to thank Bob Jeanne, Sean O'Donnell, Mary Jane West-Eberhard, and members of the Toth Lab (Iowa State University, USA) for reviewing and providing helpful feedback on the chapter. We also thank Jim Hunt and Liz Tibbetts for discussions throughout chapter preparation. Finally, we thank Amro Zayed and Clement Kent for inviting us to participate in this special issue of *Advances in Insect Physiology*. This work was supported by NSF: IOS-1146410.

REFERENCES

Agrahari, M., Gadagkar, R., 2003. Juvenile hormone accelerates ovarian development and does not affect age polyethism in the primitively eusocial wasp, *Ropalidia* marginata. J. Insect Physiol. 49, 217–222.

Amdam, G.V., Norberg, K., Fondrk, M.K., Page, R.E., 2004. Reproductive ground plan may mediate colony-level selection effects on individual foraging behavior in honey bees. Proc. Natl. Acad. Sci. U.S.A 101, 11350–11355.

Amdam, G.V., Csondes, A., Fondrk, M.K., Page, R.E., 2006. Complex social behaviour derived from maternal reproductive traits. Nature 439, 76–78.

Bengston, S.E., Jandt, J.M., 2014. The development of collective personality: the ontogenetic drivers of behavioral variation across groups. Front. Ecol. Evol. 2, 81.

Ben-Shahar, Y., Robichon, A., Sokolowski, M.B., Robinson, G.E., 2002. Influence of gene action across different time scales on behavior. Science 296, 741–744.

Berens, A.J., Hunt, J.H., Toth, A.L., Nourishment level affects caste-related gene expression in *Polistes* wasps. BMC Genomics In Review.

Berens, A.J., Hunt, J.H., Toth, A.L., 2015. Comparative transcriptomics of convergent evolution: different genes but conserved pathways underlie caste phenotypes across lineages of eusocial insects. Mol. Biol. Evol. In Press.

Bolton, A., Sumner, S., Shreeves, G., Casiraghi, M., Field, J., 2006. Colony genetic structure in a facultatively eusocial hover wasp. Behav. Ecol. 17, 873–880.

Bonckaert, W., Drijfhout, F.P., D'Ettorre, P., Billen, J., Wenseleers, T., 2012. Hydrocarbon signatures of egg maternity, caste membership and reproductive status in the common wasp. J. Chem. Ecol. 38, 42–51.

Bridge, C., Field, J., 2007. Queuing for dominance: gerontocracy and queue-jumping in the hover wasp Liostenogaster flavolineata. Behav. Ecol. Sociobiol. 61, 1253–1259.

Carpenter, J.M., 1991. Phylogenetic relationships and the origin of social behavior in the Vespidae. In: Ross, K.G., Matthews, R.W. (Eds.), Social Biology of Wasps. Cornell University Press, Ithaca.

Carroll, S.B., 2008. Evo-devo and an expanding evolutionary synthesis: a genetic theory of morphological evolution. Cell 134, 25–36.

Chapman, R.F., 2012. The Insects: Structure and Function. Cambridge University Press, Cambridge, UK.

Chapman, T.W., Stewart, S.C., 1996. Extremely high levels of inbreeding in a natural population of the free-living wasp *Ancistrocerus antilope* (Hymenoptera: Vespidae: Eumeninae). Heredity 76, 65–69.

Chilcutt, C.F., Cowan, D.P., 1992. Carnivory in adult female Eumenid wasps (Hymenoptera: Vespidae: Eumeninae) and its effect on egg production. Great Lakes Entomol. 25, 297–301.

Cowan, D.P., 1981. Parental investment in 2 solitary wasps *Ancistrocerus adiabatus* and *Euodynerus foraminatus* (Eumenidae, Hymenoptera). Behav. Ecol. Sociobiol. 9, 95–102.
Cowan, D.P., 1983. Hypotheses on cell provisioning in Eumenid wasps. Biol. J. Linn. Soc. 20, 245–247.
Daugherty, T.H.F., Toth, A.L., Robinson, G.E., 2011. Nutrition and division of labor: effects on foraging and brain gene expression in the paper wasp *Polistes metricus*. Mol. Ecol. 20, 5337–5347.
De Souza, A.R., Rodrigues, I.L., Rocha, J.V.A., Reis, W.A.A., Lopes, J.F.S., Prezoto, F., 2008. Foraging behavior and dominance hierarchy in colonies of the neotropical social wasp *Polistes ferreri* Saussure, 1853 (Hymenoptera, Vespidae) in different stages of development. Sociobiology 52, 293–303.
Desuó, I.C., Souza-Galheico, C.B., Shima, S.N., Santos, G.M.M., Cruz, J.D., Bichara, C.C., Dias, C.T.S., 2011. An adaptive view of caste differentiation in the neotropical wasp *Polybia* (*Trichothorax*) *sericea* Olivier (Hymenoptera: Vespidae). Neotrop. Entomol. 40, 653–660.
Dolezal, A.G., Brent, C.S., Hölldobler, B., Amdam, G.V., 2012. Worker division of labor and endocrine physiology are associated in the harvester ant, *Pogonomyrmex californicus*. J. Exp. Biol. 215, 454–460.
Evans, H.E., West-Eberhard, M.J., 1973. Wasps. David & Charles, Newton Abbot.
Ferreira, P.G., Patalano, S., Chauhan, R., Ffrench-Constant, R., Gabaldón, T., Guigó, R., Sumner, S., 2013. Transcriptome analyses of primitively eusocial wasps reveal novel insights into the evolution of sociality and the origin of alternative phenotypes. Genome Biol. 14, R20.
Field, J., Foster, W., 1999. Helping behaviour in facultatively eusocial hover wasps: an experimental test of the subfertility hypothesis. Anim. Behav. 57, 633–636.
Field, J., Solís, C.R., Queller, D.C., Strassmann, J.E., 1998. Social and genetic structure of paper wasp cofoundress associations: tests of reproductive skew models. Am. Nat. 151, 545–563.
Field, J., Shreeves, G., Sumner, S., 1999. Group size, queuing and helping decisions in facultatively eusocial hover wasps. Behav. Ecol. Sociobiol. 45, 378–385.
Gadagkar, R., 2009. The Social Biology of Ropalidia Marginata: Toward Understanding the Evolution of Eusociality. Harvard University Press, Cambridge, MA.
Gamboa, G.J., Heacock, B.D., Wiltjer, S.L., 1978. Division of labor and subordinate longevity in foundress associations of the paper wasp, *Polistes metricus* (Hymenoptera: Vespidae). J. Kansas Entomol. Soc. 51, 343–352.
Giray, T., Giovanetti, M., West-Eberhard, M.J., 2005. Juvenile hormone, reproduction, and worker behavior in the neotropical social wasp *Polistes canadensis*. Proc. Natl. Acad. Sci. U.S.A 102, 3330–3335.
Goodisman, M.A.D., Kovacs, J.L., Hoffman, E.A., 2007. The significance of multiple mating in the social wasp *Vespula maculifrons*. Evolution 61, 2260–2267.
Greene, A., 1991. *Dolichovespula* and *Vespula*. In: Ross, K.G., Matthews, R.W. (Eds.), The Social Biology of Wasps. Comstock Publishing Associates, Ithaca.
Harpur, B.A., Kent, C.F., Molodtsova, D., Lebon, J.M.D., Alqarni, A.S., Owayss, A.A., Zayed, A., 2014. Population genomics of the honey bee reveals strong signatures of positive selection on worker traits. Proc. Natl. Acad. Sci. U.S.A 111, 2614–2619.
Hartfelder, K., 2000. Insect juvenile hormone: from "status quo" to high society. Braz. J. Med. Biol. Res. 33, 157–177.
Hermes, M.G., Somavilla, A., Garcete-Barrett, B.R., 2013. On the nesting biology of *Pirhosigma giordani soika* (Hymenoptera, Vespidae, Eumeninae), with special reference to the use of vegetable matter. Rev. Bras. Entomol. 57, 433–436.
Hines, H.M., Hunt, J.H., O'Connor, T.K., Gillespie, J.J., Cameron, S.A., 2007. Multigene phylogeny reveals eusociality evolved twice in vespid wasps. Proc. Natl. Acad. Sci. U.S. A 104, 3295–3299.

Hoffman, E.A., Goodisman, M.A.D., 2007. Gene expression and the evolution of phenotypic diversity in social wasps. BMC Biol. 5, 23.

Hunt, J.H., 2007. The Evolution of Social Wasps. Oxford University Press, Oxford.

Hunt, J.H., Amdam, G.V., 2005. Bivoltinism as an antecedent to eusociality in the paper wasp genus Polistes. Science 308, 264–267.

Hunt, B.G., Goodisman, M.A.D., 2010. Evolutionary variation in gene expression is associated with dimorphism in eusocial vespid wasps. Insect Mol. Biol. 19, 641–652.

Hunt, J.H., Buck, N.A., Wheeler, D.E., 2003. Storage proteins in vespid wasps: characterization, developmental pattern, and occurrence in adults. J. Insect Physiol. 49, 785–794.

Hunt, J.H., Kensinger, B.J., Kossuth, J.A., Henshaw, M.T., Norberg, K., Wolschin, F., Amdam, G.V., 2007. A diapause pathway underlies the gyne phenotype in Polistes wasps, revealing an evolutionary route to caste-containing insect societies. Proc. Natl. Acad. Sci. U.S.A. 104, 14020–14025.

Hunt, J.H., Wolschin, F., Henshaw, M.T., Newman, T.C., Toth, A.L., Amdam, G.V., 2010. Differential gene expression and protein abundance evince ontogenetic bias toward castes in a primitively eusocial wasp. PLoS One 5, e10674.

Hunt, J.H., Mutti, N.S., Havukainen, H., Henshaw, M.T., Amdam, G.V., 2011. Development of an RNA interference tool, characterization of its target, and an ecological test of caste differentiation in the eusocial wasp *Polistes*. PLoS One 6, e26641.

Hurd, C.R., Jeanne, R.L., Nordheim, E.V., 2007. Temporal polyethism and worker specialization in the wasp, *Vespula germanica*. J. Insect Sci. 7, 43.

Ingram, K.K., Oefner, P., Gordon, D.M., 2005. Task-specific expression of the foraging gene in harvester ants. Mol. Ecol. 14, 813–818.

Jandt, J.M., Tibbetts, E.A., Toth, A.L., 2014. *Polistes* paper wasps: a model genus for the study of social dominance hierarchies. Insect. Soc. 61, 11–27.

Jandt, J.M., Thomson, J., Geffre, A.C., Toth, A.L., Behavior, physiology and gene expression in an ecological model species are perturbed due to lab rearing: a case study in *Polistes fuscatus* wasps. Behav. Ecol. In Review.

Jeanne, R.L., 1991. The swarm-founding Polistinae. In: Ross, K.G., Matthews, R.W. (Eds.), The Social Biology of Wasps. Cornell University, Cambridge, MA.

Jeanne, R.L., Hunt, J.H., 1992. Observations on the social wasp Ropalidia montana from peninsular India. J. Biosci. 17, 1–14.

Jeanne, R.L., Suryanarayanan, S., 2011. A new model for caste development in social wasps. Commun. Integr. Biol. 4, 373–377.

Jeanne, R.L., Downing, H.A., Post, D.C., 1988. Age polyethism and individual variation in *Polybia occidentalis,* an advanced eusocial wasp. In: Jeanne, R.L. (Ed.), Interindividual Behavioral Variability in Social Insects. Westview Press, Boulder, CO.

Johnson, B.R., Linksvayer, T.A., 2010. Deconstructing the superorganism: social physiology, groundplans, and sociogenomics. Q. Rev. Biol. 85, 57–79.

Johnson, B.R., Tsutsui, N.D., 2011. Taxonomically restricted genes are associated with the evolution of sociality in the honey bee. BMC Genomics 12, 164.

Judd, T.M., Teal, P., Hernandez, E., Choudhury, T., Hunt, J., 2015. Quantitative differences in nourishment affect caste-related physiology and development in the paper wasp *Polistes metricus*. PLoS One (in press).

Kelstrup, H.C., Hartfelder, K., Nascimento, F.S., Riddiford, L.M., 2014. Reproductive status, endocrine physiology and chemical signaling in the Neotropical, swarm-founding eusocial wasp *Polybia micans*. J. Exp. Biol. 217, 2399–2410.

Kim, B., Kim, K.W., Choe, J.C., 2012. Temporal polyethism in Korean yellowjacket foragers, *Vespula koreensis* (Hymenoptera, Vespidae). Insect. Soc. 59, 263–268.

Klahn, J.E., 1979. Philopatric and nonphilopatric foundress associations in the social wasp *Polistes fuscatus*. Behav. Ecol. Sociobiol. 5, 417–424.

Kovacs, J.L., Hoffman, E.A., Marriner, S.M., Rekau, J.A., Goodisman, M.A.D., 2010. Environmental and genetic influences on queen and worker body size in the social wasp *Vespula maculifrons*. Insect. Soc. 57, 53–65.

Linksvayer, T.A., Wade, M.J., 2005. The evolutionary origin and elaboration of sociality in the aculeate hymenoptera: maternal effects, sib-social effects, and heterochrony. Q. Rev. Biol. 80, 317–336.

Malham, J.P., Rees, J.S., Alspach, P.A., Beggs, J.R., Moller, H., 1991. Traffic rate as an index of colony size in *Vespula* wasps. N.Z. J. Zool. 18, 105–109.

Markiewicz, D.A., O'Donnell, S., 2001. Social dominance, task performance and nutrition: implications for reproduction in eusocial wasps. J. Comp. Physiol. A 187, 327–333.

Matsuura, M., Yamane, S., 1990. Biology of the Vespine Wasps. Springer-Verlag, Berlin.

Molina, Y., O'Donnell, S., 2008. Age, sex, and dominance-related mushroom body plasticity in the paperwasp *Mischocyttarus mastigophorus*. Dev. Neurobiol. 68, 950–959.

Molina, Y., O'Donnell, S., 2009. Worker reproductive competition affects division of labor in a primitively social paperwasp (*Polistes instabilis*). Insect. Soc. 56, 14–20.

Nijhout, H.F., Wheeler, D.E., 1982. Juvenile hormone and the physiological basis of insect polymorphisms. Q. Rev. Biol. 57, 109–133.

Noll, F.B., Wenzel, J.W., 2008. Caste in the swarming wasps: 'queenless' societies in highly social insects. Biol. J. Linn. Soc. 93, 509–522.

Noll, F.B., Wenzel, J.W., Zucchi, R., 2004. Evolution of caste in neotropical swarm-founding wasps (Hymenoptera: Vespidae; Epiponini). *American Museum Novitates*, COVER1-24.

O'Donnell, S., 1996. RAPD markers suggest genotypic effects on forager specialization in a eusocial wasp. Behav. Ecol. Sociobiol. 38, 83–88.

O'Donnell, S., 1998. Dominance and polyethism in the eusocial wasp *Mischocyttarus mastigophorus* (Hymenoptera: Vespidae). Behav. Ecol. Sociobiol. 43, 327–331.

O'Donnell, S., 2001. Worker age, ovary development, and temporal polyethism in the swarm-founding wasp *Polybia occidentalis* (Hymenoptera: Vespidae). J. Insect Behav. 14, 201–213.

O'Donnell, S., 2006. *Polybia* wasp biting interactions recruit foragers following experimental worker removals. Anim. Behav. 71, 709–715.

O'Donnell, S., Jeanne, R.L., 1990. Forager specialization and the control of nest repair in *Polybia occidentalis* Olivier (Hymenoptera, Vespidae). Behav. Ecol. Sociobiol. 27, 359–364.

O'Donnell, S., Jeanne, R.L., 1993. Methoprene accelerates age polyethism in workers of a social wasp (*Polybia occidentalis*). Physiol. Entomol. 18, 189–194.

O'Donnell, S., Jeanne, R.L., 1995a. The roles of body size and dominance in division of labor among workers of the eusocial wasp *Polybia occidentalis* (Olivier) (Hymenoptera, Vespidae). J. Kansas Entomol. Soc. 68, 43–50.

O'Donnell, S., Jeanne, R.L., 1995b. Worker lipid stores decrease with outside-nest task-performance in wasps: implications for the evolution of age polyethism. Experientia 51, 749–752.

O'Donnell, S., Clifford, M.R., Bulova, S.J., Deleon, S., Papa, C., Zahedi, N., 2014. A test of neuroecological predictions using paperwasp caste differences in brain structure (Hymenoptera: Vespidae). Behav. Ecol. Sociobiol. 68, 529–536.

Oldroyd, B.P., Beekman, M., 2008. Effects of selection for honey bee worker reproduction on foraging traits. PLoS Biol. 6, 0463–0470.

Oxley, P.R., Ji, L., Fetter-Pruneda, I., McKenzie, S.K., Li, C., Hu, H.F., Zhang, G.J., Kronauer, D.J.C., 2014. The genome of the clonal raider ant *Cerapachys biroi*. Curr. Biol. 24, 451–458.

Pereboom, J.J.M., Jordan, W.C., Sumner, S., Hammond, R.L., Bourke, A.F.G., 2005. Differential gene expression in queen-worker caste determination in bumble-bees. Proc. R. Soc. B Biol. Sci. 272, 1145–1152.

Pickett, K.M., Carpenter, J.M., 2010. Simultaneous analysis and the origin of eusociality in the Vespidae (Insecta: Hymenoptera). Arthropod Syst. Phylogeny 68, 3–33.

Pratte, M., 1989. Foundress association in the paper wasp *Polistes dominulus* Christ. (Hymen. Vesp.). Effects of dominance hierarchy on the division of labour. Behaviour 111, 208–219.

Queller, D.C., Zacchi, F., Cervo, R., Turillazzi, S., Henshaw, M.T., Santorelli, L.A., Strassmann, J.E., 2000. Unrelated helpers in a social insect. Nature 405, 784–787.

Reeve, H.K., 1991. Polistes. In: Ross, K.G., Matthews, R.W. (Eds.), The Social Biology of Wasps. Comstock Publishing Associates, Ithaca.

Robinson, G.E., 1987. Regulation of honey bee age polyethism by juvenile hormone. Behav. Ecol. Sociobiol. 20, 329–338.

Röseler, P.F., 1991. Reproductive competition during colony establishment. In: Ross, K.G., Matthews, R.W. (Eds.), The Social Biology of Wasps. Comstock Publishing Associates, London.

Röseler, P.-F., Röseler, I., Strambi, A., Augier, R., 1984. Influence of insect hormones on the establishment of dominance hierarchies among foundresses of the paper wasp, *Polistes gallicus*. Behav. Ecol. Sociobiol. 15, 133–142.

Ross, N.M., Gamboa, G.J., 1981. Nestmate discrimination in social wasps (*Polistes metricus*, Hymenoptera: Vespidae). Behav. Ecol. Sociobiol. 9, 163–165.

Sasaki, K., Yamasaki, K., Tsuchida, K., Nagao, T., 2009. Gonadotropic effects of dopamine in isolated workers of the primitively eusocial wasp, *Polistes chinensis*. Naturwissenschaften 96, 625–629.

Schmidt, K.C., Hunt, B.G., Smith, C.R., 2012. Queen, worker, and male yellowjacket wasps receive different nutrition during development. Insect. Soc. 59, 289–295.

Sears, A.L.W., Smiley, J.T., Hilker, M., Muller, F., Rank, N.E., 2001. Nesting behavior and prey use in two geographically separated populations of the specialist wasp *Symmorphus cristatus* (Vespidae: Eumeninae). Am. Midl. Nat. 145, 233–246.

Seppä, P., Queller, D.C., Strassmann, J.E., 2002. Reproduction in foundress associations of the social wasp, *Polistes carolina*: conventions, competition, and skew. Behav. Ecol. 13, 531–542.

Shima, S.N., Noll, F.B., Zucchi, R., 2003. Influence of the colony cycle on physiological and morphological caste variation in the perennial neotropical swarm-founding social wasp, *Protonectarina sylveirae* (Hymenoptera, Vespidae, Epiponini). Sociobiology 42, 449–466.

Shorter, J.R., Tibbetts, E.A., 2009. The effect of juvenile hormone on temporal polyethism in the paper wasp *Polistes dominulus*. Insect. Soc. 56, 7–13.

Simola, D.F., Wissler, L., Donahue, G., Waterhouse, R.M., Helmkampf, M., Roux, J., Nygaard, S., Glastad, K.M., Hagen, D.E., Viljakainen, L., Reese, J.T., Hunt, B.G., Graur, D., Elhaik, E., Kriventseva, E.V., Wen, J.Y., Parker, B.J., Cash, E., Privman, E., Childers, C.P., Munoz-Torres, M.C., Boomsma, J.J., Bornberg-Bauer, E., Currie, C.R., Elsik, C.G., Suen, G., Goodisman, M.A.D., Keller, L., Liebig, J., Rawls, A., Reinberg, D., Smith, C.D., Smith, C.R., Tsutsui, N., Wurm, Y., Zdobnov, E.M., Berger, S.L., Gadau, J., 2013. Social insect genomes exhibit dramatic evolution in gene composition and regulation while preserving regulatory features linked to sociality. Genome Res. 23, 1235–1247.

Smith, A.R., O'Donnell, S., Jeanne, R.L., 2002. Evolution of swarm communication in eusocial wasps (Hymenoptera: Vespidae). J. Insect Behav. 15, 751–764.

Smith, C.R., Toth, A.L., Suarez, A.V., Robinson, G.E., 2008. Genetic and genomic analyses of the division of labour in insect societies. Nat. Rev. Genet. 9, 735–748.

Strassmann, J.E., 1981. Wasp reproduction and kin selection: reproductive competition and dominance hierarchies among *Polistes annularis* foundresses. Florida Entomol. 64, 74–88.

Sullivan, J.P., Jassim, O., Fahrbach, S.E., Robinson, G.E., 2000. Juvenile hormone paces behavioral development in the adult worker honey bee. Horm. Behav. 37, 1–14.
Sumner, S., Pereboom, J.J.M., Jordan, W.C., 2006. Differential gene expression and phenotypic plasticity in behavioural castes of the primitively eusocial wasp, Polistes canadensis. Proc. R. Soc. B Biol. Sci. 273, 19–26.
Sumner, S., Lucas, E., Barker, J., Isaac, N., 2007. Radio-tagging technology reveals extreme nest-drifting behavior in a eusocial insect. Curr. Biol. 17, 140–145.
Suryanarayanan, S., Hermanson, J.C., Jeanne, R.L., 2011. A mechanical signal biases caste development in a social wasp. Curr. Biol. 21, 231–235.
Suzuki, T., 2003. Queen replacement without gerontocracy in the paper wasp *Parapolybia indica* in temperate Japan. Ethol. Ecol. Evol. 15, 191–196.
Tibbetts, E.A., Izzo, A.S., 2009. Endocrine mediated phenotypic plasticity: condition-dependent effects of juvenile hormone on dominance and fertility of wasp queens. Horm. Behav. 56, 527–531.
Tibbetts, E.A., Sheehan, M.J., 2012. The effect of juvenile hormone on *Polistes* wasp fertility varies with cooperative behavior. Horm. Behav. 61, 559–564.
Tibbetts, E.A., Izzo, A., Huang, Z.Y., 2011a. Behavioral and physiological factors associated with juvenile hormone in *Polistes* wasp foundresses. Behav. Ecol. Sociobiol. 65, 1123–1131.
Tibbetts, E.A., Levy, S., Donajkowski, K., 2011b. Reproductive plasticity in *Polistes* paper wasp workers and the evolutionary origins of sociality. J. Insect Physiol. 57, 995–999.
Tibbetts, E.A., Mettler, A., Donajkowski, K., 2013. Nutrition-dependent fertility response to juvenile hormone in non-social *Euodynerus foraminatus* wasps and the evolutionary origin of sociality. J. Insect Physiol. 59, 339–344.
Tobback, J., Heylen, K., Gobin, B., Wenseleers, T., Billen, J., Arckens, L., Huybrechts, R., 2008. Cloning and expression of PKG, a candidate foraging regalating gene in *Vespula vulgaris*. Anim. Biol. 58, 341–351.
Toth, A.L., Robinson, G.E., 2007. Evo-devo and the evolution of social behavior. Trends Genet. 23, 334–341.
Toth, A.L., Varala, K., Newman, T.C., Miguez, F.E., Hutchison, S.K., Willoughby, D.A., Simons, J.F., Egholm, M., Hunt, J.H., Hudson, M., Robinson, G., 2007. Wasp gene expression supports an evolutionary link between maternal behavior and eusociality. Science 318, 441–444.
Toth, A.L., Bilof, K.B.J., Henshaw, M.T., Hunt, J.H., Robinson, G.E., 2009. Lipid stores, ovary development, and brain gene expression in *Polistes metricus* females. Insect. Soc. 56, 77–84.
Toth, A.L., Varala, K., Henshaw, M.T., Rodriguez-Zas, S.L., Hudson, M.E., Robinson, G.E., 2010. Brain transcriptomic analysis in paper wasps identifies genes associated with behaviour across social insect lineages. Proc. R. Soc. B Biol. Sci. 277, 2139–2148.
Toth, A.L., Tooker, J.F., Radhakrishnan, S., Minard, R., Henshaw, M.T., Grozinger, C.M., 2014. Shared genes related to aggression, rather than chemical communication, are associated with reproductive dominance in paper wasps (*Polistes metricus*). BMC Genomics 15, 75.
Turillazzi, S., 1985. Colonial cycle of *Parischnogaster nigricans serrei* (Dubuysson) in West Java (Hymenoptera, Stenogastrinae). Insect. Soc. 32, 43–60.
Turillazzi, S., 1991. The Stenogastrinae. In: Ross, K.G., Matthews, R.W. (Eds.), The Social Biology of Wasps. Comstock Publishing Associates, Ithaca.
Turillazzi, S., 2013. The Biology of Hover Wasps. Springer, Berlin.
Weiner, S.A., Toth, A.L., 2012. Epigenetics in social insects: a new direction for understanding the evolution of castes. Genet. Res. Int. 2012, 609810.

West-Eberhard, M.J., 1987a. Flexible strategy and social evolution. In: Itô, Y., Brown, J.L., Kikkawa, J. (Eds.), Animal Societies: Theories and Facts. Japan Science Society Press, Tokyo.
West-Eberhard, M.J., 1987b. Observations of *Xenorhynchium nitidulum* (Fabricius) (Hymenoptera, Eumeninae), a Primitively Social Wasp. Psyche 94, 317–323.
West-Eberhard, M.J., 1996. Wasp societies as microcosms for the study of development and evolution. In: Turillazzi, S., West-Eberhard, M.J. (Eds.), Natural History and Evolution of Paper Wasps. Oxford University Press.
West-Eberhard, M.J., 2005. Behavior of the primitively social wasp *Montezumia cortesioides* Willink (Vespidae Eumeninae) and the origins of vespid sociality. Ethol. Ecol. Evol. 17, 201–215.
Zanette, L.R.S., Field, J., 2008. Genetic relatedness in early associations of *Polistes dominulus*: from related to unrelated helpers. Mol. Ecol. 17, 2590–2597.
Zucchi, R., Sakagami, S., Noll, F., Mechi, M., Mateus, S., Baio, M., Shima, S., 1995. *Agelaia vicina*, a swarm-founding polistine with the largest colony size among wasps and bees (Hymenoptera: Vespidae). J. N.Y. Entomol. Soc. 103, 129–137.

CHAPTER FOUR

Juvenile Hormone: A Central Regulator of Termite Caste Polyphenism

Judith Korb[1]

Department of Evolutionary Biology and Ecology, Institute of Biology I, University of Freiburg, Freiburg, Germany

[1]Corresponding author: e-mail address: judith.korb@biologie.uni-freiburg.de

Contents

Abstract

Juvenile hormone (JH) is a central regulator of insect post-embryonic development and life history traits. The foundation of termites' sociality is their unravelled developmental plasticity. Through a unique diversity of moulting types, lower termites can remain immature workers by stationary and regressive development, become sterile soldiers, or neotenic replacement reproductives that inherit the natal breeding position. How can JH, the central morphogenic hormone, regulate this diversity besides the default progressive development into a winged sexual that is common to all insects? Here, I summarize our current understanding of the interactions between socio-environmental triggers, JH and linked gene pathways that underlie termite castes and division of labour. I end with a synthetic model that may serve as a guideline to future termite research.

Advances in Insect Physiology, Volume 48
ISSN 0065-2806
http://dx.doi.org/10.1016/bs.aiip.2014.12.004

1. INTRODUCTION

Termites evolved complex colonies independently from the social Hymenoptera. They are 'social cockroaches', a monophyletic clade nested within the Blattodea (Engel et al., 2009; Inward et al., 2007a; Krishna et al., 2013; Lo et al., 2007) (Fig. 1A). Similar to most ants, social bees and wasps, their colonies are a result of polyphenism, i.e., the development of different castes from the same genetic background due to different environmental stimuli. Polyphenisms represent adaptive responses to environmental changes without disrupting successful genotypic combinations. Such environmental control of alternative phenotypes can evolve in organisms in which environmental cues can predict reliably the future selective environment (Moran, 1992). Season, food and social factors trigger different developmental pathways leading to workers, reproductives and in termites also to soldiers (e.g. Lenz, 1976; Lüscher, 1974b; Miura, 2004). This phenotypic developmental plasticity is highest in lower termites (all families except the Termitidae; Fig. 1) and especially in wood-dwelling species (Korb and Hartfelder, 2008; Noirot, 1990; Roisin, 2000; Roisin and Korb, 2011). They spend their whole colony life in a single piece of wood that serves both as food and shelter and which the termites never leave to forage outside. Here, workers are totipotent larval instars (sometimes called pseudergates, *sensu lato*), which can develop via an obligate pre-soldier instar into a sterile soldier or into two types of reproductives (Roisin and Korb, 2011) (Fig. 2): (i) via, generally, several nymphal instars (instars with wing buds) into winged sexuals (alates) that leave the nest during nuptial flights to found new colonies and (ii) via a single moult into neotenic (replacement) reproductives (hereafter also called neotenics) that reproduce within the natal nest, especially when the current reproductives die or are unhealthy. Winged sexuals reflect the default developmental pathway present in solitary insects. Soldiers and neotenics are synapomorphic termite castes and their underlying developmental pathways evolved with the transition to eusociality (Korb and Hartfelder, 2008; Noirot, 1989, 1990; Noirot and Pasteels, 1987). These progressive moults (including saltatorial ones) of workers are supplemented by two additional moulting types: (i) stationary moults where individuals moult without any change in size or morphology and (ii) regressive moults which are associated with reduction of size and/or developmental traits (especially wing buds). While stationary moults also occur in cockroaches (Bell et al., 2007), regressive moults are unique to

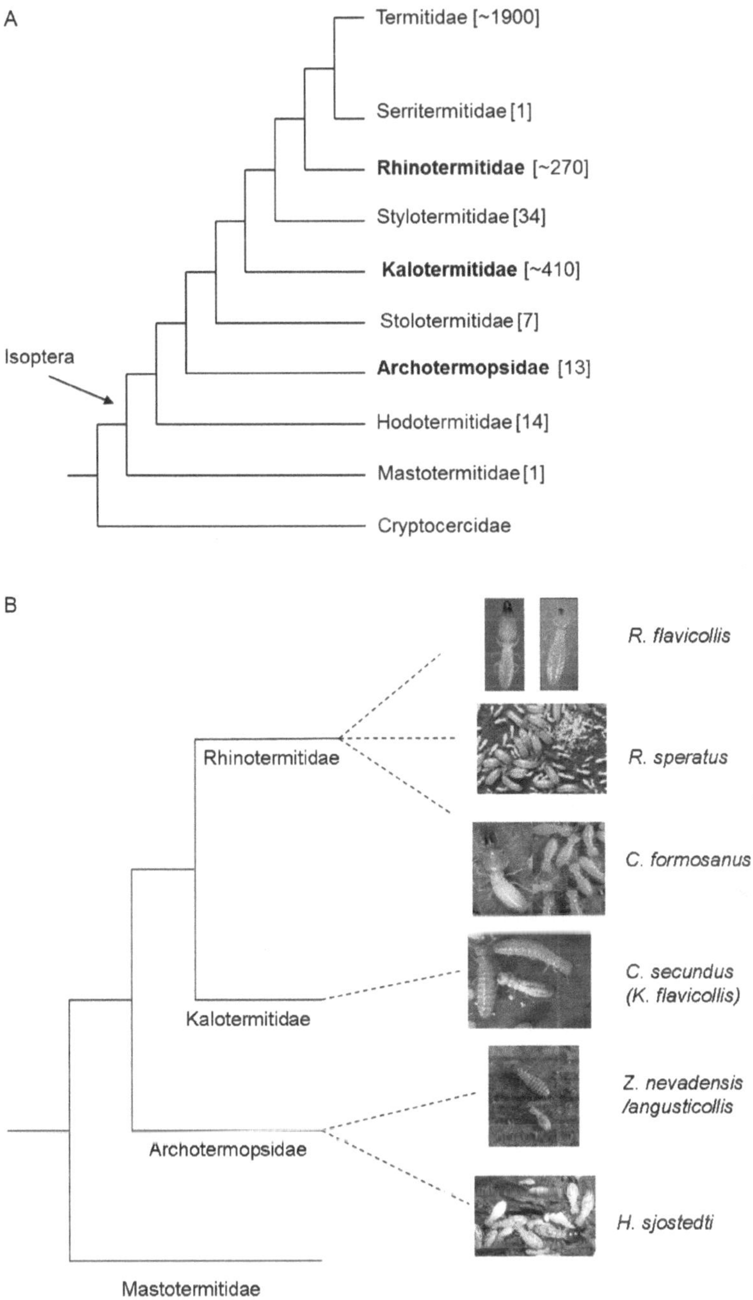

Figure 1 Termite phylogeny and model species. (A) Cladogram of termite families with the wood roach sister taxon, Cryptocercidae. []: number of species, according to Krishna et al. (2013). (B) Selected lower termite families with the model species included in this review. They are currently the most intensively studied lower termites in physiology and genetics. *Photo copyrights:* R. flavipes*: M. Scharf,* R. speratus*: K. Matsuura,* C. formosanus*: G. Henderson and* H. sjostedti*: Toru Miura.*

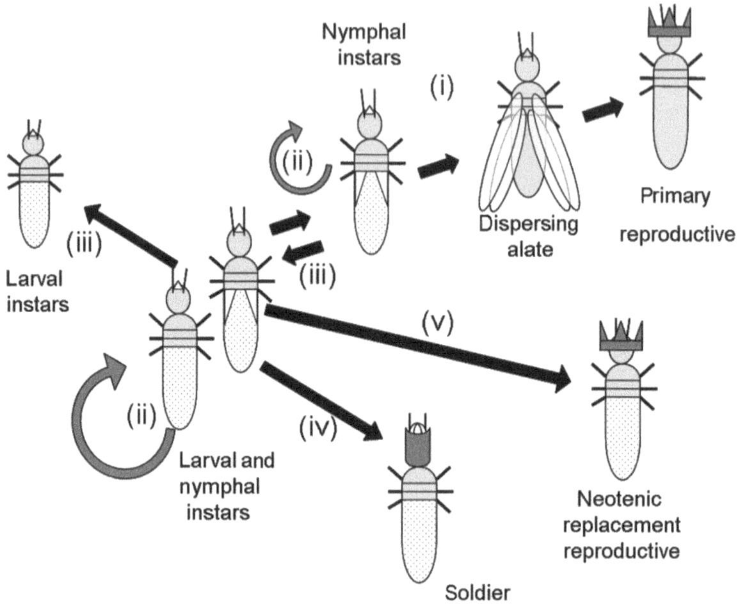

Figure 2 Developmental pathways of wood-dwelling lower termites. Workers are totipotent immatures, comprising larval (without wing buds) and nymphal (with wing buds) instars that can explore all caste options. They can develop (i) progressive via nymphal instars into winged sexuals (alates) that disperse and found a new nest; (ii) stationary, remaining in the same instar (grey semi-circle); (iii) regressive into an 'earlier' instar; (iv) via a pre-soldier into a soldier, and (v) into a neotenic replacement reproductives that reproduces within the natal nest. *Adapted from Korb et al. (2012b).*

termites. This diversity of moulting types makes termites a challenge to any endocrine model of development. How can interactions between just two morphogenic hormones, juvenile hormone (JH) and ecdysone, produce such developmental plasticity? Like in most insects, JHIII is the only JH moiety produced by termites' corpora allata and detected in hemolymph (Brent et al., 2005; Cornette et al., 2008; Greenberg and Tobe, 1985; Lanzrein et al., 1985; Park and Raina, 2004; Yagi et al., 2005).

Based on seminal termite research by Lüscher and others in the 1950s and 1960s (summarized in Lüscher, 1974b; Lenz, 1976), Nijhout and Wheeler (1982) developed a model proposing that there are three JH-sensitive periods during the inter-moult period (i.e. the time between two consecutive moults) that determine the subsequent moulting type of a lower termite worker (Fig. 3A). These three periods occur at the start, during mid-phase and during the second half of the inter-moult period, and they control, respectively, sexual reproductive traits, non-sexual adults' traits (e.g. eyes

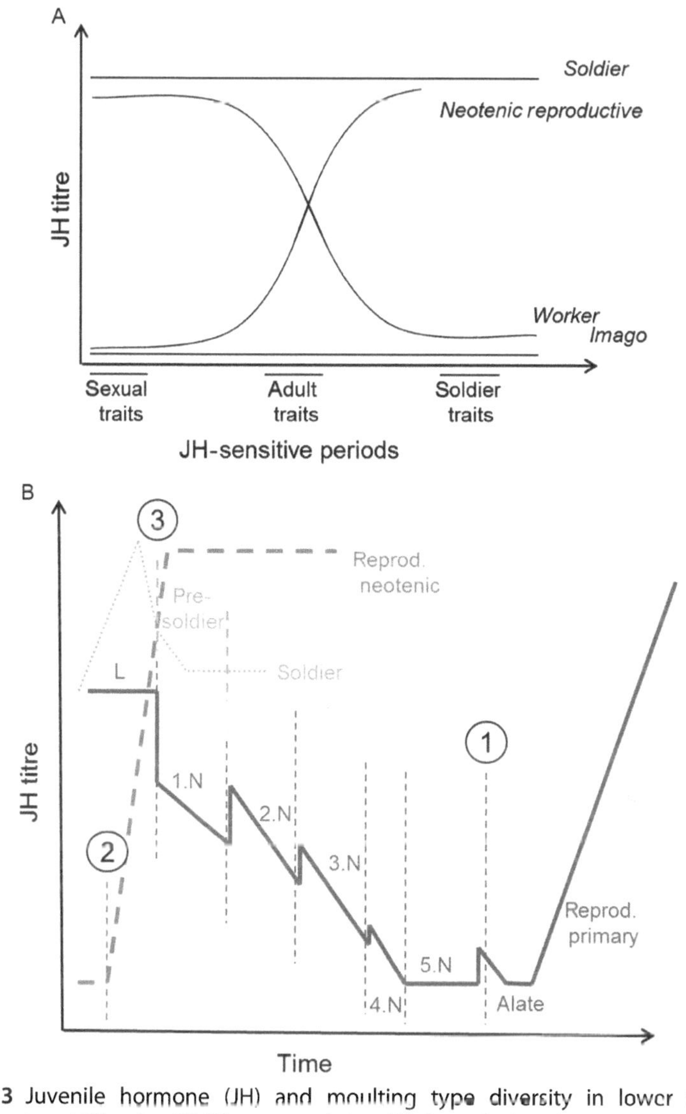

Figure 3 Juvenile hormone (JH) and moulting type diversity in lower termites. (A) Nijhout and Wheeler (1982) proposed that JH titres during three sensitive periods between two moults determine the subsequent moulting type of a worker. Continuously high or low titres result, respectively, in pre-soldier/soldier differentiation or progressive development towards winged Imagoes. Titres that vary during the inter-moult period account for neotenic and continued worker development. (B) Summary of current results on hemolymph JH titres during termite development. Blue (dark grey in the print version), solid line: progressive development towards a winged sexual; red (grey in the print version), dashed line: neotenic development; green (light grey in the print version), dotted line: soldier development; (1) imaginal moult into winged sexual (alate); (2) neotenic moult and (3) pre-soldier moult. For more information, see text. *Adapted from Korb et al. (2012b).*

and wings) and soldier traits. The sensitive periods are largely equivalent with Lüscher's phases of 'competence' (Kompetenz; Lüscher, 1952a,b). A worker developing progressively into a nymph, and finally into a winged imago, is proposed to have low JH titres throughout all three sensitive periods, while continuously high JH titres characterize pre-soldier/soldier differentiation. Development into neotenic reproductives requires low JH titres at the start but an increase during the second period. As neotenic moults have shortened inter-moult periods (Hoffmann and Korb, 2011; Lüscher, 1974b; Springhetti, 1972), they moult before the third sensitive period. Finally, the reverse pattern with first high, then low JH titres characterizes individuals remaining workers.

I will summarize our understanding of the endocrine underpinning of termite development and caste differentiation and show how recent progress fits the proposed model. I will illustrate how JH is a central mediator of lower termites' developmental plasticity and sociality and provide links to results on emerging key gene families and genetic pathways. I will concentrate on lower termites as they are most plastic. In higher termites, developmental plasticity is considerably reduced. Castes are determined in an early stage of development, even as early as in the egg by the queen (Lüscher, 1976; Okot-Kotber, 1985b). Most physiological and molecular research also focused on lower termites (noticeable exceptions, work on the higher termites *Nasutitermes takasagoensis*: e.g. Hojo et al., 2005; Toga et al., 2013, and the fungus grower *Macrotermes michaelseni*: e.g. Lanzrein et al., 1985; Okot-Kotber, 1985a,b). In this chapter, I will concentrate on the current main model species (Fig. 1B): the archotermopsids (dampwood termites) *Hodotermopsis sjostedti*, *Zootermopsis nevadensis* and *Zootermopsis angusticollis*, the kalotermitid (drywood termite) *Cryptotermes secundus* and the rhinotermitids (subterranean termites) *Reticulitermes flavipes*, *Reticulitermes speratus* and *Coptotermes formosanus*. These data will be supplemented by results for the classic model, the kalotermitid *Kalotermes flavicollis*, and for other species as appropriate and available. For recent reviews on genetic mechanisms and termite caste differentiation, I also refer the reader to Miura and Scharf (2011) and Korb and Hartfelder (2008).

I will first outline what is known about the role of JH during post-embryonic larval development in regulating progressive, stationary and regressive moults. Second, I will review results that clearly link JH to termites' caste differentiation by regulating the pre-soldier/soldier and neotenic moults. Here, most progress has been made in unravelling the mechanisms underlying soldier differentiation. Third, I will summarize results on the role of JH in adult, reproducing termites. I will show that, similar to other insects (including social Hymenoptera), JH is centrally associated with high fertility and that recent evidence suggests honest fertility signalling through a linkage between JH titres and the production of chemical scents in queens. Hence, JH seems crucial for maintenance of the reproductive monopoly by the queen and division of labour in termite colonies.

2. JH: A MAJOR REGULATOR OF TERMITES' POST-EMBRYONIC LARVAL DEVELOPMENT

Typical post-embryonic insect development is characterized by progressive moults. As a simplified description, the moulting process is initiated after a larva reaches a critical size. Then varying JH titres determine the moulting type with high titres leading to a larval moult while low titres characterize an imaginal transition (e.g. Hall and Wake, 1999; Nijhout, 1994). Recently, in *Drosophila melanogaster*, this endocrine regulation has been linked to the insulin/insulin-like growth factor signalling pathway (IIS) that regulates insect growth (e.g. Colombani et al., 2005; Mirth et al., 2005; Shingleton, 2005). How do the stationary and especially regressive moults differ from this default development?

Nijhout and Wheeler's model (1982) did not explicitly address the differences among progressive, stationary and regressive moults. They are hidden behind the imago versus worker trajectory (Fig. 3A). Lüscher proposed in the 1950s and 1960s that progressive development of immature workers is modulated by the reproductives of a colony (Lüscher, 1956, 1973, 1974b): the queen and the king prevent same sex individuals from becoming sexuals through the production of inhibitory substances (maybe directly JH) that circulate within a colony via proctodeal trophallaxis (anal feeding) and probably also through volatile compounds (Springhetti, 1969a; Stuart, 1979).

Not much progress has been made in determining the exact mechanisms underlying the various larval moulting types. The inability to predict the developmental trajectory of totipotent immatures makes tackling their molecular underpinnings extremely difficult. This especially applies for the three larval moults for which the developmental trajectories cannot be manipulated as it is possible for soldier, and partly for neotenic, development.

Recently, in the kalotermitid *C. secundus*, hemolymph JH titres have been studied during post-embryonic development using intact colonies (Korb et al., 2009a, 2012b) (Fig. 3B). The progressive transition from a larval (instar without wing buds) to a nymphal instar is characterized by a sharp drop in hemolymph JH titres supporting the idea that high JH titres hinder progressive development. During the different nymphal stages (five in the case of *C. secundus*), JH titres vary as do ecdysteroid titres, but the latter at low concentrations. Studying the time course during the inter-moult period (which lasts on average 60 days in *C. secundus*) showed significant differences in JH titres for different weeks (Korb et al., 2012b) (Fig. 3B). JH titres are elevated right after a moult. Thereafter, they remain low for varying lengths of time, before they rise again shortly before the subsequent moult. The fourth and fifth (penultimate and ultimate) nymphal instars differ from this pattern in that JH titres in the former do not drop after the moult but remain constantly at medium to high levels and that they are constantly low in the latter with a sharp peak shortly before the imaginal moult. After the imaginal moult, JH titres drop but remain higher than in the ultimate nymphal instar.

This pattern may suggest that major developmental transitions are characterized by sharp changes in JH titres (Fig. 3B): a drop during the larva–nymph transition and a rise during the nymph–imago transition. They also imply that low JH titres are associated with progressive development and that the penultimate instar is 'special': except for this instar, which is also characterized by a shortened inter-moult period, JH titres gradually decline in subsequent nymphal stages.

These data comprised workers with different developmental trajectories (progressive, stationary and regressive) as JH measurements required destructive sampling of individuals whose potential subsequent moult, hence, remained unknown. To obtain first insights into the basis underlying progressive, regressive and stationary developments, we took advantage of the fact that moulting types in *C. secundus* differ between seasons (Korb and Katrantzis, 2004). Hence, it could be shown that fourth nymphal instars, which were more likely to develop regressively, have higher JH titres than those with predominantly progressive or stationary moults (Korb et al., 2009a). Additionally, the average inter-moult period was longer for regressive moults when marked individuals were followed over two subsequent moults. This mainly seems to be due to an absence of moults with short inter-moult periods (Korb et al., 2012b).

These results support the hypothesis that high JH titres during nymphal stages prevent the development of termite workers into sexuals (Lüscher, 1956, 1974b; Springhetti, 1969a; Stuart, 1979). A parallel exists to the honeybee. JH is similarly involved in the behavioural reversion of foraging workers back to nursing tasks in the hive, for instance, after a catastrophic loss of nurses (e.g. Hartfelder et al., 2015; Robinson et al., 1992). Interestingly, this reversal has been shown to be associated with the reversal of epigenetic marks (Herb et al., 2012), which opens the question about the role of epigenetics in regulating termite development (Glastad et al., 2014; Terrapon et al., 2014; see also Glastad et al., 2015).

Similar data addressing post-embryonic larval development in termites are rare. In line with the results for *C. secundus*, Cornette et al. (2008) showed for the archotermopsid *H. sjostedti* a decline in JH titres from the larval to the nymphal instar and an increase as they passed the imaginal moult. But this species has a single nymphal instar only, and regressive moults seem to be absent. Raina et al. (2008) characterized hormone titres during moults in workers of the rhinotermitid *C. formosanus*: JH titres raised 6 days prior to ecdysis. Like in *C. secundus*, ecdysteroid titres were generally low, but they raised 3–4 days prior ecdysis.

3. JH AND CASTE DIFFERENTIATION INTO THE TWO SYNAPOMORPHIC TERMITE CASTES: THE SOLDIERS AND NEOTENIC REPRODUCTIVES

Soldiers and neotenic reproductives are the two castes that separate termites from cockroaches. Soldiers are always sterile and make termites eusocial. They can only gain indirect fitness by increasing the reproductive success of their parents, mainly through defence of siblings. They only evolved once, at the transition to a social life (Inward et al., 2007b; Noirot and Pasteels, 1987, 1988; Roisin and Korb, 2011). They are present in all species except for a few higher termites in which they have been secondarily lost. Neotenic reproductives also occur in all lower termites (Myles, 1999). The evolution of this caste allows the totipotent immature workers to gain direct fitness in the nest by inheriting the natal breeding position after the death of the current reproductives; this might have been crucial for the evolution of workers (Korb, 2007, 2009; Korb and Schneider, 2007; Myles, 1988). Depending on phylogenetic position and ecology, the degree of brood care by workers and options to inherit vary between lower termite species (Korb et al., 2012a). Even in the sister taxon of the termites, the subsocial *Cryptocercus* woodroaches, which have a similar ecology as wood-dwelling termites, the likelihood that the parents die during larval development is substantial (Nalepa, 1984), but neotenics do not exist.

How did these two key innovations, the soldiers and neotenics, evolve that distinguish cockroaches and eusocial termites? Identifying the proximate underpinnings of their development can contribute to understanding their evolution.

3.1 Soldiers

Termites' caste differentiation is best understood for soldier development. In the following, I will first outline what is known about endocrine regulation of soldier differentiation and its associated changes in gene expression. In the second part, I will show how socio-environmental effects impact soldier differentiation. Finally, I will summarize progress in our understanding of the interactions between socio-environmental and endocrine/genetic factors in regulating soldier development.

3.1.1 Endocrine regulation and associated genetic changes

The induction of soldier differentiation by JH and juvenile hormone analogues (JHAs) was one of the earliest findings in JH research (e.g. Howard and Haverty, 1979; Lüscher, 1958a,b, 1969; more recent summary: Hrdy et al., 2006). Soldier development always proceeds via a pre-soldier instar into a soldier. In a simplified view, high JH titres that are associated with an enlarged corpora allata lead to soldier differentiation by inducing a pre-soldier moult (Fig. 3B). This induction seems to depend on high JH titres, especially during the second half of the moulting interval (Lüscher, 1974a; Springhetti, 1972). Comparing the proposed model by Nijhout and Wheeler (1982) with the current results, they correspond in that soldier differentiation is associated with high JH titres (Fig. 3A and B). More specifically, recent work showed that pre-soldier differentiation is linked to a sharp JH peak during the second half of the moulting period while JH titres decline thereafter (Cornette et al., 2008) (Fig. 3B).

How are JH titres regulated? Like in other insects, the activity of the corpora allata seems to be inhibited by allatostatins (Chan et al., 2011; Yagi et al., 2005), a large family of neuropeptides present in all major insect orders (for a review, see Stay and Tobe, 2007). Allatostatins have pleiotropic effects. As brain–gut peptides they can also regulate food intake (Aguilar et al., 2003, 2006) and this may present a link to socio-environmental influences on termites' caste differentiation (Korb and Hartfelder, 2008) (see below). After production by the corpora allata, JH circulates in the hemolymph. Yet, it has a low metabolic stability as it is easy degraded by JH esterases and/or JH epoxide hydrolases (Goodman and Granger, 2005; Oakeshott et al.,

2005). To reach its targets, it must be bound to transporter proteins, which have been identified by photoaffinity labelling in *R. flavipes*, *C. formosanus* and *Z. nevadensis* (Okot-Kotber and Prestwich, 1991a,b).

Recent studies propose a further mechanism regulating JH availability and link this to pre-soldier/soldier differentiation. Hexamerins (namely the products of *Hex1* and *Hex2*), which commonly function as storage and transport proteins and which are also involved in the division of labour in the honeybee (Martins et al., 2010, 2011) and some social wasps (Hoffman and Goodisman, 2007; Hunt et al., 2003, 2007), reduce JH availability in *R. flavipes* workers, thus hindering soldier differentiation (Scharf et al., 2005a,b; Zhou et al., 2006a,b, 2007b). Fat body expression and hemolymph titres of these proteins seem to be modulated by JH, especially in the case of *Hex2*. At the same time, hexamerin reduces JH availability; this has been proposed to occur through covalent binding of Hex1 protein (Zhou et al., 2006a,b). As silencing *Hex1* expression through RNAi leads to an increased pre-soldier development, similar to an application of JHIII, it was concluded that *Hex1* and JH interact in regulation soldier differentiation. Accordingly, Hex1 is proposed to function as a *status quo* protein, its expression leads to a low availability of JH, thus preventing the transformation of workers into pre-soldiers (Zhou et al., 2006a).

There have been some discussions about these results (Hrdy et al., 2006), especially the proposed covalent binding of JH is peculiar (Korb and Hartfelder, 2008), and more work is required to unravel the details of the proposed mechanism. However, at least two recent studies provide support for a more widespread role of *Hex1* and *Hex2* in caste differentiation in termites. Cornette et al. (2013) revealed significant up-regulation of *Hex* genes 24 h after JHA application in *H. sjostedti* in a suppressive subtractive hybridisation study. Additionally, comparisons of the recently sequenced *Z. nevadensis* genome with those from several arthropods led to the identification of a total of five hexamerin genes in *Z. nevadensis*, four of which probably evolved by tandem duplication within the Blattodea/Isoptera clade, including *Hex1* and *Hex2* (Terrapon et al., 2014). Relatively high expression of both genes in worker and nymphal stages in *Z. nevadensis* is in line with the results for *R. flavipes* and supports the hypothesis that these genes have been co-opted during social evolution for caste differentiation. Hence similar to the role of hexamerins in some social wasps (Hoffman and Goodisman, 2007; Hunt et al., 2003, 2007) and the honeybee (Martins et al., 2010, 2011) as well as vitellogenin in the honeybee (e.g. Amdam et al., 2003, 2006), common storage/transport proteins which interact with JH may have been exploited during social evolution.

With regard to JH regulation, another group of genes turned up to be important during soldier differentiation: cytochrome P450 genes. These genes are known to be involved in detoxification of xenobiotics in insects, but they can also be crucial regulators of JH titres (Feyereisen, 2012). In cockroaches and locusts, the last step in the biosynthesis of JHIII from methylfarnesoate is catalysed by a P450 from the CYP15 family (Helvig et al., 2004). Also degradation of JH in cockroaches can be regulated by P450s, this time by a member from the CYP4 subfamily (Feyereisen, 2012; Sutherland et al., 1998, 2000). In *H. sjostedti* as well as in *R. flavipes*, JH/JHA-induced expression of several P450 transcripts was found (Cornette et al., 2006; Tartar et al., 2009; Zhou et al., 2007a). In addition, recently a link between a CYP15 ortholog and soldier differentiation in the latter species was established (Tarver et al., 2012) (see also below).

Strikingly, pre-soldier/soldier differentiation seems to be connected also to the IIS pathway as evidence for *H. sjostedti* suggests (Hattori et al., 2013). Similarly, in the honeybee this pathway is involved in caste differentiation and division of labour (Ament et al., 2008; Corona et al., 2007; see also Hartfelder et al., 2015). JHA-induced pre-soldier development resulted in an up-regulation of the expression of orthologs from several central IIS genes (*InR*, *FOXO* and *PKB/Akt*) during the period of soldier-specific mandible elongation. RNAi of the *InR* ortholog disrupted soldier-specific morphogenesis confirming its functional involvement during soldier differentiation. Because the IIS pathway is generally thought to act upstream of JH in insects, these results suggest that at least in *H. sjostedti* a feedback loop evolved between JH and IIS. This might enable social interactions, besides nutrition, to regulate caste determination (Hattori et al., 2013).

In addition, several more genes were identified to be differentially expressed during JHA-induced pre-soldier/soldier differentiation in *H. sjostedti* (Cornette et al., 2013; Ishikawa et al., 2010). They include not only several cuticle proteins (Koshikawa et al., 2005) but also a novel gene, *SOL1* (Miura et al., 1999). For *R. speratus*, a Laccase2 ortholog might be involved in cuticular tanning of soldiers as implied from qRT-PCR results during pre-soldier/soldier differentiation (Masuoka et al., 2013).

3.1.2 Socio-environmental effects

Environmental factors that influence JH titres and soldier differentiation include food (Liu et al., 2005b) and temperature (Lenz, 1976; Liu et al., 2005b; Scharf et al., 2007). The latter may explain seasonal cycles of soldier production in several species from temperate regions (e.g. Liu et al., 2005a; Lüscher, 1972). Waller and LaFage (1988) showed that in *C. formosanus* more soldiers are produced under higher temperatures when individuals had higher wood consumption rates and were heavier. Hence, the temperature effect on soldier production may result from better nutrition. Whether predation pressure also influences soldier numbers is still debated (Noirot and Darlington, 2000). The difficulty to link predation pressure with soldier number can have several causes, among them an inability to respond to rapidly changing predation pressure, as well as the fact that soldiers can also be involved in defending the colony against competitors (Noirot and Darlington, 2000).

Among the social factors that stimulate soldier development are the presence of reproductives and worker nestmates (e.g. Bordereau and Han, 1986; Elliott and Stay, 2008). In several species, soldiers only seem to be produced after a certain colony size is reached (Mao and Henderson, 2010). In *R. speratus*, removal of reproductives in incipient colonies prevents soldier differentiation (Maekawa et al., 2012). As the first developing soldier receives more proctodeal trophallaxis by primary reproductives than other individuals, Maekawa et al. (2012) suggested that reproductives directly favour soldier differentiation by feeding JH or other stimulatory substances to workers. This is in line with a model proposed by Lüscher (1974a) (Fig. 4A).

In contrast to reproductives, soldiers inhibit further soldier differentiation, thus regulating soldier numbers in a negative feedback process (e.g. Bordereau, 1985; Bordereau and Han, 1986; Haverty and Howard, 1981; Korb et al., 2003; Lüscher, 1969, 1974b; Mao et al., 2005; Miller, 1942; Nagin, 1972; Park and Raina, 2005; Renoux, 1976; Springhetti, 1969b; Watanabe et al., 2011). Lüscher (1974b) proposed that soldiers produce a substance that inhibits JH production in the corpora allata of workers and by doing so favour neotenic development that requires low JH titres (Fig. 4A). At least in some species, such inhibitory components seem to originate from the soldier heads as the application of soldier head extract (SHE) to the colony hinders soldier differentiation (Korb et al., 2003; Lefeuve and Bordereau, 1984; Okot-Kotber et al., 1991; Renoux, 1976).

3.1.3 How can socio-environmental factors impact endocrine regulation during soldier differentiation?

As noted above, Lüscher (1974b) suggested that the presence of soldiers impacts the physiology of workers directly by producing substances that

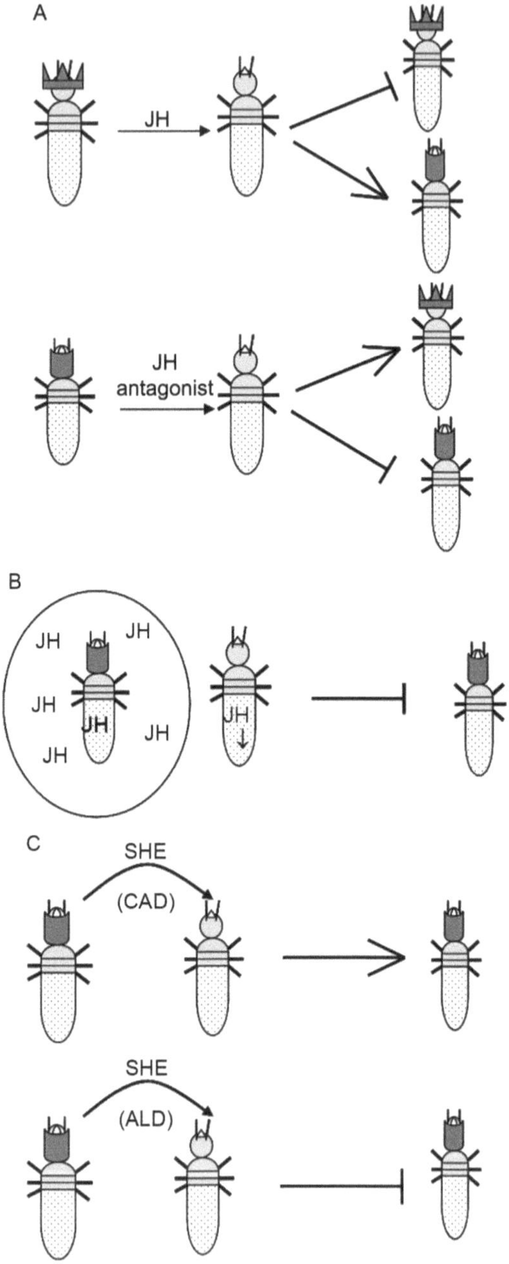

Figure 4 Models that were proposed to explain the social effects of nestmates on soldier differentiation. (A) Lüscher proposed that the queen produces a pheromone (maybe JH) that inhibits the development of reproductives while favouring soldier differentiation, whereas soldiers have the opposite effect. (B) Henderson suggested that soldiers function as JH sink for workers. Their presence reduces JH availability within the colony, hence inhibiting soldier development. (C) Scharf, Tarver and co-workers proposed that in *R. flavipes* γ-cadinene (CAD) and its aldehyde, γ-cardinenal (ALD) are the active compounds of soldier head extracts (SHE) with the former simulating soldier differentiation while the latter inhibits it.

inhibit JH production in workers (Fig 4A). Henderson (1998) extended this idea and proposed that soldiers act as a 'JH-sponge' that reduces JH titres in workers (Fig. 4B). Evidence supporting this hypothesis comes from research on rhinotermitids: JH titres in workers rise after removal from the colony (Mao et al., 2005; Okot-Kotber et al., 1993), which can lead to pre-soldier differentiation (Mao et al., 2005). Yet when workers are kept with soldiers, JH titres of the former are reduced and no pre-soldiers develop (Mao et al., 2005; Park and Raina, 2005). Similarly, JH titres in workers increase with colony size, but the addition of soldiers reduces worker's JH titres (Mao and Henderson, 2010). Also after application of JHIII, JH titres of workers (comprising endogenous and applied JHIII) are reduced when soldiers are added (Watanabe et al., 2011). This implies that soldiers may act as a direct JH sink for workers. Thus, they may complement the JH-reducing effect that hexamerins seem to have at the sub-organismic level.

In addition to soldiers functioning as JH sinks, compounds from the soldier's heads also inhibit soldier differentiation, at least in some species. How SHE impact development has been studied in *R. flavipes* (Fig. 4C): although SHE alone inhibits soldier differentiation in this species (Okot-Kotber et al., 1991), the combined application of JHIII and SHE to groups of workers results synergistically in increased pre-soldier differentiation (Tarver et al., 2009). Along with this combined application, the gene expression profile in workers changes in a pattern that is similar to the application of JHIII alone (Tarver et al., 2010). Testing 49 candidate genes revealed a differential expression of *Hex1* and *Hex2* as well as two vitellogenins and several genes known from other insects to function in JH biosynthesis or degradation (among them P450s). These results suggest a link between the social effects of SHE and endocrine regulation of soldier differentiation (Fig. 4C). A few questions remain: the similar expression profile of JHIII alone and JHIII combined with SHE made it difficult to separate a JHIII effect from an additional SHE effect. Furthermore, in all other studies, including one on *R. flavipes* (Okot-Kotber et al., 1991), SHE suppressed soldier development, in contrast to the enhancing effect found in this study. An answer to the last question may be the composition of SHE. Active head extract components were identified to be γ-cadinene (CAD) and its aldehyde γ-cardinenal (ALD) (Tarver et al., 2009). The former seems to stimulate soldier development, while the latter inhibits it (Tarver et al., 2011). Hence, different ALD to CAD ratios might explain an inhibitory versus enhancing effect of SHE on soldier differentiation (Fig. 4C). Yet, this mechanism might be restricted as these candidate primer pheromones were neither found in other *R. flavipes* populations (Perdereau et al., 2010) nor in *C. formosanus* (M. Tarver, personal communication). Recently, a follow-up study (Sen et al., 2013) showed that JHIII and SHE have opposite impacts on gene expression in the termites' gut and that combined treatments result in intermediate effects, supporting a general inhibitory function of SHE. Furthermore, this study also suggests an important role of the termite gut in JH-dependent caste development.

The effect of CAD, ALD and JHIII on one P450, *CYP15F1*, was studied in more detail (Tarver et al., 2012). In cockroaches and locusts, *CYP15A1* orthologs catalyse the final step in JH biosynthesis (Helvig et al., 2004). *CYP15F1* was rapidly and sustainably induced after application of JHIII, CAD or ALD. The latter is surprising given its supposed inhibitory function. In addition, RNAi of *CYP15F1* led to a reduction in pre-soldier development showing the functional significance of JH in soldier induction.

3.2 Neotenic reproductives

3.2.1 Endocrine regulation

In contrast to extensive research on soldiers, less work is done on the regulation of the second synapomorphic termite caste, the neotenics. Neotenic development requires low JH titres (e.g. Light, 1944; Lüscher, 1952a, 1974b). In *K. flavicollis*, the sensitive period seems to be during the first half of the inter-moult period as the competence of workers to become neotenics is highest directly after a moult and decreases gradually (Lüscher, 1952a, 1974b) (Fig. 3). Neotenic development is also characterized by a shortened inter-moult period (Hoffmann and Korb, 2011; Lüscher, 1952a,b) (Fig. 3B). This may imply an interaction between JH and ecdysone: low JH titres post-moulting may trigger a pre-mature ecdysone peak and a precocious subsequent moult. This hypothesis allows a separation between neotenic and progressive development without invoking a second, mid-phase JH-sensitive period proposed by Nijhout and Wheeler (1982) (Fig. 3).

3.2.2 Socio-environmental effects

Equivalent to soldiers, it is known since a long time that neotenic development is generally inhibited by the presence of reproductives (e.g. Grasse and Noirot, 1946; Lüscher, 1952a, 1974b; Shimizu, 1970; Springhetti, 1969a, 1985). Such a negative feedback is especially pronounced in the Kalotermitidae in which colonies are headed by a single pair of reproductives (Lenz, 1976), whereas it is less strong in species with multiple reproductives (Lenz, 1985). No inhibiting effect is known in species with asexual queen succession, such as *R. speratus*. Here, queens produce many neotenic reproductives parthenogenetically as future replacements of themselves (Matsuura, 2011). In contrast to the generally inhibitory effect of reproductives, soldiers can stimulate neotenic differentiation in some (e.g. *K. flavicollis*) but probably not all species (e.g. *Z. nevadensis*, *Z. angusticollis*) (Lüscher, 1974b).

Based mainly on results for *K. flavicollis* and *Zootermopsis*, Lüscher (1974b) proposed that fertile reproductives inhibit further reproductive differentiation through propagating JH (or similar compounds) within the colony, which at the same time favours soldier development (Fig. 4A). These substances may circulate within a colony through proctodeal feeding among nestmates. Behavioural experiments in *C. secundus* support this hypothesis. Individuals which receive most proctodeal feeding are less likely to develop progressively into nymphal instars (Korb and Schmidinger, 2004). Furthermore, in queenless colonies, workers which will successfully develop into neotenics act more often as donors than others during proctodeal interactions (Hoffmann and Korb, 2011). In addition to this contact-requiring feeding mechanism, a volatile compound is necessary to maintain reproductive monopoly (Himuro et al., 2011; Lüscher, 1974b; Matsuura et al., 2010).

4. JH, FECUNDITY AND REPRODUCTIVE DIVISION OF LABOUR

JH also plays a pivotal role in adult insects where it affects fundamental life history traits such as ageing and fecundity (e.g. Flatt et al., 2005). High JH titres are essential for egg production; they influence vitellogenin (Vg) availability, a precursor to the yolk protein vitellin that forms the egg yolk. In social Hymenoptera, JH was identified as a major player in caste differentiation as well as maintenance of division of labour (e.g. Ament et al., 2008; Corona et al., 2007; Libbrecht et al., 2013; Rascon et al., 2011).

4.1 JH and fecundity

Similarly, reproducing termite queens are characterized by high JH titres, and a link between JH and fecundity seems to exist in all major termite study systems. In the termopsid *Z. nevadensis*, a large corpora allata of queens indicates high JH activity (summarized in Lüscher, 1974b) and Brent et al. (2005) showed that JH stimulates oogenesis of mated *Z. angusticollis* queens. In *H. sjostedti*, JH titres increase during alate maturation of females (Cornette et al., 2008). Among the kalotermitids, *K. flavicollis* queens have enlarged corpora allata and measurements of *C. secundus* reproductives revealed high hemolymph JH titres (Korb et al., 2009a, 2012b). In the rhinotermitid *R. flavipes*, an increase in corpora allata activity correlates with an increase in the number of vitellogenic ovarioles in neotenics, and reproducing neotenic queens have elevated JH titres (Elliott and Stay, 2007, 2008). Finally, in *R. speratus* JH titres rise from nymphal instar to reproductives and the cyclical egg production in founding queens correlates with JH titres and *Vg* expression (Maekawa et al., 2010). JH titres of neotenics are continuously high even when no vitellogenic eggs are present (Maekawa et al., 2010).

An exception to that rule that reproductives have high JH titres seems to be the neotenics from *H. sjostedti* (Cornette et al., 2008). This could have several explanations: first, vitellogenesis occurs cyclically. However, Maekawa et al. (2010) found continuously high JH titres in cyclically reproducing *R. speratus* neotenics. Second, neotenic *H. sjostedti* queens have a very low fecundity. This requires further testing; do JH titres correlate with fecundity or with the number of co-occurring neotenics across taxa? Third, hemolymph JH titres do not necessarily reflect JH synthesis (Treiblmayr et al., 2006), which is certainly true.

4.2 Genes associated with reproduction: 'Queen genes'

Neotenic queens of totipotent lower termites differ from the workers from which they develop mainly in that they reproduce and maintain reproductive division of labour. Confounding traits which are linked to dispersal and nest foundation (e.g. wings or compound eyes) are absent. This offers the possibility to identify 'queen genes' (i.e. genes linked to reproduction and reproductive dominance of the queen) and the underlying molecular mechanisms of caste differentiation.

In *C. secundus*, five genes were identified to be specifically over-expressed in (neotenic) queens (neofem) using cDNA representational difference analysis (Weil et al., 2007): an esterase-lipase (*Neofem1*), a β-glucosidase (*Neofem2*), a vitellogenin (*Neofem3*), a *Cyp4* P450 gene (*Neofem4*) and a 'novel' gene (*Neofem5*) with no orthologs. A comparative study in *Cryptotermes cynocephalus* identified 13 genes that are highly expressed in neotenic queens (Weil et al., 2009b). Strikingly, in most cases the caste-specific expression pattern was not conserved, despite the close phylogenetic relationship of both Cryptotermes species. Only four genes overlapped that were over-expressed in neotenic females: *Neofem1*, *Neofem2*, *Neofem3* and *Neofem6*. The last gene was not identified in the *C. secundus* study (Weil et al., 2007), but its differential expression was confirmed in the latter study (Weil et al., 2009b). *Neofem6* shows closest similarity to *takeout*. The products of these *Neofem* genes may play important roles in female reproductives, in particular in controlling caste determination and reproductive division of labour.

For the β-glycosidase gene, *Neofem2*, as well as for the P450 gene, *Neofem4*, their functional significance in maintaining reproductive division of labour was shown in RNAi experiments (Hoffmann et al., 2014; Korb et al., 2009b) (for more details to *Neofem4* see below): down-regulation of these genes in queens resulted in behavioural changes of workers typical for colonies lacking reproductives; after silencing, the workers did not recognize the neotenic queen although she was present and did not change her behaviour. β-glucosidases similar to *Neofem2* may play a role in egg pheromones in *R. speratus* (Matsuura et al., 2009) and in sex-specific signalling in the Madeiran cockroach (Cornette et al., 2003), implicating that *Neofem2* may be involved in communication. The exact mechanism by which *Neofem2* functions remains to be discovered. Yet it seems not to be involved in the production of the cuticular royalty scent, as *Neofem4* is (Hoffmann et al., 2014) (see below).

Genes implicated in female reproduction and reproductive dominance were also targeted in *C. formosanus*. Here, a microarray study compared differential expression between non-reproducing and reproducing females using dealates (i.e. alates that had shed their wings but had not mated yet) and reproducing primary queens after colony foundation (Husseneder et al., 2011). Strikingly, among the identified genes was an ortholog of *Neofem2*, one of the 'queen genes' in *C. secundus* and *C. cynocephalus*. In total, about seven times more genes were over-expressed in reproducing queens than dealates. These genes commonly related to immunity (*GNBP2*), nutrition (especially several *endo*-beta-1, 4-glucanases), protein storage, regulation of caste differentiation and reproduction (e.g. *Hex1*, *Hex2*; JH-binding proteins; two CYP4 orthologs: *CYP4C9*, *CYPU3*). Notably, *Neofem4* from *C. secundus* also belongs to the CYP4 family and another gene, *CYP4C7*, from this subfamily metabolizes JH in a cockroach and is linked to ovulation (Sutherland et al., 1998, 2000). This makes CYP4 genes strong candidates for being linked with fecundity in termites and potentially being co-opted during social evolution in maintaining reproductive division of labour. Further support for this hypothesis is shown in the following.

4.3 JH, fertility signalling and reproductive division of labour

The fertility of a queen, as the sole reproducer within a colony, is critical for workers and soldiers that can gain indirect fitness benefits by helping to raise her offspring, i.e., their siblings. 'Fertility monitoring' systems evolved in social insects that reveal the queen's fertility to the altruistic castes. Hydrocarbons on the cuticle (cuticular hydrocarbons, CHCs) of queens, and in some species also on her eggs, seem to function as such fertility indicators (summarized in Liebig, 2010; Peeters and Liebig, 2009; Van Oystaeyen et al., 2014; van Zweden and d'Ettorre, 2010). Although other substances can also play a role (e.g. Hanus et al., 2009; Himuro et al., 2011; Le Conte and Hefetz, 2008; Matsuura et al., 2010), changes in CHC profiles between reproducing and non-reproducing individuals have been identified in all major clades of social insects, including termites (Liebig, 2010; Peeters and Liebig, 2009; Van Oystaeyen et al., 2014). In *Z. nevadensis* and *C. secundus*, fertile queens are characterized by long-chained CHCs (Liebig et al., 2009; Weil et al., 2009a). A change of the CHC profile in *C. secundus* queens results in a breakdown of reproductive division of labour in that workers behave like in queenless colonies (Hoffmann et al., 2014). *Neofem4*, a member of cytochrome CYP4 family, seems crucially involved in producing this queen scent (Hoffmann et al., 2014): it is over-expressed in queens. Silencing *Neofem4* with RNAi changes a queen's CHC profile to become more worker-like, and this results in workers behaving like in queenless colonies (Hoffmann et al., 2014). Strikingly, the expression of *Neofem4* is contingent on high JH titres, which implies that it may function

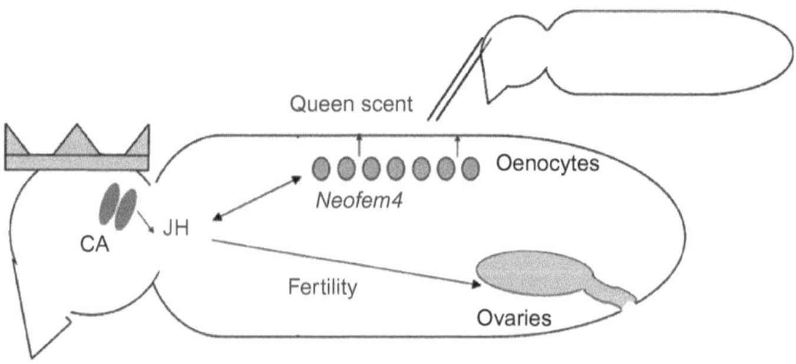

Figure 5 Model to explain the maintenance of reproductive division of labour in *C. secundus* colonies. Reproductive queens are characterized by high hemolymph JH titres which are associated with an over-expression of the P450 gene, *Neofem4*. A high expression of this gene is crucial for the production of the royalty scent. Hence, workers can smell the fertility of the queen and reproductive division of labour within a colony is maintained.

as an honest signal of fecundity (Fig. 5): only highly fecund queens can produce the signal. An ortholog of *Neofem4* was also identified in the genome of *Z. nevadensis*, and notably it was also differently expressed between castes with highest expression levels in female reproductives and eggs (Terrapon et al., 2014). This would make *Neofem4* a strong candidate for fertility signalling across termites, but note that it seems not differentially expressed in *C. cynocephalus* (Weil et al., 2009b).

5. CONCLUSIONS

During the last decades, we made considerable progress in our understanding of the role of JH in underlying the developmental plasticity and high diversity of moulting types in termites, especially with regard to soldier differentiation. These results overall are in line with the model proposed by Nijhout and Wheeler (1982) (Fig. 3). Beyond this, they imply that (i) soldier differentiation is initiated by a sharp JH rise above background levels rather than continuously high JH titres (Fig. 3B), (ii) the second, mid-phase sensitive period may not be necessary to account for neotenic development, the shortened inter-moult period triggered by low post-moult JH titres may explain it (Fig. 3B) and (iii) regressive development, not addressed by the model, might be due to high JH titres associated with long inter-moult periods (Korb et al., 2009a). Based on results for all model species, we can generate the following synthetic model that aims at summarizing our current understanding of the interaction between JH and socio-environmental as well as endogenous factors in termites' caste differentiation (Fig. 6): hemolymph JH titres in totipotent worker seem to be reduced by hexamerins preventing soldier differentiation. At the colony level, soldiers may act as a further JH sink. As is common for insects, JH production seems to be under the control of allatostatins which due to their pleiotropic function as brain–gut peptides may also function as socio-environmental sensors. Additional feedback from the socio-environment may occur through interactions of JH with the IIS pathway. Hence, similar to what was proposed for *D. melanogaster* (Shingleton, 2005), growth rate and development can be linked. In line, storage proteins such as hexamerins and vitellogenins as well as nutrition-related proteins and 'JH-associated' P450 genes repeatedly emerged across species as major candidates of caste differentiation. This model is still based on few data and is rather hypothetical but it may guide future research to finally understand the proximate underpinning of termite sociality.

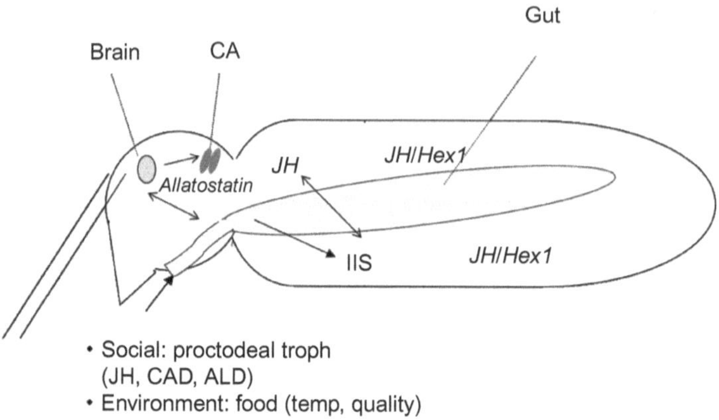

Figure 6 Synthetic model that aims at summarizing our current understanding of the interaction between juvenile hormone (JH) and socio-environmental as well as endogenous factors in termite caste differentiation, based on results for several study species. Endogenous JH titres are critical in determining the developmental trajectory of a totipotent lower termite worker. According to current results, their availability in the hemolymph seems to be regulated by Hexamerin storage/transport proteins. Endogenous feedback mechanisms between the gut, the corpora allata (CA) and genes from the insulin/insulin-like growth factor signalling pathway (IIS) may interact with socio-environmental factors in regulating hemolymph JH titres and hence a workers developmental trajectory. For further information, see text.

ACKNOWLEDGEMENTS

I thank M. Scharf and M. Tarver for discussions of their work, Clement Kent & Amro Zayed for making this special volume possible and inviting me to contribute, and G. Henderson, K. Matsuura, T. Miura and M. Scharf for providing photos for Fig. 1.

REFERENCES

Aguilar, R., Maestro, J.L., Vilaplan, L., Pascual, N., Piulachs, M.D., Belles, X., 2003. Allatostatin gene expression in brain and midgut, and activity of synthetic allatostatins on feeding-related processes in the cockroach *Blattella germanica*. Regul. Pept. 115, 171–177.

Aguilar, R., Maestro, J.L., Belles, X., 2006. Effects of myoinhibitory peptides on food intake in the German cockroach. Physiol. Entomol. 31, 257–261.

Amdam, G.V., Norberg, K., Hagen, A., Omholt, S.W., 2003. Social exploitation of vitellogenin. Proc. Natl. Acad. Sci. U.S.A. 100, 1799–1802.

Amdam, G.V., Csondes, A., Fondrk, M.K., Page, R.E., 2006. Complex social behaviour derived from maternal reproductive traits. Nature 439, 76–78.

Ament, S.A., Corona, M., Pollock, H.S., Robinson, G.E., 2008. Insulin signaling is involved in the regulation of worker division of labor in honey bee colonies. Proc. Natl. Acad. Sci. U.S.A. 105, 4226–4231.

Bell, W.J., Roth, L.M., Nalepa, C.A., 2007. Cockroaches: Ecology, Behavior, and Natural History. The Johns Hopkins University Press, Baltimore, MD.

Bordereau, C., 1985. The role of pheromones in termite caste differentiation. In: Watson, J.A.L., Okot-Kotber, B.M., Noirot, C. (Eds.), Caste Differentiation in Social Insects. Pergamon Press, Oxford, pp. 221–226.

Bordereau, C., Han, S.H., 1986. Stimulatory influence of the queen and king on soldier differentiation in the higher termites *Nasutitermes lujae* and *Cubitermes fungifaber*. Insect. Soc. 33, 296–305.

Brent, C.S., Schal, C., Vargo, E.L., 2005. Endocrine changes in maturing primary queens of *Zootermopsis angusticollis*. J. Insect Physiol. 51, 1200–1209.

Chan, K.K., Abel, D.S., Stay, B., 2011. Fine structure of corpora allata of castes with different rates of juvenile hormone production in the termite *Reticulitermes flavipes*. Arthropod Struct. Dev. 40, 26–38.

Colombani, J., Bianchini, L., Layalle, S., Pondeville, E., Dauphin-Villemant, C., Antoniewski, C., Carre, C., Noselli, S., Leopold, P., 2005. Antagonistic actions of ecdysone and insulins determine final size in Drosophila. Science 310, 667–670.

Cornette, R., Farine, J.P., Abed-Viellard, D., Quennedey, B., Brossut, R., 2003. Molecular characterization of a male-specific glycosyl hydrolase, Lma-p72, secreted on to the abdominal surface of the Madeira cockroach *Leucophaea maderae* (Blaberidae, Oxyhaloinae). Biochem. J. 372, 535–541.

Cornette, R., Koshikawa, S., Hojo, M., Matsumoto, T., Miura, T., 2006. Caste-specific cytochrome P450 in the damp-wood termite *Hodotermopsis sjostedti* (Isoptera, Termopsidae). Insect Mol. Biol. 15, 235–244.

Cornette, R., Gotoh, H., Koshikawa, S., Miura, T., 2008. Juvenile hormone titers and caste differentiation in the damp-wood termite *Hodotermopsis sjostedti* (Isoptera, Termopsidae). J. Insect Physiol. 54, 922–930.

Cornette, R., Hayashi, Y., Koshikawa, S., Miura, T., 2013. Differential gene expression in response to juvenile hormone analog treatment in the damp-wood termite *Hodotermopsis sjostedti* (Isoptera, Archotermopsidae). J. Insect Physiol. 59, 509–518.

Corona, M., Velarde, R.A., Remolina, S., Moran-Lauter, A., Wang, Y., Hughes, K.A., Robinson, G.E., 2007. Vitellogenin, juvenile hormone, insulin signaling, and queen honey bee longevity. Proc. Natl. Acad. Sci. U.S.A. 104, 7128–7133.

Elliott, K.L., Stay, B., 2007. Juvenile hormone synthesis as related to egg development in neotenic reproductives of the termite *Reticulitermes flavipes*, with observations on urates in the fat body. Gen. Comp. Endocrinol. 152, 102–110.

Elliott, K.L., Stay, B., 2008. Changes in juvenile hormone synthesis in the termite *Reticulitermes flavipes* during development of soldiers and neotenic reproductives from groups of isolated workers. J. Insect Physiol. 54, 492–500.

Engel, M.S., Grimaldi, D.A., Krishna, K., 2009. Termites (Isoptera): their phylogeny, classification, and rise to ecological dominance. Am. Mus. Novit. 3650, 1–27.

Feyereisen, R., 2012. Insect CYP genes and P450 enzymes. In: Gilbert, L.I. (Ed.), Insect Molecular Biology and Biochemistry. Academic Press, London, pp. 236–316.

Flatt, T., Tu, M.P., Tatar, M., 2005. Hormonal pleiotropy and the juvenile hormone regulation of *Drosophila* development and life history. Bioessays 27, 999–1010.

Glastad, K.M., Hunt, B.G., Goodisman, M.A.D., 2014. Evolutionary insights into DNA methylation in insects. Curr. Opin. Insect Sci. 1, 25–30.

Glastad, K.M., Chau, L.M., Goodisman, M.A.D., 2015. Epigenetics in Social Insects. In: Zayed, A., Kent, C.F. (Eds.), Genomics, Physiology and Behavior of Social Insects (Vol. 48). Academic Press, London, pp. 227–270.

Goodman, W.G., Granger, N.A., 2005. The juvenile hormones. In: Gilbert, L.I., Iatrou, K., Gill, S.S. (Eds.), Comprehensive Insect Molecular Science, vol. 3. Elsevier, Oxford, pp. 319–408.

Grasse, P.P., Noirot, C., 1946. La production des sexués néoteniques chez le termite á cou jaune (*Calotermes flavicollis*): inhibition germinale et inhibition somatique. C. R. Acad. Sci. 223, 869–871.

Greenberg, S., Tobe, S.S., 1985. Adaptation of a radiochemical assay for juvenile hormone biosynthesis to study caste differentiation in a primitive termite. J. Insect Physiol. 31, 347–352.
Hall, B.K., Wake, M.H., 1999. The Origin and Evolution of Larval Forms. Academic Press, San Diego, CA.
Hanus, R., et al., 2009. Beyond cuticular hydrocarbons: evidence of proteinaceous secretion specific to termite kings and queens. Proc. R. Soc. London B. 277, 995–1002.
Hartfelder, K., Guidugli-Lazzarini, K.R., Cervoni, M.S., Santos, D.E., Humann, F.C., 2015. Old Threads Make New Tapestry – Rewiring of Signalling Pathways Underlies Caste Phenotypic Plasticity in the Honey Bee, *Apis mellifera* L. In: Zayed, A., Kent, C.F. (Eds.), Genomics, Physiology and Behavior of Social Insects (Vol. 48). Academic Press, London, pp. 1–36.
Hattori, A., Sugima, Y., Sasa, C., Miyakawa, H., Ishikawa, Y., Miyazaki, S., Okada, Y., Cornette, R., Lavine, L.C., Emlen, D.J., Koshikawa, S., Miura, T., 2013. Soldier morphogenesis in the damp-wood termite is regulated by the insulin signaling pathway. J. Exp. Biol. 320, 295–306.
Haverty, M.I., Howard, R.W., 1981. Production of soldiers and maintenance of soldier proportions by laboratory experimental groups of *Reticulitermes flavipes* (Kollar) and *Reticulitermes virginicus* (Banks) (Isoptera: Rhinotermitidae). Insect. Soc. 28, 32–39.
Helvig, C., Koener, J.F., Unnihathan, G.C., Feyereisen, R., 2004. CYP15A1, the cytochrome P450 that catalyzes epoxidation of methylfarnesoate to juvenile hormone III in cockroach corpora allata. Proc. Natl. Acad. Sci. U.S.A. 101, 4024–4029.
Henderson, G., 1998. Primer pheromones and possible soldier caste influence on the evolution of sociality in lower termites. In: Vander Meer, R.K., Breed, M.D., Espelie, K.E., Winston, M.L. (Eds.), Phermone Communication in Social Insects. Westview Press, Boulder, CO, pp. 314–330.
Herb, B.R., Wolschin, F., Hansen, K.D., Aryee, M.J., Langmead, B., Irizarry, R., Amdam, G.V., Feinberg, A.P., 2012. Reversible switching between epigenetic states in honeybee behavioral subcastes. Nat. Neurosci. 15, 1371–1373.
Himuro, C., Yokoi, T., Matsuura, K., 2011. Queen-specific volatile in a higher termite *Nasutitermes takasagoensis* (Isoptera: Termitidae). J. Insect Physiol. 57, 962–965.
Hoffman, E.A., Goodisman, M.A.D., 2007. Gene expression and the evolution of phenotypic diversity in social wasps. BMC Biol. 5, e23.
Hoffmann, K., Korb, J., 2011. Is there conflict over direct reproduction in lower termite colonies? Anim. Behav. 81, 265–274.
Hoffmann, K., Gowin, J., Hartfelder, K., Korb, J., 2014. The scent of royalty: a P450 gene signals reproductive status in a social insect. Mol. Biol. Evol. 31, 2689–2696.
Hojo, M., Morioka, M., Matsumoto, T., Miura, T., 2005. Identification of soldier caste-specific protein in the frontal gland of nasute termite *Nasutitermes takasagoensis* (Isoptera: Termitidae). Insect Biochem. Mol. Biol. 35, 347–354.
Howard, R.W., Haverty, M.I., 1979. Termites and juvenile hormone analogues: a review of methodology and observed effects. Sociobiology 4, 269–278.
Hrdy, I., Kuldova, J., Hanus, R., Wimmer, Z., 2006. Juvenile hormone III, hydroprene and a juvenogen as soldier caste differentiation regulators in three *Reticulitermes* species: potential of juvenile hormone analogues in termite control. Pest Manag. Sci. 62, 848–854.
Hunt, J.H., Buck, N.A., Wheeler, D.E., 2003. Storage proteins in vespid wasps: characterization, developmental pattern, and occurrence in adults. J. Insect Physiol. 49, 785–794.
Hunt, J.H., Kensinger, B.J., Kossuth, J.A., Henshaw, M.T., Norberg, K., Wolschin, F., Amdam, G.V., 2007. A diapause pathway underlies the gyne phenotype in *Polistes* wasps, revealing an evolutionary route to caste-containing insect societies. Proc. Natl. Acad. Sci. 104, 14020–14025.
Husseneder, C., McGregor, C., Lang, R.P., Collier, R., Delatte, J., 2011. Transcriptome profiling of female alates and egg-laying queens of the Formosan subterranean termite. Comp. Biochem. Physiol. D 7, 14–27.

Inward, D., Beccaloni, G., Eggleton, P., 2007a. Death of an order: a comprehensive molecular phylogenetic study confirms that termites are eusocial cockroaches. Biol. Lett. 3, 331–335.

Inward, D.J.G., Vogler, A.P., Eggleton, P., 2007b. A comprehensive phylogenetic analysis of termites (Isoptera) illuminates key aspects of their evolutionary biology. Mol. Phylogenet. Evol. 44, 953–967.

Ishikawa, Y., Okada, Y., Ishikawa, A., Miyakawa, H., Koshikawa, S., Miura, T., 2010. Gene expression changes during caste-specific neuronal development in the damp-wood termite *Hodotermopsis sjostedti*. BMC Genomics 11, e314.

Korb, J., 2007. Workers of a drywood termite do not work. Front. Zool. 4, e7.

Korb, J., 2009. Termites: an alternative road to eusociality and the importance of group benefits in social insects. In: Gadau, J., Fewell, J.H. (Eds.), Organization of Insect Societies. From Genome to Sociocomplexity. Harvard University Press, Cambridge, pp. 128–147.

Korb, J., Hartfelder, K., 2008. Life history and development—a framework for understanding the ample developmental plasticity in lower termites. Biol. Rev. 83, 295–313.

Korb, J., Katrantzis, S., 2004. Influence of environmental conditions on the expression of the sexual dispersal phenotype in a lower termite: implications for the evolution of workers in termites. Evol. Dev. 6, 342–352.

Korb, J., Schmidinger, S., 2004. Help or disperse? Cooperation in termites influenced by food conditions. Behav. Ecol. Sociobiol. 56, 89–95.

Korb, J., Schneider, K., 2007. Does kin structure explain the occurrence of workers in a lower termite? Evol. Ecol. 21, 817–828.

Korb, J., Roux, E., Lenz, M., 2003. Proximate factors influencing soldier development in the basal termite *Cryptotermes secundus* (Hill). Insect. Soc. 50, 299–303.

Korb, J., Hoffmann, K., Hartfelder, K., 2009a. Endocrine signatures underlying plasticity in postembryonic development of a lower termite, *Cryptotermes secundus* (Kalotermitidae). Evol. Dev. 11, 269–277.

Korb, J., Weil, T., Hoffmann, K., Foster, K.R., Rehli, M., 2009b. A gene necessary for reproductive suppression in termites. Science 324, 758.

Korb, J., Buschmann, M., Schafberg, S., Liebig, J., Bagnères, A.G., 2012a. Brood care and social evolution in termites. Proc. R. Soc. B 279, 2662–2671.

Korb, J., Hoffmann, K., Hartfelder, K., 2012b. Molting dynamics and juvenile hormone titer profiles in the nymphal stages of a lower termite, *Cryptotermes secundus* (Kalotermitidae)—signatures of developmental plasticity. J. Insect Physiol. 58, 376–383.

Koshikawa, S., Cornette, R., Hojo, M., Maekawa, K., Matsumoto, T., Miura, T., 2005. Screening of genes expressed in developing mandibles during soldier differentiation in the termite *Hodotermopsis sjostedti*. FEBS Lett. 579, 1365–1370.

Krishna, K., Grimaldi, D.A., Krishna, V., Engel, M.S., 2013. Treatise on the Isoptera of the world. Bull. Am. Mus. Nat. Hist. 377, 1–2704.

Lanzrein, B., Gentinetta, V., Fehr, R., 1985. Titres of juvenile hormone and ecdysteroids in reproductives and eggs of *Macrotermes michaelseni*: relation to caste determination? In: Watson, J.A.L., Okot-Kotber, B.M., Noirot, C. (Eds.), Caste Differentiation in Social Insects. Pergamon Press, Oxford, pp. 307–328.

Le Conte, Y., Hefetz, A., 2008. Primer pheromones in social Hymenoptera. Annu. Rev. Entomol. 53, 26.1–26.20.

Lefeuve, P., Bordereau, C., 1984. Soldier formation regulated by a primer pheromone from the soldier frontal gland in a higher termite, *Nasutitermes lujae*. Proc. Natl. Acad. Sci. U.S.A. 81, 7665–7668.

Lenz, M., 1976. The dependence of hormone effects in termite caste determination on external factors. In: Lüscher, M. (Ed.), Phase and Caste Determination in Insects—Endocrine Aspects. Pergamon Press, Oxford, pp. 73–89.

Lenz, M., 1985. Is inter- and intraspecific variability of lower termite neotenic number due to adaptive thresholds for neotenic elimination?—considerations from studies on *Porotermes adamsoni* (Froggatt) (Isoptera: Termopsidae). In: Watson, J.A.L., Okot-Kotber, B.M., Noirot, C. (Eds.), Caste Differentiation in Social Insects. Pergamon Press, Oxford, pp. 125–146.

Libbrecht, R., Corona, M., Wende, F., Azevedo, D.O., Serrao, J.E., Keller, L., 2013. Interplay between insulin signaling, juvenile hormone, and vitellogenin regulates maternal effects on polyphenism in ants. Proc. Natl. Acad. Sci. U.S.A. 110, 11050–11055.

Liebig, J., 2010. Hydrocarbon profiles indicate fertility and dominance status in ant, bee, and wasp colonies. In: Blomquist, G.J., Bagneres, A.G. (Eds.), Insect Hydrocarbons. Biology, Biochemistry, and Chemical Ecology. Cambridge University Press, Cambridge, pp. 254–281.

Liebig, J., Eliyahu, D., Brent, C.S., 2009. Cuticular hydrocarbon profiles indicate reproductive status in the termite *Zootermopsis nevadensis*. Behav Ecol Sociobiol. 63, 1799–1807.

Light, S.F., 1944. Experimental studies on ectohormonal control of the development of supplementary reproductives in the termite genus *Zootermopsis* [formerly Termopsis]. In: University of California Publications in Zoology, vol. 43. University of California Press, Berkeley, CA, pp. 413–454.

Liu, Y.X., Henderson, G., Mao, L., Laine, R.A., 2005a. Seasonal variation of juvenile hormone titers of the Formosan subterranean termite, *Coptotermes formosanus* (Rhinotermitidae). Environ. Entomol. 34, 557–562.

Liu, Y.X., Henderson, G., Mao, L.X., Laine, R.A., 2005b. Effects of temperature and nutrition on juvenile hormone titers of *Coptotermes formosanus* (Isoptera: Rhinotermitidae). Ann. Entomol. Soc. Am. 98, 732–737.

Lo, N., Engel, M.S., Cameron, S., Nalepa, C.A., Tokuda, G., Grimaldi, D., Kitade, O., Krishna, K., Klaas, K.D., Maekawa, K., Miura, T., Thompson, G.J., 2007. Save Isoptera: a comment on Inward et al. Biol. Lett. 3, 564–565.

Lüscher, M., 1952a. Die Produktion und Elimination von Ersatzgeschlechtstieren bei der Termite *Kalotermes flavicollis* (Fabr.). Z. Vgl. Physiol. 34, 123–141.

Lüscher, M., 1952b. Untersuchungen über das individuelle Wachstum bei der Termite *Kalotermes flavicollis* Fabr. (Ein Beitrag zum Kastenbildungsproblem). Biol. Zentralbl. 71, 529–543.

Lüscher, M., 1956. Die Entstehung von Ersatzgeschlechtstieren bei der Termite *Kalotermes flavicollis* (Fabr.). Insect. Soc 3, 119–128.

Lüscher, M., 1958a. Experimentelle Erzeugung von Soldaten bei der Termite *Kalotermes flavicollis* (Fabr). Naturwissenschaften 45, 69–70.

Lüscher, M., 1958b. Über die Entstehung der Soldaten bei Termiten. Rev. Suisse Zool. 65, 372–377.

Lüscher, M., 1969. Die Bedeutung des Juvenilhormons für die Differenzierung der Soldaten bei der Termite *Kalotermes flavicollis*. In: Proceedings of the VI Congress IUSSI, pp. 165–170.

Lüscher, M., 1972. Environmental control of juvenile hormone (JH) secretion and caste differentiation in termites. Gen. Comp. Endocrinol 3 (Suppl.), 504–509.

Lüscher, M., 1973. The influence of the composition of experimental groups on caste development in *Zootermopsis* (Isoptera). In: Proceedings of the VII Congress IUSSI, pp. 253–255.

Lüscher, M., 1974a. Die Kompetenz zur Soldatenbildung bei Larven (Pseudergaten) der Termite *Zootermopsis angusticollis*. Rev. Suisse Zool. 81, 711–714.

Lüscher, M., 1974b. Kasten und Kastendifferenzierung bei niederen Termiten. In: Schmidt, G.H. (Ed.), Sozialpolymorphismus bei Insekten. Wissenschaftliche Verlagsgesellschaft, Stuttgart, pp. 694–739.

Lüscher, M., 1976. Evidence for an endocrine control of caste determination in higher termites. In: Lüscher, M. (Ed.), Phase and Caste Determination in Insects: Endocrine Aspects. Pergamon Press, Oxford, pp. 91–103.

Maekawa, K., Ishitani, K., Gotoh, H., Cornette, R., Miura, T., 2010. Juvenile hormone titre and vitellogenin gene expression related to ovarian development in primary reproductives compared with nymphs and nymphoid reproductives of the termite *Reticulitermes speratus*. Physiol. Entomol. 35, 52–58.

Maekawa, K., Nakamura, S., Watanabe, D., 2012. Termite soldier differentiation in incipient colonies is related to parental proctodeal trophallactic behavior. Zool. Sci. 29, 213–217.

Mao, L.X., Henderson, G., 2010. Group size effect on worker juvenile hormone titers and soldier differentiation in Formosan subterranean termite. J. Insect Physiol. 56, 725–730.

Mao, L., Henderson, G., Liu, Y., Laine, R.A., 2005. Formosan subterranean termite (Isoptera: Rhinotermitidae) soldiers regulate juvenile hormone levels and caste differentiation in workers. Ann. Entomol. Soc. Am. 98, 340–345.

Martins, J.R., Nunes, F.M.F., Cristino, A.S., Simoes, Z.L.P., Bitondi, M.M.G., 2010. The four hexamerin genes in the honey bee: structure, molecular evolution and function deduced from expression patterns in queens, workers and drones. BMC Mol. Biol. 11, e23.

Martins, J.R., Anhezini, L., Dalacqua, R.P., Simoes, Z.L.P., Bitondi, M.M.G., 2011. A honey bee hexamerin, HEX 70a, is likely to play an intranuclear role in developing and mature ovarioles and testioles. PLoS 6, 329006.

Masuoka, Y., Miyazaki, S., Saiki, R., Tsuchida, T., Maekawa, K., 2013. High *Laccase2* expression is likely involved in the formation of specific cuticular structures during soldier differentiation of the termite *Reticulitermes speratus*. Arthropod Struct. Dev. 42, 469–475.

Matsuura, K., 2011. Sexual and asexual reproduction in termites. In: Bignell, D.E., Roisin, Y., Lo, N. (Eds.), Biology of Termites: A Modern Synthesis. Springer, Dordrecht, Heidelberg, London, New York, pp. 255–278.

Matsuura, K., Yashiro, T., Shimizu, K., Tatsumi, S., Tamura, T., 2009. Cuckoo fungus mimics termite eggs by producing the cellulose-digesting enzyme beta-glucosidase. Curr. Biol. 19, 30–36.

Matsuura, K., Himuro, C., Yokoi, T., Yamamoto, Y., Vargo, E.L., Keller, L., 2010. Identification of a pheromone regulating caste differentiation in termites. Proc. Natl. Acad. Sci. U.S.A. 107, 12963–12968.

Miller, E.M., 1942. The problem of castes and caste differentiation in *Prorhinotermes simplex* (Hagen). Bull. Univ. Miami Sch. Med. Jackson Meml. Hosp. 15, 3–27.

Mirth, C., Truman, J.W., Riddiford, L.M., 2005. The role of the prothoracic gland in determining critical weight to metamorphosis in *Drosophila melanogaster*. Curr. Biol. 15, 1796–1807.

Miura, T., 2004. Proximate mechanisms and evolution of caste polyphenism in social insects: from sociality to genes. Ecol. Res. 19, 141–148.

Miura, T., Scharf, M.E., 2011. Molecular basis underlying caste differentiation in termites. In: Bignell, D.E., Roisin, Y., Lo, N. (Eds.), Biology of Termites: A Modern Synthesis. Springer, Dordrecht, Heidelberg, London, New York, pp. 211–254.

Miura, T., Kamikouchi, A., Sawata, M., Takeuchi, H., Natori, S., Kubo, T., Matsumoto, T., 1999. Soldier caste-specific gene expression in the mandibular glands of *Hodotermopsis japonica* (Isoptera: Termopsidae). Proc. Natl. Acad. Sci. U.S.A. 96, 13874–13879.

Moran, N.A., 1992. The evolutionary maintenance of alternative phenotypes. Am. Nat. 139, 971–989.

Myles, T., 1988. Resource inheritance in social evolution from termite to man. In: Sobotnikoff, S. (Ed.), The Ecology of Social Behavior. Academic Press, New York, pp. 379–423.

Myles, T.G., 1999. Review of secondary reproduction in termites (Insecta: Isoptera) with comments on its role in termite ecology and social evolution. Sociobiology 33, 1–91.

Nagin, R., 1972. Caste determination in *Neotermes jouteli* (Banks). Insect. Soc. 1, 39–61.

Nalepa, C.A., 1984. Colony composition, protozoan transfer and some life history characteristics of the woodroach *Cryptocercus punctulatus* Scudder (Dictyoptera: Cryptocercidae). Behav. Ecol. Sociobiol. 14, 273–279.

Nijhout, H.F., 1994. Insect Hormones. Princeton University Press, Princeton, NJ.

Nijhout, F.H., Wheeler, D.E., 1982. Juvenile hormone and the physiological basis of insect polyphenisms. Q. Rev. Biol. 57, 109–133.

Noirot, C., 1989. Social structure in termite societies. Ethol. Ecol. Evol. 1, 1–17.

Noirot, C., 1990. Sexual castes and reproductive strategies in termites. In: Engels, W. (Ed.), An Evolutionary Approach to Castes and Reproduction. Springer Verlag, Berlin, pp. 5–35.

Noirot, C., Darlington, J.P.E.C., 2000. Termite nests: architecture, regulation and defence. In: Abe, T., Bignell, D.E., Higashi, M. (Eds.), Termites: Evolution, Sociality, Symbiosis and Ecology. Kluwer Academic Publishers, Dordrecht, pp. 121–139.

Noirot, C., Pasteels, J.M., 1987. Ontogenic development and evolution of the worker caste in termites. Experientia 43, 851–860.

Noirot, C., Pasteels, J.M., 1988. The worker caste is polyphyletic in termites. Sociobiology 14, 15–20.

Oakeshott, J.G., Claudianos, C., Campbell, P.M., Newcomb, R.D., Russell, R.J., 2005. Biochemical genetics and genomics of insect esterases. In: Gilbert, L.I., Iatrou, K., Gill, S.S. (Eds.), Comprehensive Insect Molecular Science vol. 5. Elsevier, San Diego, CA, pp. 309–381.

Okot-Kotber, B.M., 1985a. Caste polymorphism in a higher termite, *Macrotermes michaelseni* (Termitidae: Macrotermitinae). In: Watson, J.A.L., Okot-Kotber, B.M., Noirot, C. (Eds.), Caste Differentiation in Social Insects. Pergamon Press, Oxford, pp. 87–106.

Okot-Kotber, B.M., 1985b. Mechanisms of caste determination in a higher termite, *Macrotermes michaelseni* (Isoptera, Macrotermitidae). In: Watson, J.A.L., Okot-Kotber, B.M., Noirot, C. (Eds.), Caste Differentiation in Social Insects. Pergamon Press, Oxford, pp. 267–306.

Okot-Kotber, B.M., Prestwich, G.D., 1991a. Identification of a juvenile-hormone binding protein in the castes of the termite, *Reticulitermes flavipes*, by photoaffinity labeling. Insect Biochem. 21, 775–784.

Okot-Kotber, B.M., Prestwich, G.D., 1991b. Juvenile hormone binding proteins of termites detected by photoaffinity labeling: comparison of *Zootermopsis newadensis* with two rhinotermitids *Coptotermes formosanus* and *Reticulitermes flavipes*. Arch. Insect Biochem. Physiol. 17, 119–128.

Okot-Kotber, B.M., Ujváry, I., Mollaaghababa, R., Szurdoki, F., Matolcsy, G., Prestwich, G.D., 1991. Physiological influence of fenoxycarb pro-insecticides and soldier head extracts of various termite species on soldier differentiation in *Reticulitermes flavipes* (Isoptera). Sociobiology 19, 77–90.

Okot-Kotber, B.M., Prestwich, G.D., Strambi, A., Strambi, C., 1993. Changes in morphogenetic hormone titers in isolated workers of the termite *Reticulitermes flavipes* (Kollar). Gen. Comp. Endocrinol. 90, 290–295.

Park, Y.I., Raina, A.K., 2004. Juvenile hormone III titers and regulation of soldier caste in *Coptotermes formosanus* (Isoptera: Rhinotermitidae). J. Insect Physiol. 50, 561–566.

Park, Y.I., Raina, A.K., 2005. Regulation of juvenile hormone titers by soldiers in the Formosan subterranean termite *Coptotermes formosanus*. J. Insect Physiol. 51, 385–391.

Peeters, C., Liebig, J., 2009. Fertility signaling as a general mechanism of regulating reproductive division of labor in ants. In: Gadau, J., Fewell, J.H. (Eds.), Organization of Insect Societies. From Genome to Sociocomplexity. Harvard University Press, Cambridge, pp. 220–242.

Perdereau, E., Dedeine, F., Christides, J.P., Bagneres, A.G., 2010. Variations in worker cuticular hydrocarbons and soldier isoprenoid defensive secretions within and among introduced and native populations of the subterranean termite, *Reticulitermes flavipes*. J. Chem. Ecol. 36, 1189–1198.

Raina, A., Park, Y.I., Gelman, D., 2008. Molting in workers of the Formosan subterranean termite *Coptotermes formosanus*. J. Insect Physiol. 54, 155–161.

Rascon, B., Mutti, N.S., Tolfsen, C., Amdam, G., 2011. Hony bee life history plasticity: development, behavior, and aging. In: Flatt, T., Heyland, A. (Eds.), Mechanisms of Life History Evolution. Oxford University Press, Oxford, pp. 253–266.

Renoux, J., 1976. Le polymorphisme de *Schedorhinotermes lamanianus* (Sjöstedt) (Isoptera: Rhinotermitidae). Essai d'interprétation. Insectes Soc. 23, 279–494.

Robinson, G.E., Page, R.E., Strambi, C., Strambi, A., 1992. Colony integration in honey bees: mechanisms of behavioral reversion. Ethology 90, 336–348.

Roisin, Y., 2000. Diversity and evolution of caste patterns. In: Abe, T., Bignell, D.E., Higashi, M. (Eds.), Diversity and Evolution of Caste Patterns. Kluwer Academic Publishers, Dordrecht, Netherlands, pp. 95–119.

Roisin, Y., Korb, J., 2011. Social organisation and the status of workers in termites. In: Bignell, D.E., Roisin, Y., Lo, N. (Eds.), Biology of Termites: A Modern Synthesis. Springer, Dordrecht, Heidelberg, London, New York, pp. 133–164.

Scharf, M.E., Ratliff, C.R., Wu-Scharf, D., Zhou, X., Pittendrigh, B.R., Bennett, G.W., 2005a. Effects of juvenile hormone III on *Reticulitermes flavipes*: changes in hemolymph protein composition and gene expression. Insect Biochem. Mol. Biol. 35, 207–215.

Scharf, M.E., Wu-Scharf, D., Zhou, X., Pittendrigh, B.R., Bennett, G.W., 2005b. Gene expression profiles among immature and adult reproductive castes of the termite *Reticulitermes flavipes*. Insect Mol. Biol. 14, 31–44.

Scharf, M.E., Buckspan, C.E., Grzymala, T.L., Zhou, X., 2007. Regulation of polyphenic caste differentiation in the termite *Reticulitermes flavipes* by interaction of intrinsic and extrinsic factors. J. Exp. Biol. 210, 4390–4398.

Sen, R., Raychoudhury, R., Cai, Y., Sun, Y., Lietze, V.-U., Boucias, D.G., Scharf, M.E., 2013. Differential impacts of juvenile hormone, soldier head extract and alternate caste phenotypes on host and symbiont transcriptome composition in the gut of the termite *Reticulitermes flavipes*. BMC Genomics 14, e491.

Shimizu, K., 1970. Studies on the caste differentiation of the supplementary reproductives of the Japanese termite, *Reticulitermes speratus* (KOLBE). Agr. Bull. Miyazahi Univ. 17, 1–46.

Shingleton, A.W., 2005. Body-size regulation: combining genetics and physiology. Curr. Biol. 15, R825–R827.

Springhetti, A., 1969a. Il controllo sociale della differenziazione degli alati in *Kalotermes flavicollis* Fabr. (Isoptera). Ann. Univ. Ferrara Sez. III Biol. Anim. 3, 73–96.

Springhetti, A., 1969b. Influenza dei reali sulla differenziazione dei soldati di *Kalotermes flavicollis* Fabr. (Isoptera). In: Proceedings of the VI Congress IUSSI, pp. 267–273.

Springhetti, A., 1972. The competence of *Kalotermes flavicollis* F. (Isoptera) pseudergates to differentiate into soldiers. Monit. Zool. Ital. 6, 97–111.

Springhetti, A., 1985. The function of the royal pair in the society of *Kalotermes flavicollis* (Fabr.) (Isoptera: Kalotermitidae). In: Watson, J.A.L., Okot-Kotber, B.M., Noirot, C. (Eds.), Caste Differentiation in Social Insects. Pergamon Press, Oxford, pp. 165–176.

Stay, B., Tobe, S.S., 2007. The role of allatostatins in juvenile hormone synthesis in insects and crustaceans. Annu. Rev. Entomol. 52, 277–299.

Stuart, A.M., 1979. The determination and regulation of the neotenic reproductive caste in the lower termites (Isoptera): with special reference to the genus *Zootermopsis* (Hagen). Sociobiology 4, 223–237.

Sutherland, T.D., Unnithan, G.C., Andersen, J.F., Evans, P.H., Murataliev, M.B., Szabo, L.Z., Mash, E.A., Bowers, W.S., Feyereisen, R., 1998. A cytochrome P450 terpenoid hydroxylase linked to the suppression of insect juvenile hormone synthesis. Proc. Natl. Acad. Sci. U.S.A. 95, 12884–12889.

Sutherland, T.D., Unnithan, G.C., Feyereisen, R., 2000. Terpenoid omega-hydroxylase (CYP4C7) messenger RNA levels in the corpora allata: a marker for ovarian control of juvenile hormone synthesis in *Diploptera punctata*. J. Insect Physiol. 46, 1219–1227.

Tartar, A., Wheeler, M.M., Zhou, X., Coy, M.R., Boucias, D.G., Scharf, M.E., 2009. Parallel metatranscriptome analyses of host and symbiont gene expression in the gut of the termite *Reticulitermes flavipes*. Biotechnol. Biofuels 2, e25.

Tarver, M.R., Schmelz, E.A., Rocca, J.R., Scharf, M.E., 2009. Effects of soldier-derived terpenes on soldier caste differentiation in the termite *Reticulitermes flavipes*. J. Chem. Ecol. 35, 256–264.

Tarver, M.R., Zhou, X.G., Scharf, M.E., 2010. Socio-environmental and endocrine influences on developmental and caste-regulatory gene expression in the eusocial termite *Reticulitermes flavipes*. BMC Mol. Biol. 11, e28.

Tarver, M.R., Schmelz, E.A., Scharf, M.E., 2011. Soldier caste influences on candidate primer pheromone levels and juvenile hormone-dependent caste differentiation in workers of the termite *Reticulitermes flavipes*. J. Insect Physiol. 57, 771–777.

Tarver, M.R., Coy, M.R., Scharf, M.E., 2012. Cyp15F1—a novel cytochrome P450 gene linked to juvenile hormone-dependent caste differentiation in the termite *Reticulitermes flavipes*. Arch. Insect Biochem. Physiol. 80, 92–108.

Terrapon, N., Li, C., Robertson, H.M., Ji, L., Meng, X., Booth, W., Chen, Z., Childers, C.P., Glastad, K.M., Gokhale, K., Gowin, J., Gronenberg, W., Hermansen, R.A., Hu, H., Hunt, B.G., Huylmans, A.K., Khalil, S.M.S., Mitchell, R.D., Munoz-Torres, M.C., Mustard, J.A., Pan, H., Reese, J.T., Scharf, M.E., Sun, F., Vogel, H., Xiao, J., Yang, W., Yang, Z., Yang, Z., Zhou, Y., Zhu, J., Brent, C.S., Elsik, C.G., Goodisman,, M.a.D., Liberles, D.A., Roe, R.M., Vargo, E.L., Vilcinskas, A., Wang, J., Bornberg-Bauer, E., Korb, J., Zhang, G., Liebig, L., 2014. Molecular traces of alternative social organization in a termite genome. Nat Commun 5, e3636.

Toga, K., Saiki, R., Maekawa, K., 2013. Hox gene *deformed* is likely involved in mandibular regression during presoldier differentiation in the nasute termite *Nasutitermes takasagoensis*. J. Exp. Zool. 320B, 385–392.

Treiblmayr, K., Pascual, N., Piulachs, M.D., Keller, T., Belles, X., 2006. Juvenile hormone titer versus juvenile hormone synthesis in female nymphs and adults of the German cockroach, *Blattella germanica*. J. Insect Sci. 6, e43.

Van Oystaeyen, A., Oliveira, R.C., Holman, L., Van Zweden, J.S., Romero, C., Oi, C.A., D'ettorre, P., Khalesi, M., Billen, J., Wäckers, F., Millar, J.G., Wenseleers, T., 2014. Conserved class of queen pheromones stops social insect workers from reproducing. Science 343, 287–290.

Van Zweden, J.S., D'Ettorre, P., 2010. Nestmate recognition in social insects and the role of hydrocarbons. In: Blomquist, G.J., Bagneres, A.G. (Eds.), Insect Hydrocarbons. Biology, Biochemistry, and Chemical Ecology. Cambridge University Press, Cambridge, pp. 222–243.

Waller, D.A., La Fage, J.P., 1988. Environmental influence on soldier differentiation in *Coptotermes formosanus* Shiraki (Rhinotermitidae). Insect. Soc. 35, 144–152.

Watanabe, D., Gotoh, H., Miura, T., Maekawa, K., 2011. Soldier presence suppresses presoldier differentiation through a rapid decrease of JH in the termite *Reticulitermes speratus*. J. Insect Physiol. 57, 791–795.

Weil, T., Rehli, M., Korb, J., 2007. Molecular basis for the reproductive division of labour in a lower termite. BMC Genomics 8, e198.
Weil, T., Hoffmann, K., Kroiss, J., Strohm, E., Korb, J., 2009a. Scent of a queen - cuticular hydrocarbons specific for female reproductives in lower termites. Naturwissenschaften 96, 315–319.
Weil, T., Korb, J., Rehli, M., 2009b. Comparison of queen-specific gene expression in related lower termite species. Mol. Biol. Evol. 26, 1841–1850.
Yagi, K.J., Kwok, R., Chan, K.K., Setter, R.R., Myles, T.G., Tobe, S.S., Stay, B., 2005. Phe-Gly-Leu-amide allatostatin in the termite *Reticulitermes flavipes*: content in brain and corpus allatum and effect on juvenile hormone synthesis. J. Insect Physiol. 51, 357–365.
Zhou, X., Oi, F.M., Scharf, M.E., 2006a. Social exploitation of hexamerin: RNAi reveals a major caste-regulatory factor in termites. Proc. Natl. Acad. Sci. U.S.A. 103, 4499–4504.
Zhou, X., Tarver, M.R., Bennett, G.W., Oi, F.M., Scharf, M.E., 2006b. Two hexamerin genes from the termite *Reticulitermes flavipes*: sequence, expression, and proposed functions in caste regulation. Gene 376, 47–58.
Zhou, X., Song, C., Grzymala, T.L., Oi, F.M., Scharf, M.E., 2007a. Juvenile hormone and colony conditions differentially influence cytochrome P450 gene expression in the termite *Reticulitermes flavipes*. Insect Mol. Biol. 15, 749–761.
Zhou, X., Tarver, M.R., Scharf, M.E., 2007b. Hexamerin-based regulation of juvenile hormone-dependent gene expression underlies phenotypic plasticity in a social insect. Development 134, 601–610.

CHAPTER FIVE

The Physiology and Genomics of Social Transitions in Aphids

Patrick Abbot[1]
Department of Biological Sciences, Vanderbilt University, Nashville, Tennessee, USA
[1]Corresponding author: e-mail address: patrick.abbot@vanderbilt.edu

Contents

Abstract

Aphids are models for studies of plant/insect interactions, microbial symbioses and insect polyphenisms. Some aphids express an unusual social polyphenism by producing morphologically and behaviourally specialized juvenile nymphs (soldiers) that defend their family group from predators. To date, there have been few functional studies of social aphids. The recent publication of the pea aphid genome offers new opportunities for such work. I describe two directions for functional studies of social aphids that capture unique features of their social biology. Although social aphids are plant herbivores, soldiers attack other insects, implying unique chemosensory adaptations not seen in aphids generally. Secondly, social aphids produce castes of aggressive soldiers, by expressing unique patterns of development and photoperiodic responses to seasonal cues. I provide examples that illustrate the unusual chemosensory and developmental transitions that social aphids exhibit. Functional studies of social aphid chemosensory and developmental biology will provide new insights into sociality in aphids and provide opportunities to integrate social and functional biology in hemimetabolous insects.

Advances in Insect Physiology, Volume 48
ISSN 0065-2806
http://dx.doi.org/10.1016/bs.aiip.2014.12.005

1. INTRODUCTION

In a volume devoted to social insects, a chapter on aphids may be easily overlooked, like that part of a newspaper you thumb past quickly. But take a moment and consider aphids. In addition to being serious pests with sizable economic impacts, aphids are models in the study of insect development, insect–plant interactions and insect symbioses (Brisson and Sterns, 2006). Understanding how sociality evolves in such organisms can therefore lead in directions that are unusual for social insect biology, offering the possibility of conceptual linkages between social evolution and disparate areas of insect biology. There is some evidence, for example, that aphid 'queens' regulate caste development by withholding nutritional symbionts, thereby consigning their soldier-like daughters to a sterile dead end (Fukatsu and Ishikawa, 1992). There is evidence that soldiers in some species have transformed proteolytic enzymes related to feeding on plants into secreted animal-killing venoms (Kutsukake et al., 2004). Aphids also offer key experimental advantages for the study of social behaviour. For example, aphid clonality permits genotype-specific, common garden manipulations familiar to plant biologists—but in a social insect. And, the transitional, on-the-cusp nature of aphid sociality allows for much-needed reconstruction of the decisive factors that govern the tipping points back and forth between sociality and alternative life history strategies.

It is worthwhile to ask what can aphids teach us about the mechanisms underlying social evolution? The short answer is that it remains to be seen. Since their discovery in the 1970s, the focus of much of the research on social aphids has been behavioural, taxonomic and evolutionary, with an emphasis on natural history. Functional studies are only recently becoming common (Ijichi et al., 2005b; Shibao et al., 2010). There are several reasons for this. Many species have complex life cycles and are difficult to rear in the lab. Many are found in far-flung regions globally, making comprehensive studies of diversity challenging. And we do not yet have a genome for a social aphid, which is the starting point for functional research.

Thus, the following is less a review about what is known about social aphids than a prospectus for functional approaches to aphid sociality. First, I provide an overview of sociality in aphids, followed by brief introductions to aphid genomics and development, as gleaned from the work on non-social aphids. I then use these overviews to describe four general questions about social aphids that are promising directions for social aphid research in relation to mechanistic perspectives on insect sociality. They are: (1) What is the basis of soldier morphogenesis and caste regulation in aphids? (2) What are the seasonal and photoperiodic correlates of social life styles in aphids? (3) What mechanisms underlie the positive chemotactic response that aphid soldiers have to natural enemies? (4) How do aphids defend themselves, and what are the roles of salivary and secretory molecules in aphid defence?

2. BACKGROUND ON SOCIAL APHIDS

True aphids in the family Aphididae, comprising about 4000 species, are small, soft-bodied herbivorous insects. They are hemipterans, and they feed on plant phloem sap. All are obligately associated with nutritional endosymbiotic bacteria. One of the remarkable features of their complex life cycles is that nearly every task in their life style, whether it is dispersal or reproduction, is associated with a discrete, morphologically specialized adult phenotype, which is why they are models for studying insect polyphenisms (Hartfelder and Emlen, 2005; Srinivasan and Brisson, 2012; Tagu et al., 2005). Most aphids are monophagous and specialize on a taxonomically restricted set of host plants. About 10% of species seasonally alternate between distinct host plant species (known as heteroecy). Aphids express cyclical parthenogenesis, in which annual cycles of asexual reproduction are punctuated by a sexual generation. Many species are agricultural pests in temperate regions, particularly in the Northern Hemisphere, where they are more diverse than in lower or southern latitudes.

Depending on how sociality is defined in aphids (Costa, 2006), all 'eusocial' species occur in only two subfamilies (Hormaphidinae and Eriosomatinae; formally Pemphiginae). Within these two subfamilies, there are about 50 eusocial species, and eusociality has arisen or been lost independently multiple times. No eusocial species have been described from the Aphidinae, the large subfamily that includes species such as the pea aphid (*Acyrthosiphon pisum*). Social aphids form family groups composed of a reproductive female (a foundress or fundatrix) and one or more generations of morphologically and/or behaviourally specialized soldier-like nymphs, produced asexually by the foundress (Fig. 1). Often soldiers are associated with plant galls, but many species form galls and do not produce soldiers, or soldiers are produced at other life cycle stages in the absence of galls. As the name implies, the primary function of soldiers is defence against natural

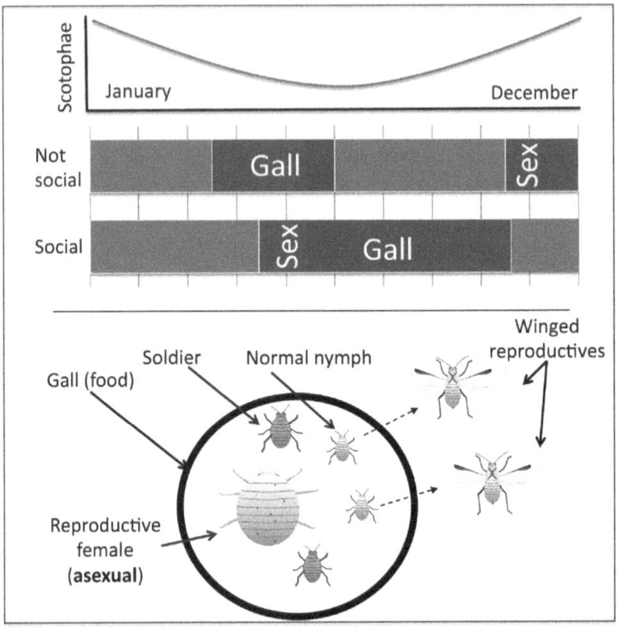

Figure 1 An example of how the life cycle of a social aphid in the genus *Pemphigus* can differ from that of a non-social species, and an example of a generalized structure of a social aphid group. Most social aphids alternate between two unrelated host plants (top panel). Galls are formed on the primary host, and while soldiers are produced on the secondary host in some species, many social aphids produce soldiers on the primary host. Typically, host alternating aphids fly from their galls during the short nights of summer and produce the sexual generation as nights get longer in the fall (top panel, scotophase). A typical, non-social species alternates between two hosts (light and dark grey shading). A sexual phase occurs in autumn, and the gall foundress hatches from a diapausing egg in spring (dark grey). However, a feature of social aphids is that on whichever host the soldiers are formed, residency on that host plant tends to be prolonged. The social *P. obesinymphae* also host alternates (top panel). However, the galling phase is much longer (dark grey), and consequently, the sexual generation occurs in the spring, not the autumn. In the bottom panel, a typical social group is illustrated. The gall-forming foundress asexually produces monomorphic offspring, which are behaviourally specialized for defence. In some species, soldiers are sterile, or additional generations of females produce soldiers. Eventually however, some nymphs become winged adults, and in most cases, disperse.

enemies, although many exhibit other unusual adaptations (Table 1). Some species have castes of juveniles that are distinctly dimorphic, and produce sterile soldiers with enlarged forelimbs for grasping their enemies, and/or sharp pronotal horns that they use to stab their enemies. Others do have

Table 1 A sample of aphid species with soldiers, various adaptations they exhibit and possible functional implications (development or chemosensory) associated with those adaptations

Species	Observed adaptation	Putative functional implication	Representative citations
Hormaphidinae			
Astegopteryx bambucifoliae	Attack eggs of predators; fighting apterous adults	Chemosensory	Aoki and Kurosu (2010)
Astegopteryx spinocephala	Guard eggs in gall	Development	Kurosu et al. (2006)
Ceratoglyphina bambusae	Soldiers attack conspecific non-soldiers; gall repair; invade and transform galls of another species	Chemosensory	Aoki and Kurosu (2010)
Ceratovacuna japonica	Seasonal variation in soldier armature; soldiers attack ants	Development	Aoki and Kurosu (2010) and Kurosu et al. (1990)
Ceratovacuna lanigera	Attack eggs of predators; mark predators for attack	Chemosensory	Aoki and Kurosu (2010)
Hamamelistes miyabei	First-instar soldiers with delayed moult	Development	Shibao and Fukatsu (2003)
Nipponaphis monzeni	Altruistic repair of galls by soldiers; galls absorb water	Chemosensory	Kurosu et al. (2003) and Kutsukake et al. (2012)
Pseudoregma bambucicola	Patrolling soldiers attack eggs of predators; soldier embryogenesis	Chemosensory/ development	Ijichi et al. (2004) and Ohara (1985)
Pseudoregma nicolaiae	Ants affect caste differences	Development	Shingleton and Foster (2000)
Quadrartus yoshinomiyai	Post-reproductive adults attack predators with gluc	Chemosensory	Uematsu et al. (2010)
Tuberaphis styraci	Venomous soldiers; soldiers regulated by density, contact	Chemosensory	Aoki and Kurosu (2010), Kutsukake et al. (2004) and Shibao et al. (2004a)

Continued

Table 1 A sample of aphid species with soldiers, various adaptations they exhibit and possible functional implications (development or chemosensory) associated with those adaptations—cont'd

Species	Observed adaptation	Putative functional implication	Representative citations
Eriosomatinae			
Colophina clematicola	Attack predator eggs and other aphid species	Chemosensory	Kurosu and Aoki (1988)
Epipemphigus niisimae	Fighting fundatrices and gall usurpation	Chemosensory	Aoki and Makino (1982)
Grylloprociphilus imbricator	Older nymphs are soldiers	Development	Aoki et al. (2001)
Pemphigus dorocola	Chimeric offspring	Development	Aoki and Kurosu (1991)
Pemphigus obesinymphae	Foundress presence required to maintain soldiers; conditional development by soldiers	Chemosensory/development	Withgott et al., (1997) and Abbot et al. (2001)
Pemphigus spyrothecae	Caste formation by life cycle compression	Development	Foster (1990) and Rhoden and Foster (2002)

dimorphic castes, and juveniles are uniformly aggressive without morphological specializations. In many species, soldiers pierce their enemies with the needle-like mouthparts (stylets; Kurosu et al., 1994; Fig. 2), and in some, the mouthparts themselves can be dimorphic. (Note that in the aphid literature, distinctions are made about what does or does not constitute a soldier. For simplicity, I will use 'soldier' and omit other caveats regarding aphid sociality. Interested readers who wish to sample the diversity of social aphid biology (Table 1) are directed to the many enlightening papers from Shigeyuki Aoki and Utako Kurosu (e.g., Aoki and Kurosu, 2010; and references therein), and Takema Fukatsu and colleagues (e.g., Kutsukake et al., 2012; Shibao et al., 2010; and references therein). General reviews and guides can be found in Blackman and Eastop (1994), Stern and Foster (1996), Costa (2006), Pike et al., 2007; Pike and Foster (2008)).

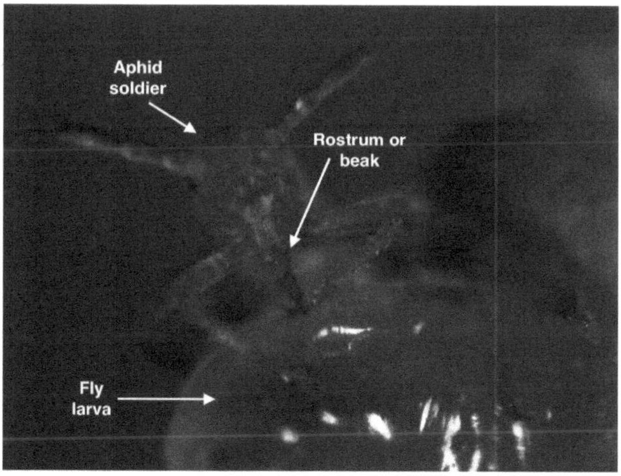

Figure 2 A first-instar soldier of *Pemphigus obesinymphae* attacking fly larvae with its mandibular stylets. *Photo credit: P. Abbot.*

3. AN APHID GENOME

In discussing the prospects for functional studies of social aphids, it is worthwhile to keep in mind that there is no published genome of a social aphid. Moreover, there has been about 75 million years of evolution and the accumulation of key differences between the solitary pea aphid *A. pisum*, which currently provides the reference genome for aphids, and eusocial aphid species themselves. That said, many features of aphids and their genomes are functionally conserved, and inevitably, the pea aphid genome will be the pole star that will guide genomic research on social aphids. The initial release revealed a low GC content genome with a haploid size of 517 Mb (The International Aphid Genomics Consortium, 2010). The current release includes more than 23,000 coding sequences and about 19,000 predicted genes, a large fraction of which lack functional annotation. The pea aphid has three autosomes and one sex chromosome, but the nearly 24,000 genomic scaffolds have not yet been assigned to chromosomes because a sufficient linkage map does not yet exist. Perhaps the most surprising feature of the pea aphid genome is the extensive degree of gene duplication, with over 2000 families exhibiting copy number variation, more than any other sequenced insect genome (Godfray, 2010). This result alone provides a clue to the genomic flexibility that may underlie the unusual phenotypes exhibited by social aphids. Included in the families exhibiting extensive duplication are microRNA, siRNA and piRNA genes, many of which are orphans. Also significant is what is absent from the genome, including genes involved in the IMD immune signalling pathway, as well as other lineage-specific gene losses (Gerardo et al., 2010).

In general, the pea aphid genome reveals the potential for epigenetic regulation of aphid development and phenotypic plasticity (see the chapter on "Epigenetics in Social Insects" by Glastad et al., this volume). A large fraction (about 38%) of the genome, for example, is composed of transposable and other repetitive elements, many of which likely are involved in regulatory functions. Aphids have a functional DNA methylation system, including copies of each of the three canonical classes of DNA methyltransferases in animals, as well as additional aphid-restricted copies of genes involved in methylation (Srinivasan and Brisson, 2012; Walsh et al., 2010). Aspects of DNA methylation in aphids are comparable to honey bees, reflecting the conservation of ancestral patterns of methylation systems in insects (Hunt et al., 2010). Like other herbivorous insects, aphids must find and identify their host plants, while avoiding predators and finding mates. Genes involved in chemoreception exhibit patterns similar to other insects, including extensive duplication of odorant receptor and gustatory receptor gene families, with many orphans (Zhou et al., 2010). The genome exhibits the signatures of the sap-feeding aphid diet, and present are genes involved in sugar transport, as well as a reduced but functional set of P450 and other genes with plant detoxification functions (Price et al., 2010; Ramsey et al., 2010). One expectation about the aphid genome was that the complexity of their life cycles would be encoded in the genome itself. However, relative to *Drosophila*, the life cycle complexity seems to be encoded largely by differences in regulatory networks involved in development. Additionally, while there is overall conservation of genes involved in insect embryogenesis, missing are developmental genes known from *Drosophila* such as *bicoid* and *oskar* (Shigenobu et al., 2010). All of the signalling pathways involved in insect development are intact, but modified with the expansions and losses typical of the rest of the genome (Brisson et al., 2010). Genes involved in juvenile hormone (JH) are present, although some such as the JH-binding hexamerins, known to be important in caste regulation in lower termites, are absent. Similarly, genes orthologous to the core circadian machinery in *Drosophila* are present, but the *period/timeless* machinery is generally not well conserved (Cortés et al., 2010). Finally, there is clear signature of bacterial symbiosis in the aphid genome, with complementarity with and gene transfer from the primary aphid symbionts

Buchnera aphidicola, and the expansion of amino acid transporters and gene families involved in maintaining a functional symbiosis (Ramsey et al., 2010; Wilson et al., 2010).

4. TWO AXES OF SOCIAL TRANSITIONS IN APHIDS

Traditionally, most approaches to understanding the evolution of sociality in aphids have centred on galls as the localizing feature that establishes the genetic cohesion and vulnerability to predators that favour kin-selected evolution of defence (Abbot, 2009). This behavioural ecological approach has been well suited to integrating aphids into the larger field of social evolution (Stern and Foster, 1996). However, this approach has been less successful in drawing attention to the mechanisms that underlie the adaptive features of social aphids, especially those that are not obviously analogous to those of other social insects (West Eberhard, 2003). A focus on more mechanistic aspects of sociality in aphids can provide insights into social traits themselves and can generate hypotheses for the evolutionary routes to complex sociality in aphids (Shibao et al., 2010; Table 1).

What are examples of mechanistic approaches of aphid sociality? As suggested above, there are two axes or 'transitions' that capture distinctive features of social aphids relative to non-social aphids and other social insects:

1. *A developmental transition*. Social aphids require the buildup of aggressive, specialized nymphs in order to effectively defend their colonies (Stern and Foster, 1996). To do so, typically some or all of the first- or second-instar nymphs in their clonal groups either do not moult or exhibit some degree of developmental inhibition that prolongs early juvenile stadia. In many species, these nymphs exhibit maternally determined or environmentally cued morphological dimorphisms (Akimoto, 1992; Pike et al., 2004; Sakata and Itô, 1991; Shibao and Fukatsu, 2003; Shibao et al., 2010). The soldier stage of social aphids is a polyphenism that involves a change in the normal genetic and hormonal regulatory networks that govern juvenile aphid development (Ijichi et al., 2005b). In addition, the groups formed by social aphids tend to be longer-lived than the aggregations of non-social aphids (Akimoto, 1992; Aoki and Moran, 1994; Kurosu and Aoki, 2009; Pike et al., 2004). The extension of the social group implies a fundamental difference between social and non-social aphids in the photoperiodic response to seasonal and environmental cues that govern life cycle

events such as reproduction and dispersal, which I characterize as a *developmental transition* (Simon et al., 2010).

2. *A chemosensory transition*. Social aphids express physiological or morphological adaptations for sensing and manipulating their environment that are not observed elsewhere among aphids generally. They exhibit strongly aggressive responses to threats from natural enemies, and many even insert their feeding mouthparts into predators (Aoki and Kurosu, 2010; Stern and Foster, 1996). Often predators die, likely because soldiers secrete salivary or midgut-derived factors that act as venoms (Kutsukake et al., 2004; Lawson et al., 2014). Such traits are unusual, given that aphids are prey for many predators and tend to avoid them. As well, aphids are not predators, and thus ancestrally, aphids lack adaptations for injuring other animals. All aphids use an array of chemosensory modalities and effector molecules to locate and to feed on their host plants, and to respond to their environment (Döring, 2014; Powell et al., 2006; Webster, 2012; Will, 2013). The manner which aphid soldiers respond to and deter predators must be co-opted from these plant-associated chemosensory and secretory modalities, which I characterize as a *chemosensory transition*.

4.1 Social aphids are characterized by unique developmental regulation of nymphal morphology and physiology and exhibit altered photoperiodic responses to seasonal and environmental cues

4.1.1 Overview of the general problem

Ceratovacuna japonica (Aphididae: Hormaphidinae) is a social aphid that illustrates the distinctive features of the soldier caste in aphids (Aoki and Kurosu, 2010; Aoki et al., 1981). *C. japonica* occurs in southeast Asia and Japan and seasonally alternates between *Styrax* spp. and bamboo (*Sasa* spp.). On bamboo, which is the secondary host, open colonies occur with castes of enlarged, sterile first-instar 'pseudoscorpion-like' soldiers. These soldiers have a number of unusual morphological traits, including enlarged and more sclerotized tergites, limbs and frontal horns, which they use to pierce their enemies. Oddly, the morphology of soldiers can change over the course of a season (Hattori and Itino, 2008).

Understanding caste regulation has been a central challenge of the study of social insects since Darwin (Hölldobler and Wilson, 1990; Ratnieks et al., 2010; Schwander et al., 2010). How are castes regulated in social aphids? The mechanisms and seasonal cues involved in soldier production in aphids are only poorly understood. Soldier phenotypes in aphids are unusually plastic for social insects and respond to a diversity of cues with potentially synergistic and antagonistic interactions (Shibao et al., 2010), including crowding or size (Sakata et al., 1991; Schütze and Maschwitz, 1991; Shibao, 1999; Shibao et al., 2010; Shingleton and Foster, 2001; Stern, 1994), tactile and possibly non-volatile stimulation associated with density (Shibao et al., 2003, 2004a), season or colony age (Akimoto, 1992; Ijichi et al., 2005a; Stern et al., 1997), host plant characteristics (Sakata et al., 1991), the presence of predators (Shibao, 1998) and the presence of tending ants (Shingleton and Foster, 2000). How such cues are integrated or expressed within and between species is unclear. Given our current understanding of the molecular and physiological basis of aphid polyphenisms (Goggins, 2007), it is likely that mechanisms expressed by non-social aphids in the context of herbivory have been redeployed by social aphids in the context of sociality (Hartfelder and Emlen, 2005).

4.1.2 What the pea aphid has to say about social aphid caste regulation

Like termites, aphids are remarkable for the degree to which they exhibit both genetically encoded polymorphisms and environmentally induced discrete phenotypes, or polyphenisms (Hartfelder and Emlen, 2005; Srinivasan and Brisson, 2012). But, unlike termites, most aphid species are not social, and where they contrast with other social insects is that the social polyphenisms they exhibit are a subset of a generalized genomic and developmental capacity for phenotypic plasticity, common to the family. While the various aphid polyphenisms differ in the specific mechanisms at work, the general features must be the same, in that a parthenogenetic female receives a cue from the environment that determines alternative developmental programmes, and that cue either directly affects development or acts maternally on unborn offspring (Brisson, 2010). However, it is becoming increasingly evident that key differences exist between the developmental programmes that determine the various aphid polyphenisms, and it may prove that social polyphenisms share common components of some developmental programmes, and not others (Duncan et al., 2013). Thus, in order to understand the development and regulation of castes in aphids, it is worthwhile to consider briefly the general features of polyphenisms in aphids.

The principle polyphenisms in aphids are related to reproduction, dispersal and colour. The dispersal or winged polyphenism has received the most study, for reasons of pest management. What these studies have revealed is that the wing polyphenism responds to a complex, multifactorial set of interacting cues that can vary from species to species (Mehrparvar et al., 2013). There are two types of winged female aphids, one produced in autumn related to sexual reproduction and the other produced in spring and summer solely for dispersal to new host plants. The winged forms that will produce the sexual generation are typically under the control of photoperiod and temperature, and insensitive to ecological and tactile cues such as crowding or predation. The winged aphids produced for dispersal to new host plants (where asexual reproduction will occur) are primarily influenced by variation in host plant quality, tactile cues such as crowding, predation and pathogen pressure, and mutualists (Müller et al., 2001). In pea aphids, there is accumulating evidence that sexual and asexual generations differ in early development, implying only partially overlapping sets of genes and regulatory networks governing the output of otherwise phenotypically similar sexual and asexual adults (Duncan et al., 2013; Gallot et al., 2012; Miura et al., 2003).

In insects, development is regulated by ecdysteroids, the neurosecretory prothoracicotropic hormones (PTTH), and JHs. For many years, juvenile hormone (JHII in aphids) was thought to inhibit the production of winged aphids, and thereby promote the development of wingless nymphs and adults. In pea aphids, morphs are not conditional, but are determined maternally and expressed post-natally after several rounds of ecdysis (Braendle et al., 2006; Ogawa and Miura, 2014). This is not true of all species however, and in some, winged forms can be cued post-natally. For the short-day, sexual generation produced in autumn, this appears to be the case, and reduction in JH levels generally correspond to the production of winged forms (Braendle et al., 2006). However, the picture is less clear for the long-day spring- and summer-winged forms produced by the aphid foundress that in social aphids also produces the soldier morphs. A number of studies have shown that JH titres remain high in the long-day, summer-winged polyphenism (Ishikawa et al., 2011). JH thus has contrasting effects, depending on environmental cues and morph specification. Less is known currently about the translation and integration of endocrine cues neuronally, or the genetic basis of winged polyphenisms (Brisson, 2010; Le Trionnaire et al., 2013). Both JH and melatonin are thought to be involved in the transduction of photoperiodic signals (Gao and Hardie, 1997). The genes known to underlie wing development in *Drosophila* are present in the pea aphid genome, and some exhibit differential expression across development. One gene in particular, *apterous*, a transcription factor which is known to be involved in wing morphogenesis in *Drosophila*, exhibits strong patterns of differential expression between winged and wingless morphs (Brisson et al., 2010). Other functionally validated candidates in flies (Bloch et al., 2013), for example, those involved in insulin signalling, are present in aphids and exhibit tissue-specific and photoperiodic patterns that indicate likely roles in life cycle scheduling (Le Trionnaire et al., 2013). Generally, the implication is that, despite the evolutionary distance between flies and aphids, core genes and regulatory units may be sufficiently conserved, and at least some functional insights from *Drosophila* can be leveraged to gain insight into aphid polyphenisms.

At the moment, no study comparable to that of Brisson et al. (2010) has been undertaken for social aphids. However, functional studies of aphid caste regulation have been pioneered in the lab of Takema Fukatsu at the National Institute of Advanced Industrial Science and Technology, Japan. Using field and laboratory studies using artificially reared aphids, Fukatsu and colleagues have begun to characterize the proximate mechanisms involved in caste regulation in a social aphid. *Tuberaphis styraci* (Aphididae: Hormaphidinae) forms coral-like galls on the tree *Styrax obassia* in Japan and mainland China, where a soldier polyphenism occurs among second-instar juveniles. Generally in social aphids, the first-instar soldier morphs are determined prenatally and their phenotypes are expressed at birth (Ijichi et al., 2005b). However, this is not always the case, and soldier polyphenisms can be conditional. In *T. styraci*, for example, maternal cues during embryogenesis determine soldier specification in the second-instar nymphs (Shibao et al., 2003). However, there is some flexibility in developmental trajectories. Normal first-instar nymphs can be induced to develop into second-instar soldiers when conditions become crowded. Remarkably, they are insensitive to soldier density, and only contact with normal nymphs promotes post-natal differentiation into soldiers (Shibao et al., 2004b). The specific cues themselves and how they are translated remain unknown, but one intriguing possibility is that the signal involves non volatile chemicals transmitted both by contact and through the host plant itself (Shibao et al., 2010). Why some species express post-natal plasticity in soldier development and others do not is not clear, but it is worth noting that no other species has been studied functionally to the extent of *T. styraci*. Evidence from the North American species, *Pemphigus obesinymphae* (Aphididae: Eriosomatinae), suggests that conditionality in response to non-volatile chemical cues may be common in social aphids. Withgott et al. (1997) showed that *P. obesinymphae* soldiers respond developmentally to colony status, and that they foundress supplies some cue that regulates post-natal development, not unlike primer pheromones in bees (Grozinger et al., 2003). And in the same species, Abbot et al. (2001) demonstrated that soldier phenotypes can be conditionally expressed in the presence of unrelated foundresses.

4.1.3 Seasonality in social aphid biology

Photoperiodism is the tracking of day (photophase) or night length (scotophase), coupled with the storage, integration and physiological deployment of these phase ratios in the service of seasonal life history schedules. As herbivorous insects with complex life cycles in temperate regions, aphids express strong patterns of photoperiodic timing (Hardie and Nunes, 2001). One of the major life history correlates of social evolution in aphids is that the life cycle phase in which sociality is expressed (the galls of species with primary host sociality, or the colonies of species with secondary host sociality) tends to be extended or prolonged relative to the equivalent phase in non-social species. The extension of the 'social' life cycle stage implies an altered response to the seasonal and photoperiodic cues that orchestrate aphid life cycles. As well, photoperiodic cues are likely important in regulation of the hormones and neuroendocrine effectors that govern soldier development (Hartfelder and Emlen, 2005).

Aphid life cycles can be notoriously difficult to study (Moran, 1992). Most of our current understanding of photoperiodism in aphids has been derived from studies on species in the Aphidinae (Simon et al., 2010). Much less is known about photoperiodic regulation of host alternation in the subfamilies in which social aphids are found (Dixon, 1998, 2005). And few studies have explored evolutionary or functional aspects of aphid life cycles in both social and related non-social species (Stern, 1994, 1998). Generally in aphids, in the fall months, a minimum number of shorter days and longer autumnal nights (scotophase) trigger a switch from viviparous to oviparous (sexual) reproduction that ultimately results in the production of a cold-hardy, yolk-filled, overwintering egg (Gao et al., 1999; Le Trionnaire et al., 2013; Shingleton et al., 2003). However, in at least one social species, there has been a change in the response to temperature-modulated photoperiodic input. The details reveal how aphid life cycles may be intimately related to the origins of social behaviour in aphids.

Abbot and Withgott (2004) characterized the life cycles of two species in North America in the genus *Pemphigus*, one of which has aggressive monomorphic soldiers and an extended gall-forming stage on its primary host

(Moran, 1992; Fig. 2). Aphids in the genus *Pemphigus* form galls on poplars (*Populus* spp.), and most seasonally alternate between poplars in the fall, winter and spring, and the roots of herbaceous crucifers and composites in the summer. Poplars exhibit heterophyly (discrete variation in leaves on a shoot; Eckenwalder, 1996). The early, spring flush leaves expand from buds that were preformed the previous autumn. The sexually produced foundresses form galls on these leaves upon hatching. Weeks later, a summer or late flush of leaves expand from neoformed buds. The spring flush leaves senesce first at the height of summer, along with any galls that they bear. Only the summer or late leaves persist into autumn, sometimes 4 or 5 months after their initial flush. Poplar heterophyly is key to understanding the extension of the galling phase and its relationship to the social polyphenisms. The social *P. obesinymphae* forms galls on the summer flush, and as consequence, its galling phase is greatly prolonged relative to other North American aphids in the genus. *P. obesinymphae* overwinters on its secondary host, and only returns to poplars in the spring, when they bear the spring leaves. In contrast to most aphids, therefore, the sexual generation in *P. obesinymphae* is initiated by unknown photoperiodic cues in the spring, and it is the longevity of the summer flush leaves themselves that facilitate the association between aphid sociality and long-lived galls. Possibly, soldiers evolved in *P. obesinymphae* because of the shift to summer flush leaves, which either extended the vulnerability inherent to galls, or because predators are more abundant in late versus early summer (Wilch, 1999).

4.1.4 Promises of a functional biology of social aphid caste regulation

Even in the absence of a fully sequenced social aphid genome, studies on species such as *T. styraci* illustrate the promise of future comparative, functional studies (Ijichi et al., 2005b) Lawson, S.G. (personal communication), for example, compared the pattern of gene expression across development between two species in the genus *Pemphigus* that co-occur on poplars in North America, but differ in the presence or absence of monomorphic soldiers. Unsurprisingly, a large fraction of genes were differentially expressed between first and fourth-instar nymphs in the species without soldiers. However, relatively fewer genes were differentially expressed across development in the social species. Although these results remain to be validated, possibly the difference reflects the fact that first-instar soldiers, which grow slowly, do not upregulate genes normally expressed during early juvenile growth.

Functional studies of social aphids will provide insight into whether there are differential patterns of expression in genes underlying hormonal and neurogenic effects on development in social aphids (Brisson, 2010). Are there differential patterns of expression of JH and other neuroendocrine or signalling factors such as ecdysteroids, PTTHs and insulin signalling between social aphids and their non-social relatives? Do JH or JH-degrading factors play roles similar to those they play in termite soldier differentiation (see the chapter on "Juvenile Hormone, a Central Regulator of Termite Caste Polyphenism" by Korb, this volume; Korb et al., 2003), or do they operate in a manner similar to what is known about JH activity during aphid viviparous reproduction (Ishikawa et al., 2011)? How are maternally expressed genes regulated during soldier embryogenesis, relative to normal nymphs and winged adults (Ijichi et al., 2004)? Answering such questions will provide fundamental insights into social aphid biology.

4.2 Social aphids exhibit altered gustatory and chemosensory responses and exhibit altered secretory biology

4.2.1 Overview of the general problem

An aggressive or predaceous wasp is not surprising, whether it is social or solitary. The same is not true of aphids. The plant-parasitic Sternorrhyncha (the aphids, white flies and scale insects) express various competitive, defensive and mutualistic traits in response to predators and competitors (Aoki and Makino, 1982; Dixon, 1998; Hartbauer, 2010; Whitham, 1979). However, functionally, aphid soldiers act as predators, even though they often die in the process (Foster and Rhoden, 1998). In a manner opposite to what all other aphids do, aphid soldiers are strongly and positively stimulated by natural enemies (Arakaki, 1992; Fig. 2). For example, the syrphid fly, *Metasyrphus confrator*, is a predator of aphids in Japan. *M. confrator* lays its eggs in or near aphid colonies, and once hatched, the relatively immobile larva prey upon aphids nearby. But like *C. japonica*, the pseudoscorpion-like soldiers of *Pseudoregma bambucicola* attack the syrphid larva with their frontal horns and their eggs with their feeding stylets. The attacks on eggs can be so frequent that when soldier density is high, *M. confrator* lay their eggs in the relative safety of spider webs (Aoki et al., 1984; Ohara, 1985). Despite appearances, soldier attacks are not nutritional (Rhoden and Foster, 2002). Like honey bees defending their colonies, soldiers of species in the genus *Pseudoregma* die within a few minutes of attacking *M. confrator* larva (Ohara, 1985), and even die when attacking non-predaceous insects (Arakaki, 1992). Where it has been examined, the rapid death of soldiers when they attack is a general feature of social aphids, for reasons that are unclear.

Social aphids are among the 'fortress defenders', in which the primary expression of altruistic behaviours involves defence of siblings rather than brood care (Queller and Strassmann, 1998). Understanding how social aphids express defence is tantamount to understanding how they are social. In other words, the challenge in social aphids is to characterize the chemosensory basis of soldier response to potential predators, and the insecticidal basis of their behavioural expression of altruism. To appreciate how unusual it is for an aphid soldier to attack another insect, a brief overview of how aphids feed is useful.

4.2.2 What the pea aphid has to say about the chemical ecology of social aphids

The coordination and integration of the colonies of social insects involve the use of pheromones and other semiochemicals. Most of an aphid's life is spent on the leaves and stems of plants, and chemoreception is the principle modality through which aphids, social or non-social, perceive their environment (Powell et al., 2006; Will, 2013). Like other insects, aphids use peripheral sense organs (mostly hair or peg-like sensilla) for olfaction, contact and taste, aggregated in sensory fields located mostly on the antenna, mouthparts and legs. Upon the first contact with the leaf surface, aphids use the tip of their proboscis to sample their host, and process information regarding host features using both volatile and non-volatile semiochemicals. Next, they penetrate the plant epidermis with their stylets. Initially during this sampling period, they secrete a gel-like saliva that will stabilize the delicate stylet as it penetrates the plant (Tjallingii, 2006). They next secrete and re-ingest a small amount of watery saliva that is mixed with plant intra- and intercellular fluids, which provides post-ingestive feedback via mandibular and alimentary-localized sense organs. Ultimately, they pinpoint and puncture the phloem-containing sieve elements, which induces the production of large amounts of watery saliva that contains various effector molecules and proteins that manipulate or block plant defences and/or promote feeding success. The information that aphids receive during this process is encoded in chemical signals in the form of odorant or taste ligands that ultimately couple to receptors on the dendrites innervating the sensilla. It was thought that, relative to the holometabolous insects, aphids and other hemipterans have fewer odorant and chemosensory receptors. Smadja et al. (2009), however, identified 79 ORs and 77 GRs in the pea aphid genome, more than *Nasonia*, and consistent with *Drosophila* and *Anopheles*. Cao et al. (2014) identified only 45 candidate odorant receptors in the cotton aphid, *Aphis gossypii*, fewer than the pea aphid. *A. gossypii* is a highly polyphagous aphid, and the authors speculated that reduced number of ORs in *A. gossypii* reflected its indiscriminate feeding habits.

At present, nothing is known about chemoreception in social aphids. In the context of how aphids respond to their host plants, it is difficult to speculate how aphid soldiers perceive their natural enemies. The response of aphid soldiers to other insects like *M. confrator* must be mediated in part by non-volatile chemoreception (Schütze and Maschwitz, 1991). The chemical ecology of aphids is a broad topic and encompasses both volatile and non-volatile semiochemistry (Pickett and Glinwood, 2005), however, and only a handful of studies have investigated the chemical ecology of aphid societies. Aphids produce sex pheromones during their sexual generation, but sexual reproduction is not immediately relevant to aphid sociality. Many species are known to emit warning pheromones (mostly the sesquiterpene (*E*)-β-farnesene) in response to predators from specialized structures on their abdomens known as cornicles, but there is little known about the use of warning pheromones by social aphids. Shibao et al. (2005) report that the social aphid *T. styraci* (Aphididae: Hormaphidinae) emit (*E*)-β-farnesene when threatened (cited in Aoki and Kurosu, 2010). There are a few examples in which aphid soldiers attack other aphids or can be stimulated to attack inanimate objects (Shingleton and Foster, 2001). But most evidence indicates that they do not or rarely attack clone mates, conspecifics, other aphids or inanimate objects (Aoki et al., 1991; Schütze and Maschwitz, 1991), suggesting that simply mechanical stimulation is not sufficient in most species to provoke sustained, purposeful defence behaviour. Schütze and Maschwitz (1991) used paper soaked with the haemolymph of crushed insects to demonstrate that the horned soldiers of *Pseudoregma sundanica* (Aphididae: Hormaphidinae) respond to specific contact semiochemicals in haemolymph, rather than non-specific or volatile stimuli. Aphid soldiers, they suggested, are tuned to respond to the wounds of predators.

4.2.3 The secretory biology of social aphids

The defining feature of sociality in aphids is defence, but it is unclear how soldiers effectively deter, attack or kill natural enemies. Kutsukake et al. (2004) provided the first clue and showed that *T. styraci* soldiers orally inject a midgut-expressed, proteolytic enzyme into wax moth larva, indicating that venom use may act as a key innovation in aphid social evolution. *T. styraci* soldiers, but not normal nymphs, express a copy of the proteolytic enzyme that exhibits evidence of strong positive selection, typical of venom genes generally (Fry et al., 2009). However, a subsequent study of species related to *T. styraci* did not find evidence of similar soldier-specific expression or rate acceleration by *catB* orthologs (Kutsukake et al., 2008). Leaving aside the species with soldiers that attack with frontal horns, and not stylets, aphidologists would have been dubious that aphids possess adaptations for purposely secreting anything from midguts (or as Kutsukake et al. (2004) described it, 'vomit'). To understand why, it necessary to describe the pea aphid alimentary canal.

Aphids feed on phloem contained within the sieve cells of their host plants, and they can feed both passively or actively, determined by how they orchestrate a valve and pump system in their anterior alimentary canal (McLean and Kinsey, 1984; Prado and Tjallingii, 1994; Will, 2013). The aphid pharynx is divided into two chambers, separated by a piston-like value called the precibarial valve. Food enters a pharyngeal canal anterior to the valve from the maxillary stylets, called the precibarium, where it is held until the precibarial valve opens. Behind the valve, a chamber called the cibarial pump functions as a conduit between the anterior alimentary canal and oesophagus, connected to the midgut at the cardiac valve. Aphids mostly feed passively, by way of the turgor pressure of sieve cells. However, in the process of sampling their host plants, or when turgor pressure drops, then muscles connected to the cibarial pump contract, enlarging the canal posterior to the closed cibarial valve and creating a pressure differential between the anterior and posterior canals. The pressure differential also causes the oesophagus and cardiac valve to collapse, sealing the midgut. Opening the cibarial valve allows food to enter the space of the cibarial pump. Next, the valve closes, sealing the food in the dilated pump, and relaxation of the dilator muscles forces food into the midgut through the cardiac valve, which lacks independent muscular control. Sensilla on both sides of the cibarial valve allow aphids to judge whether to ingest food fully, or to expel food that has been sampled in the stylets and anterior alimentary canal. Overall, because of leakage across valves and food that inevitably remains in the canals connecting each major unit of the feeding channel, active feeding results in food moving in a two steps forward and one step back manner as valves and muscles build and relax pressure differentials.

If the pea aphid is a guide, then the oesophagus and cardiac valve of species like *T. styraci* should prevent anything more than leakage from the midgut. In one of the more visually striking results on social aphids in recent years, however, Kurosu et al. (2003) and Kutsukake et al. (2009) showed that in addition to attacking predators, *Nipponaphis monzeni* (Aphididae: Hormaphidinae) soldiers can repair damage to their galls by secreting massive amounts of a rapidly coagulating substance that seals the damage and spurs plant tissue regeneration. They speculate that *T. styraci* has co-opted aspects of their circulating immunity for social functions. The coagulating substance may be derived from specialized hemocytes that contain the components of the immunity-associated prophenoloxidase system (PPO) for wound healing and encapsulation. How the PPO cascade is cued, and how its products are secreted is unclear. But it is clear that social aphids are likely to possess unique soft tissue adaptations that facilitate such things as midgut evacuation, and thus defence. If so, then whatever soldiers ingest, sequester or metabolize become candidates for insecticidal activity. Lawson (personal communication), for example, has evidence that *P. obesinymphae* translocate both plant metabolites and aphid-borne viruses when they attack other insects. It would be useful to know more about the alimentary canals of aphids in the Eriosomatinae and Hormaphidinae, and if aphid soldiers have mechanisms for active secretion from more than their salivary glands.

4.2.4 Promises of a functional biology of social aphid chemosensory and secretory biology

The clues from the chemosensory and secretory biology of *T. styraci* and *N. monzeni* suggest possibilities for integrating facets of aphid functional biology with social evolution. A social aphid genome will be immensely valuable in cataloguing chemosensory receptors and will shed light on the unique roles that semiochemicals play in aphid soldier behaviour. Prior to that, an obvious first study would be to use *de novo* sequencing approaches to compare within and between species patterns of differential, tissue-specific expression of chemosensory transcripts in soldiers and normal aphids. And if active midgut secretion is a general feature of aphid soldiers that attack with their stylets, then it implies a bottom-up role of plants in social aphid biology. Plants have various means of chemical defence, and thus plant defences become candidates for either aphid defences. Or, as Kutsukake et al. (2008) suggested, plants may act as evolutionary drivers of the chemical weaponry of social aphids. Finally, the scab repair exhibited by *T. styraci* suggests interesting angles on the roles of galls in aphid social evolution that complement the traditional 'defendable resource' treatment. Galls may act as agents of clonal cohesion and vulnerability that may favour altruistic acts of defence (Abbot, 2009; Stern and Foster, 1996), but the capacity for defence itself by gall-forming aphids may have its roots in the intricacies of manipulation of plant biology that accompanies the gall-forming habit (Kutsukake et al., 2009).

5. CONCLUSIONS

Functional studies of social aphids are not common at the moment. But aphids are unusual in that they are models for functional studies of topics ranging from plant–insect interactions to insect–microbial symbioses. Ironically, functional research on *non-social* aphids exceeds that of many insects, including eusocial Hymenoptera. What this means is that once the tools for studying social aphids are available, and as functional work on pea aphids and its relatives become more refined, a number of opportunities will immediately present themselves for experimental, functional studies of social evolution in aphids. As well, given the number of independent origins of sociality in aphids, comparative genomic and transcriptomic studies of closely related social and non-social species should offer unique insights into the functional adaptations that characterize sociality in aphids (Table 1).

ACKNOWLEDGEMENTS

The author thanks Stephen Ferguson, Amro Zayed, Clement Kent and one anonymous reviewer for comments on the manuscript, and the National Science Foundation for the support (NSF IOS 1147033).

REFERENCES

Abbot, P., 2009. On the evolution of dispersal and altruism in aphids. Evolution 63, 2687–2696.

Abbot, P., Withgott, J.H., 2004. Phylogenetic and molecular evidence for allochronic speciation in gall-forming aphids (*Pemphigus*). Evolution 58, 539–553.

Abbot, P., Withgott, J.H., Moran, N.A., 2001. Genetic conflict and conditional altruism in social aphid colonies. Proc. Natl. Acad. Sci. U.S.A. 98, 12068–12071.

Akimoto, S., 1992. Shift in life-history strategy from reproduction to defense with colony age in the galling aphid *Hemipodaphis persimilis* producing defensive first instar larvae. Res. Popul. Ecol. 34, 359–372.

Aoki, S., Kurosu, U., 1991. An aptera and its abnormal offspring found in a gall of *Pemphigus dorocola* (Homoptera, Aphidoidea). Jpn. J. Entomol. 59, 555–563.

Aoki, S., Kurosu, U., 2010. A review of the biology of Cerataphidini (Hemiptera, Aphididae, Hormaphidinae), focusing mainly on their life cycles, gall formation, and soldiers. Psyche 2010, 1–34, Article ID 380351.

Aoki, S., Makino, S., 1982. Gall usurpation and lethal fighting among fundatrices of the aphid. Kontyû 50, 365–376.

Aoki, S., Moran, N.A., 1994. *Pemphigus obesinymphae*, a new American aphid species with defenders and swollen nymphs (Homoptera: Aphidoidea: Pemphigidae). J. N. Y. Entomol. Soc. 102, 251–260.

Aoki, S., Akimoto, S., Yamane, S., 1981. Observations *on Pseudoregma alexanderi* (Homoptera, Pemphigidae), an aphid species producing pseudoscorpion-like soldiers on bamboos. Kontyû 49, 355–366.

Aoki, S., Kurosu, U., Usuba, S., 1984. First instar larvae of the sugar-cane wooly aphid, *Ceratovacuna lanigera* (Homoptera, Pemphigidae), attack its predators. Kontyû 52, 458–460.

Aoki, S., Kurosu, U., Stern, D.L., 1991. Aphid soldiers discriminate between soldiers and non-soldiers, rather than between kin and non-kin, in *Ceratoglyphina bambusae*. Anim. Behav. 42, 865–866.

Aoki, S., Kurosu, U., Von Dohlen, C.D., 2001. Colony defense by wingpadded nymphs in *Grylloprociphilus imbricator* (Hemiptera: Aphididae). Fla. Entomol. 84, 431–434.

Arakaki, N., 1992. Predators of the sugar cane woolly aphid, *Ceratovacuna lanigera* (Homoptera: Aphididae) in Okinawa and predator avoidance of defensive attack by the aphid. Appl. Entomol. Zool. 27, 159–161.

Blackman, R.L., Eastop, V.F., 1994. Aphids on the World's Trees: An Identification and Information Guide. University Press, London, UK, 987 pp.

Bloch, G., Hazan, E., Rafaeli, A., 2013. Circadian rhythms and endocrine functions in adult insects. J. Insect Physiol. 59, 56–69.

Braendle, C., Davis, G.K., Brisson, J.A., Stern, D.L., 2006. Wing dimorphism in aphids. Heredity 97, 192–199.

Brisson, J.A., 2010. Aphid wing dimorphisms: linking environmental and genetic control of trait variation. Philos. Trans. R. Soc. Lond. B Biol. Sci. 365, 605–616.

Brisson, J.A., Stern, D.L., 2006. The pea aphid, *Acyrthosiphon pisum*: an emerging genomic model system for ecological, developmental and evolutionary studies. BioEssays 28, 747–755.

Brisson, J.A., Ishikawa, A., Miura, T., 2010. Wing development genes of the pea aphid and differential gene expression between winged and unwinged morphs. Insect Mol. Biol. 19, 63–73.

Cao, D., Liu, Y., Walker, W.B., Li, J., Wang, G., 2014. Molecular characterization of the *Aphis gossypii* olfactory receptor gene families. PLoS One 9, e101187.

Cortés, T., Ortiz-Rivas, B., Martínez-Torres, D., 2010. Identification and characterization of circadian clock genes in the pea aphid *Acyrthosiphon pisum*. Insect Mol. Biol. 19, 123–139.

Costa, J.T., 2006. The Other Social Insect Societies. Harvard University Press, Cambridge, MA.

Dixon, A.F.G., 1998. Aphid Ecology, second ed. Chapman & Hall, London, 300 pp.

Dixon, A.F.G., 2005. Insect Herbivore-Host Dynamics: Tree-Dwelling Aphids. Cambridge University Press, Cambridge, 199 p.

Döring, T.F., 2014. How aphids find their host plants, and how they don't. Ann. Appl. Biol. 165, 3–26.

Duncan, E.J., Leask, M.P., Dearden, P.K., 2013. The pea aphid (*Acyrthosiphon pisum*) genome encodes two divergent early developmental programs. Dev. Biol. 377, 262–274.

Eckenwalder, J.E., 1996. Systematics and evolution of *Populus*. In: Stettler, R.F., Bradshaw Jr., H.D., Heilman, P.E., Hinckley, T.M. (Eds.), Biology of *Populus* and Its Implications for Management and Conservation. NRC Research Press, Ottawa, pp. 7–32.

Foster, W.A., 1990. Experimental evidence for effective and altruistic colony defence against natural predators by soldiers of the gall-forming aphid *Pemphigus spyrothecae* (Hemiptera: Pemphigidae). Behav. Ecol. Sociobiol. 27, 421–430.

Foster, W.A., Rhoden, P.K., 1998. Soldiers effectively defend aphid colonies against predators in the field. Anim. Behav. 55, 761–765.

Fry, B.G., Roelants, K., Champagne, D.E., et al., 2009. The toxicogenomic multiverse: convergent recruitment of proteins into animal venoms. Annu. Rev. Genomics Hum. Genet. 9, 483–511.

Fukatsu, T., Ishikawa, H., 1992. Soldier and male of an eusocial aphid *Colophina arma* lack endosymbiont: implications for physiological and evolutionary interaction between host and symbiont. J. Insect Physiol. 38, 1033–1042.
Gallot, A., Shigenobu, S., Hashiyama, T., Jaubert-Possamai, S., Tagu, D., 2012. Sexual and asexual oogenesis require the expression of unique and shared sets of genes in the insect *Acyrthosiphon pisum*. BMC Genomics 13, 76.
Gao, N., Hardie, J., 1997. Melatonin and the pea aphid, *Acyrthosiphon pisum*. J. Insect Physiol. 43, 615–620.
Gao, N., von Schantz, M., Foster, F.G., Hardie, J., 1999. The putative brain photoperiodic photoreceptors in the vetch aphid, *Megoura viciae*. J. Insect Physiol. 45, 1011–1019.
Gerardo, N.M., Altincicek, B., Anselme, C., Atamian, H., Barribeau, S.M., de Vos, M., et al., 2010. Immunity and other defenses in pea aphids, *Acyrthosiphon pisum*. Genome Biol. 11, R21.
Godfray, H.C.J., 2010. The pea aphid genome. Insect Mol. Biol. 19, 1–4.
Goggin, F.L., 2007. Plant–aphid interactions: molecular and ecological perspectives. Curr. Opin. Plant Biol. 10, 399–408.
Grozinger, C.M., Sharabash, N.M., Whitfield, C.W., Robinson, G.E., 2003. Pheromone-mediated gene expression in the honey bee brain. Proc. Natl. Acad. Sci. U.S.A. 100 (Suppl. 2), 14519–14525.
Hardie, J., Nunes, M.V., 2001. Aphid photoperiodic clocks. J. Insect Physiol. 47, 821–832.
Hartbauer, M., 2010. Collective defense of *Aphis nerii* and *Uroleucon hypochoeridis* (Homoptera, Aphididae) against natural enemies. PLoS One 5, e10417.
Hartfelder, K., Emlen, D.J., 2005. Endocrine control of insect polyphenism. In: Gilbert, L.I., Iatrou, K., Gill, S. (Eds.), Comprehensive Molecular Insect Science, Vol. 3, Endochrinology. Elsevier, Amsterdam, pp. 652–702.
Hattori, M., Itino, T., 2008. Soldiers' armature changes seasonally and locally in an eusocial aphid (Homoptera: Aphididae). Sociobiology 52, 429–436.
Hölldobler, B., Wilson, E.O., 1990. The Ants. Harvard University Press, Cambridge, 732 pp.
Hunt, B.G., Brisson, J.A., Yi, S.V., Goodisman, M.A.D., 2010. Functional conservation of DNA methylation in the pea aphid and the honeybee. Genome Biol. Evol. 2, 719–728.
Ijichi, N., Shibao, H., Miura, T., Matsumoto, T., 2004. Soldier differentiation during embryogenesis of a social aphid, *Pseudoregma bambucicola*. Entomol. Sci. 7, 143–155.
Ijichi, N., Shibao, H., Miura, T., Matsumot, T., Fukatsu, T., 2005a. Analysis of natural colonies of a social aphid *Colophina arma*: population dynamics, reproductive schedule, and survey for ecological correlates with soldier production. Appl. Entomol. Zool. 40, 239–245.
Ijichi, N., Shibao, H., Miura, T., Matsumoto, T., Fukatsu, T., 2005b. Comparative analysis of caste differentiation during embryogenesis of social aphids whose soldier castes evolved independently. Insect. Soc. 52, 177–185.
Ishikawa, A., Ogawa, K., Gotoh, H., Walsh, T.K., Tagu, D., Brisson, J.A., Rispe, C., Jaubert-Possamai, S., Kanbe, T., Tsubota, T., Shiotsuki, T., Miura, T., 2011. Juvenile hormone titre and related gene expression during the change of reproductive modes in the pea aphid. Insect Mol. Biol. 21, 49–60.
Korb, J., Roux, E.A., Lenz, M., 2003. Proximate factors influencing soldier development in the basal termite *Cryptotermes secundus* (Hill). Insect. Soc. 50, 299–303.
Kurosu, U., Aoki, A., 1988. Monomorphic first instar larvae of *Colophina clematicola* (Homoptera, Aphidoidea) attack predators. Kontyû 56, 867–871.
Kurosu, U., Aoki, S., 2009. Extremely long-closed galls of a social aphid. Psyche 4, 1–9.
Kurosu, U., Stern, D.L., Aoki, S., 1990. Agonistic interactions between ants and gall-living soldier aphids. J. Ethol. 8, 139–141.

Kurosu, U., Aoki, S., Fukatsu, T., 1994. Taxonomic position of the aphid "*Astegopteryx*" *takenouchii* (Homoptera), with notes on its defenders on the secondary host. Jpn. J. Entomol. 62, 363–368.

Kurosu, U., Aoki, S., Fukatsu, T., 2003. Self-sacrificing gall repair by aphid nymphs. Proc. Biol. Sci. 270, S12–S14.

Kurosu, U., Buranapanichpan, S., Aoki, S., 2006. *Astegopteryx spinocephala* (Hemiptera: Aphididae), a new aphid species producing sterile soldiers that guard eggs laid in their gall. Entomol. Sci. 9, 181–190.

Kutsukake, M., Shibao, H., Nikoh, N., 2004. Venomous protease of aphid soldier for colony defense. Proc. Natl. Acad. Sci. U.S.A. 101, 11338–11343.

Kutsukake, M., Nikoh, N., Shibao, H., Rispe, C., Simon, J.C., Fukatsu, T., 2008. Evolution of soldier-specific venomous protease in social aphids. Mol. Biol. Evol. 25, 2627–2641.

Kutsukake, M., Shibao, H., Uematsu, K., Fukatsu, T., 2009. Scab formation and wound healing of plant tissue by soldier aphid. Proc. Biol. Sci. 276, 1555–1563.

Kutsukake, M., Meng, X.-Y., Katayama, N., Nikoh, N., Shibao, S., Fukatsu, T., 2012. An insect-induced novel plant phenotype for sustaining social life in a closed system. Nat. Commun. 3, 1187.

Lawson, S.P., Graham, C., Legan, A., Abbot, P., 2014. Comparative phenotyping across a social transition in aphids. Anim. Behav. 96, 117–125.

Le Trionnaire, G., Wucher, V., Tagu, D., 2013. Genome expression control during the photoperiodic response of aphids. Physiol. Entomol. 38, 117–125.

McLean, D.L., Kinsey, M.G., 1984. The precibarial valve and its role in the feeding behavior of the pea aphid, *Acyrthosiphon pisum*. Bull. Entomol. Soc. Am. 30, 26–31.

Mehrparvar, M., Zytynska, S.E., Weisser, W.W., 2013. Multiple cues for winged morph production in an aphid metacommunity. PLoS One 8, e58323.

Miura, T., Braendle, C., Shingleton, A., 2003. A comparison of parthenogenetic and sexual embryogenesis of the pea aphid *Acyrthosiphon pisum* (Hemiptera: Aphidoidea). J. Exp. Zool. 295, 59–81.

Moran, N.A., 1992. The evolution of aphid life cycles. Annu. Rev. Entomol. 37, 321–348.

Müller, C.B., Williams, I.S., Hardie, J., 2001. The role of nutrition, crowding and interspecific interactions in the development of winged aphids. Ecol. Entomol. 26, 330–340.

Ogawa, K., Miura, T., 2014. Aphid polyphenisms: trans-generational developmental regulation through viviparity. Front. Physiol. 5, 1–11.

Ohara, K., 1985. Observations on the prey-predator relationship between *Pseudoregma bambucicola* (Homoptera, Pemphigidae) and *Metasyrphus confrater* (Diptera, Syrphidae), with special reference to the behaviour of the aphid soldiers. Esakia 23, 107–110.

Pickett, J.A., Glinwood, R.T., 2005. Chemical ecology. In: van Emden, H., Harrington, R. (Eds.), Aphids as Crop Pests. CABI International, Wallingford, Oxon, GBR.

Pike, N., Foster, W.A., 2008. The ecology of altruism in a clonal insect. In: Korb, J., Heinze, J. (Eds.), Ecology of Social Evolution. Springer-Verlag, Berlin, pp. 37–56.

Pike, N., Braendle, C., Foster, W.A., 2004. Seasonal extension of the soldier instar as a route to increased defence investment in the social aphid *Pemphigus spyrothecae*. Ecol. Entomol. 29, 89–95.

Pike, N., Whitfield, J.A., Foster, W.A., 2007. Ecological correlates of sociality in *Pemphigus* aphids, with a partial phylogeny of the genus. BMC Evol. Biol. 7, 185.

Powell, G., Tosh, C.R., Hardie, J., 2006. Host plant selection by aphids: behavioral, evolutionary, and applied perspectives. Annu. Rev. Entomol. 51, 309–330.

Prado, E., Tjallingii, W.F., 1994. Aphid activities during sieve element punctures. Entomol. Exp. Appl. 72, 157–165.

Price, D.R.G., Tibbles, K., Shigenobu, S., Smertenko, A., Russell, C.W., Douglas, A.E., Fitches, E., Gatehouse, A.M.R., Gatehouse, J.A., 2010. Sugar transporters of the major

facilitator superfamily in aphids; from gene prediction to functional characterization. Insect Mol. Biol. 19, 97–112.
Queller, D.C., Strassmann, J.E., 1998. Kin selection and social insects. Bioscience 48, 165–175.
Ramsey, J.S., Rider, D.S., Walsh, T.K., De Vos, M., Gordon, K.H.J., Ponnala, L., Macmil, S.L., Roe, B.A., Jander, G., 2010. Comparative analysis of detoxification enzymes in *Acyrthosiphon pisum* and *Myzus persicae*. Insect Mol. Biol. 19, 155–164.
Ratnieks, F.L.W., Foster, K.R., Wenseleers, T., 2010. Darwin's special difficulty: the evolution of 'neuter insects' and current theory. Behav. Ecol. Sociobiol. 65, 481–492.
Rhoden, P.K., Foster, W.A., 2002. Soldier behaviour and division of labour in the aphid genus *Pemphigus* (Hemiptera, Aphididae). Insect. Soc. 49, 257–263.
Sakata, K., Itô, Y., 1991. Life history characteristics and behaviour of the bamboo aphid, *Pseudoregma bambucicola* (Hemiptera: Pemphigidae), having sterile soldiers. Insectes Soc. 38, 317–326.
Sakata, K., Itô, Y., Yukawa, J., Yamane, S., 1991. Ratio of sterile soldiers in the bamboo aphid, *Pseudoregma bambucicola* (Homoptera: Aphididae) colonies in relation to social and habitat conditions. Appl. Entomol. Zool. 26, 463–468.
Schütze, M., Maschwitz, U., 1991. Enemy recognition and defence within trophobiotic associations with ants by the soldier caste of *Pseudoregma sundanica* (Homoptera: Aphidoidea). Entomol. Gen. 16, 1–12.
Schwander, T., Lo, N., Beekman, N., Oldroyd, B.P., Keller, L., 2010. Nature versus nurture in social insect caste differentiation. Trends Ecol. Evol. 25, 275–282.
Shibao, H., 1998. Social structure and the defensive role of soldiers in a eusocial bamboo aphid, *Pseudoregma bambucicola* (Homoptera: Aphididae): a test of the defence-optimization hypothesis. Res. Popul. Ecol. 40, 325–333.
Shibao, H., 1999. Quantitative genetic analysis of fitness-related traits in the soldier-producing aphid, *Pseudoregma bambucicola* (Homoptera: Aphididae): genetic variation and costs of soldier production. Appl. Entomol. Zool. 34, 57–68.
Shibao, H., Fukatsu, T., 2003. Altruistic defenders in a gall-forming aphid of the tribe Hormaphidini (Homoptera, Aphididae, Hormaphidinae) on its primary host. Insect. Soc. 50, 167–173.
Shibao, H., Lee, J.-M., Kutsukake, M., Fukatsu, T., 2003. Aphid soldier differentiation: density acts on both embryos and newborn nymphs. Naturwissenschaften 90, 501–504.
Shibao, H., Kutsukake, M., Fukatsu, T., 2004a. Density triggers soldier production in a social aphid. Proc. R. Soc. Lond. B 271, S71–S74.
Shibao, H., Kutsukake, M., Fukatsu, T., 2004b. The proximate cue of density-dependent soldier production in a social aphid. J. Insect Physiol. 50, 143–147.
Shibao, H., Kutsukake, M., Matsuyama, S., Suzuki, T., Fukatsu, T., 2005. Warning pheromone excreted for reinforcement—soldier deployment system of eusocial aphid by a chemical substance. Kagaku to Seibutsu 43, 4–6.
Shibao, H., Kutsukake, M., Matsuyama, S., Fukatsu, T., Shimada, M., 2010. Mechanisms regulating caste differentiation in an aphid social system. Commun. Integr. Biol. 3, 1–5.
Shigenobu, S., Bickel, R.D., Brisson, J.A., Butts, T., Chang, C.-C., Christiaens, O., Davis, G.K., Duncan, E.J., Ferrier, D.E.K., Iga, M., Janssen, R., Lin, G.-W., Lu, H.-L., McGregor, A.P., Miura, T., Smagghe, G., Smith, J.M., van der Zee, M., Velarde, R.A., Wilson, M.J., Dearden, P.K., Stern, D.L., 2010. Comprehensive survey of developmental genes in the pea aphid, *Acyrthosiphon pisum*: frequent lineage-specific duplications and losses of developmental genes. Insect Mol. Biol. 19, 47–62.
Shingleton, A.W., Foster, W.A., 2000. Ant tending influences soldier production in a social aphid. Proc. R. Soc. Lond. B 267, 1863–1868.
Shingleton, A.W., Foster, W.A., 2001. Behaviour, morphology and the division of labour in two soldier-producing aphids. Anim. Behav. 62, 671–679.

Shingleton, A.W., Sisk, G.C., Stern, D.L., 2003. Diapause in the pea aphid (*Acyrthosiphon pisum*) is a slowing but not a cessation of development. BMC Dev. Biol. 3, 1–12.

Simon, J.-C., Stoeckel, S., Tagu, D., 2010. Evolutionary and functional insights into reproductive strategies of aphids. C. R. Biol. 333, 488–496.

Smadja, C., Shi, P., Butlin, R.K., Robertson, H.M., 2009. Large gene family expansions and adaptive evolution for odorant and gustatory receptors in the pea aphid, *Acyrthosiphon pisum*. Mol. Biol. Evol. 26, 2073–2086.

Srinivasan, D.G., Brisson, J.A., 2012. Aphids: a model for polyphenism and epigenetics. Gen. Res. Int. 2012, 1–12.

Stern, D.L., 1994. A phylogenetic analysis of soldier evolution in the aphid family Hormaphididae. Proc. Biol. Sci. 256, 203–209.

Stern, D.L., 1998. Phylogeny of the tribe Cerataphidini (Homoptera) and the evolution of the horned soldier aphids. Evolution 52, 155–165.

Stern, D.L., Foster, W.A., 1996. The evolution of soldiers in aphids. Biol. Rev. 71, 27–79.

Stern, D.L., Whitfield, J.A., Foster, W.A., 1997. Behavior and morphology of monomorphic soldiers from the aphid genus *Pseudoregma* (Cerataphidini, Hormaphididae): implications for the evolution of morphological castes in social aphids. Insect. Soc. 44, 379–392.

Tagu, D., Sabater-Muñoz, B., Simon, J.-C., 2005. Deciphering reproductive polyphenism in aphids. Invertebr. Reprod. Dev. 48, 71–80.

The International Aphid Genomics Consortium, 2010. Genome sequence of the pea aphid *Acyrthosiphon pisum*. PLoS Biol. 8, e1000313. http://dx.doi.org/10.1371/journal.pbio.1000313.

Tjallingii, W.F., 2006. Salivary secretions by aphids interacting with proteins of phloem wound responses. J. Exp. Bot. 57, 739–745.

Uematsu, K., Fukatsu, T., Shimada, M., 2010. Altruistic colony defense by menopausal female insects. Curr. Biol. 20, 1182–1186.

Walsh, T.K., Brisson, J.A., Robertson, H.M., Gordon, K., Jaubert-Possamai, S., Tagu, D., Edwards, O.R., 2010. A functional DNA methylation system in the pea aphid, *Acyrthosiphon pisum*. Insect Mol. Biol. 19, 215–228.

Webster, B., 2012. The role of olfaction in aphid host location. Physiol. Entomol. 37, 10–18.

West Eberhard, M.J., 2003. Developmental Plasticity and Evolution. Oxford University Press, Oxford, 794 pp.

Whitham, T.G., 1979. Territorial behaviour of *Pemphigus* gall aphids. Nature 279, 324–325.

Wilch, M.H., 1999. Predation and prey response in the galls of *Pemphigus populiramulorum*. Masters thesis, University of Arizona, Tucson, AZ.

Will, T., 2013. How phloem-feeding insects face the challenge of phloem-located defenses. Front. Plant Sci. 4, 336.

Wilson, A.C.C., Ashton, P.D., Calevro, F., Charles, H., Colella, S., Febvay, G., Jander, G., Kushlan, P.F., MacDonald, S.J., Schwarz, J.F., Thomas, G.H., Douglas, A.E., 2010. Genomic insight into the amino acid relations of the pea aphid, *Acyrthosiphon pisum*, with its symbiotic bacterium Buchnera aphidicola. Insect Mol. Biol. 19, 249–258.

Withgott, J.H., Abbot, D.K., Moran, N.A., 1997. Maternal death relaxes developmental inhibition in nymphal aphid defenders. Proc. R. Soc. Lond. B 264, 1197–1202.

Zhou, J.J., Vieira, F.G., He, X.L., Smadja, C., Liu, R., Rozas, J., Field, L.M., 2010. Genome annotation and comparative analyses of the odorant-binding proteins and chemosensory proteins in the pea aphid *Acyrthosiphon pisum*. Insect Mol. Biol. 19, 113–122.

CHAPTER SIX

Insect Reward Systems: Comparing Flies and Bees

Eirik Søvik*,[1], Clint J. Perry†, Andrew B. Barron‡
*Department of Biology, Washington University in St. Louis, St. Louis, Missouri, USA
†School of Chemical and Biological Sciences, Queen Mary University, London, United Kingdom
‡Department of Biological Sciences, Macquarie University, Sydney, New South Wales, Australia
[1]Corresponding author: e-mail address: eirik.sovik@gmail.com

Contents

Abstract

Many elements of animal behaviour are organised by an innate reward-seeking drive stemming from neurobiological reward systems. The behavioural concept of reward and its neurobiological substrates was initially developed in mammalian systems, and there it has become clear that several novel social behaviours evolved through the co-option of reward pathways. Only more recently has reward been explored in insects. In this review, we consider current knowledge about reward pathways in the two predominant insect models: *Drosophila melanogaster* and the honey bee *Apis mellifera*. These two models are phylogenetically distantly related and have vastly different ecologies: fruit flies are mostly solitary while honey bees live in complex societies involving social foraging and brood care. Initially, it was assumed the reward system was essentially similar between these two organisms, but more recent studies have appeared to highlight quite significant differences. Here, we critically evaluate apparent differences in the neurobiology of the reward system between these organisms. We discuss which differences may be real and which may be reflective of the very different

Advances in Insect Physiology, Volume 48
ISSN 0065-2806
http://dx.doi.org/10.1016/bs.aiip.2014.12.006

modes of analysis applied in these two models. Finally, we discuss how modification of reward systems might have contributed to social evolution in insects.

1. INTRODUCTION

In behavioural neuroscience, reward has an operational definition; stimuli that elicit approach or seeking behaviour from animals (e.g. sugars or other nutrients) are considered rewards, and the attainment of these stimuli is considered to be rewarding (Loeb, 1918). Reward and its neurobiological correlates have been the focus of study for almost a century (Loeb, 1918), but until the 1990s the majority of this work was conducted in mammalian model systems (Schultz, 2010). This rich research tradition (predominantly with mammals) has resulted in a nuanced understanding of the neural substrates that are involved in signalling different conceptual aspects of reward (Berridge et al., 2009), and an understanding of how reward pathways can become the driving force of the evolution of novel social behaviours (O'Connell and Hofmann, 2011). However, this dominant focus on mammalian systems has done little to advance our knowledge of how reward systems operate in other animals, such as the insects.

The study of reward systems in insects was inspired by a landmark study (Hammer, 1993), demonstrating that the activation of a single identified neuron could replace sucrose reward during conditioning in honey bees (*Apis mellifera*). Since this pivotal study, a range of studies have examined different aspects of rewards in insects and have attempted to pinpoint the neural substrates responsible (Perry and Barron, 2013). From these studies, there appeared to be quite significant differences between the results from insects and the more established understanding of the physiology of reward in mammals. In mammals, dopamine (DA) systems emerged as key neurochemical signals for rewards (Schultz, 1997, 2002, 2007), while in insects octopamine (OA) systems were considered to signal rewards (Hammer and Menzel, 1998). OA and DA are chemically quite similar but OA is only a trace amine in mammalian nervous systems.

Recently, this view has changed following a number of studies using the fruit fly (*Drosophila melanogaster*) that have demonstrated a role for specific DA circuits in the fly reward system (Burke et al., 2012; Kim et al., 2007; Krashes et al., 2009; Selcho et al., 2009). However, so far no evidence has emerged that DA plays a role in reward signalling in honey bees. Some have suggested that this reflects neurobiological differences between the reward systems of honey bees and fruit flies, while others have suggested that the difference might reflect differences in the experimental methods applied to these two organisms (Barron et al., 2010; Perry and Barron, 2013). In this review, we carefully compare and contrast the current status of knowledge in the two systems. We consider the real and apparent differences between these two influential models. Finally, we speculate how modification of the reward system by the evolutionary process might have contributed to the evolution of social behaviour in insects.

2. WHAT DO WE MEAN BY REWARD?

The idea that behaviour is guided by reward and punishment goes back at least to the ancient Greeks (e.g. Epicurus, translated in Bailey, 1926). Reward and reward seeking is now recognised to be fundamental to the organisation of behaviour (Dreher and Tremblay, 2009). Reward not only supports elementary processes such as drinking, eating and reproduction but also encompasses a broad range of behaviour from arousal and attention to foraging and decision making to gambling and social relationships. Dysfunctions in the reward system can cause disruptions in learning, planning, social interactions and general evaluation of our surroundings (Søvik and Barron, 2013). Understanding the reward system has become vital for understanding all goal-directed animal behaviour, from the very basic to the most complex. Understandably, the focus of research has been on mammalian models, and here we briefly review how analyses of reward in mammals have helped develop our understanding of biological reward systems.

What do we mean by reward? One of the earliest, and perhaps still most useful, experimental definitions of reward began with Loeb (1918) who defined rewards as stimuli that elicit approach and punishment as stimuli that elicit avoidance. Pavlov (1927) conceptualised rewards as objects that bring about a change in behaviour, or rather that caused learning. These definitions are operational in that the behaviour that the stimulus induces qualifies it as a reward or punishment. In common usage, reward is often synonymous with the word pleasure. In general writing and many sociological disciplines, it is common to consider anything that gives us the conscious experience of pleasure as a reward. Scientific investigation of reward, however, has shown that reward involves multiple mechanisms and psychological components.

The major components of reward include liking: the conscious and unconscious pleasurable impact of reward; wanting: the conscious and unconscious desires for reward; and learning: associations, representations and predictions about future rewards based on past experiences (Berridge and Kringelbach, 2008). Extensive research in mammalian and human systems has demonstrated that these different psychological components are mediated by partly dissociable brain substrates (Berridge et al., 2009).

Direct brain stimulation, functional neuroimaging, neural recording and tracing studies have identified several overlapping areas in the mammalian brain that respond to reward, including areas within the brain stem and midbrain up into the frontal cortex, with the strongest response to reward occurring along the medial forebrain bundle (Leknes and Tracey, 2008; Wise and Rompre, 1989). Lesion and psychopharmacological studies suggest that the midbrain DA systems are vital to reward processing in the mammalian brain (Schultz, 2010). Early studies on electrical self-stimulation and drug addiction showed that stimulation of midbrain areas with a strong dopaminergic innervation would elicit approach behaviour, and dopaminergic projections within the midbrain play a central role in reward processing (Ikemoto, 2010). One of the major mammalian DA pathways, known as the mesocorticolimbic DA pathway, connects dopaminergic neurons in the ventral tegmentum to the nucleus accumbens and separately to the frontal cortex (Carr and Sesack, 2000; Omelchenko and Sesack, 2009; Van Bockstaele and Pickel, 1995). Two areas in this pathway, the ventral pallidum and nucleus accumbens, have been found to be major foci for reward learning and pleasure (Smith et al., 2007). Although the exact function of DA is still to be determined, we now know that DA plays a much more complicated role in reward than as simply a pleasure transmitter. Different DA circuits are involved in the different liking, wanting and learning elements of the reward response, and they interact with a range of other neurochemical systems. Opioids, serotonin, endocannabinoids and γ-aminobutyric acid (GABA) also play a role in different components of reward (Gardner, 2005; Kranz et al., 2010; Van Ree et al., 2000; Vlachou and Markou, 2010).

Mammalian studies have emphasised the distinction between wanting and liking of reward. In rodent models, the latter is indicated by distinctive orofacial expressions of the rodents that are similar to the innate expressions of humans to food reward and are considered indicative of the affective state of the animal. Affective neuroscience studies of rodents have indicated hedonic (liking) reactions, orofacial reactions to food stimuli, to be coordinated by a network of hedonic hotspots distributed within the nucleus accumbens and ventral pallidum but also in other forebrain and deep brainstem regions. Microinjections of drugs that activate neuronal opioid, endocannabinoid or related neurochemical receptors in these hedonic hotspots significantly increase the normal number of "liking" reactions to a sucrose taste (Mahler et al., 2007; Peciña and Berridge, 2005; Smith and Berridge, 2005; Smith et al., 2009). Berridge and colleagues blocked DA systems via antagonists in rats and showed that hedonic reactions to food reward were as strong as normal but the "wanting" reactions (e.g. eating more voraciously) had changed. Similarly, stimulation of anatomically separate regions within the same midbrain areas have been shown to increase "wanting," but not affect the rats' "liking" of the food through their hedonic responses (Berridge and Valenstein, 1991). These among many other experiments have shown that the liking and wanting psychological components of reward have distinct neuroanatomical and neurochemical pathways.

Assessment of liking in insects is difficult. For obvious reasons, insects do not display the orofacial reactions that are considered indicative of liking in mammals. Consequently, the insect literature tends to discuss not whether a reward is liked or sought, but simply the valance of a reward (Galizia, 2014; Knaden et al., 2012; Parnas et al., 2013). Valence describes stimuli simply by the extent to which they solicit a positive or negative response from an animal. That response might be approach versus avoidance behaviour, or an appetitive versus a rejection response. The valence concept is aligned well with the simple bioassays that have dominated much of the *Drosophila* literature, and is entirely operational in definition with no assumption of the internal mental processes of the animal. It may, however, incorporate both liking and wanting in the observed behavioural responses. The term valence as applied in the invertebrate literature is very different from the concept of emotional valence that is sometimes used in human and comparative psychological literature to describe positive and negative emotional reactions.

Our objective below is to review what is currently known about the reward systems of honey bees and fruit flies to assess the degree to which they might be comparable. This issue is important because honey bees and fruit flies are the two dominant insect neuroscience models, and it is important to consider to what degree findings might translate across these systems despite the significant phylogenetic distance between them. The literature is currently a little confusing: for a long time, neurobiological similarities were assumed across insect orders; however, new findings highlight an increasing divergence in our understanding of honey bee and fruit fly reward systems.

3. REWARD SYSTEMS IN HONEY BEES AND FLIES

The two major insect models for comparative neuroscience are the fruit fly and the European honey bee. Since Benzer established *Drosophila* as the model organism for behavioural neurogenetics (Weiner, 1999), the fly has been widely used to study the molecular underpinnings of memory (Davis, 2005). Early investigations did not aim at understanding the fruit fly reward system *per se*, but rather to identify molecular components involved in memory processing. This research initially focused on aversive conditioning (Quinn et al., 1974), but subsequent investigators modified the assays used to enable the investigation of reward conditioning (Tempel et al., 1983), which facilitated study of the fruit fly reward system. The honey bee has arguably been as important a model as *Drosophila* for studies of insect learning. Appetitive behaviours in honey bees were first studied by von Frisch at the turn of the twentieth century (von Frisch, 1914); however, mechanistic studies of the neurobiology of reward did not begin in earnest until studies of reward learning using the proboscis extension reflex (Kuwabara, 1957).

3.1 Experimental approaches

3.1.1 Behavioural paradigms utilised to study reward systems

The most widely used experimental approach to dissect reward systems in *Drosophila* is the t-maze. In this apparatus, groups of fruit flies are typically trained to associate a sucrose reward with an odour stimulus (Tempel et al., 1983). Afterwards, the valence of the rewarded odour is assessed, by examining whether the group of fruit flies distributes themselves closer to the reward-associated odour than a control odour. This is a very different approach from the most common assay used to assess reward learning with honey bees: proboscis extension response (PER) conditioning. In PER experiments, individual honey bees are restrained so that only the antennae and proboscis are free to move (Bitterman et al., 1983; Kuwabara, 1957). In this position, honey bees reflexively extend their proboscis if sucrose is touched to the antenna. During conditioning, honey bees are exposed to an odour paired with sucrose exposure to the antenna. Post-conditioning, odour exposure becomes sufficient to elicit proboscis extension, and this response is considered indicative of reward learning.

3.1.2 Fruit fly experiments: Advanced genetic tools allow for careful investigation of the role of specific neural circuits in reward

The power of *Drosophila* lies in the advanced genetic tools that have been developed for this model system. Early fruit fly studies relied on discovering mutants with interesting defects. This process more or less relied on chance, as investigators induced random mutations and carefully studied mutants' behaviour with the hope of discovering phenotypes of interest (this approach is known as forward genetics). However, this all changed when the advent of the genomic era allowed *Drosophila* researchers to investigate the functions of variants of known genes or gene knockouts in different experimental paradigms (this is known as reverse genetics). Here, we give a brief summary of the techniques used in some of the most revealing experiments on reward.

Perhaps the most important single genetic tool has been the incorporation of the GAL4-UAS driver system in fruit flies (Brand and Perrimon, 1993). This is a powerful method that allows for manipulation of expression of a gene of interest in selected tissue only. For the GAL4-UAS system to work, two transgenes are necessary: a GAL4 transcription factor and a GAL4 upstream activating sequence (UAS). Typically, two constructs (one containing the promoter region of a gene of interest upstream from GAL4, and another with the gene of interest downstream from UAS) are introduced into two separate fly lines. When the gene upstream of GAL4 is transcribed, so is GAL4. The resulting GAL4 protein binds to UAS and initiates transcription of the transgene and any gene linked to UAS. The beauty of this system is that GAL4 is only expressed in cells where the upstream gene is normally expressed. Thus, if GAL4 is downstream of a particular enzyme or transcription factor, the UAS-linked transgene will only be expressed in cells naturally expressing the enzyme or transcription factor, while leaving all other cells unaffected. Since these are two separate transgenes, it is possible to maintain separate lines containing GAL4 and UAS without the transgene being expressed because there is no UAS sequence for the GAL4 protein to bind to in lines containing GAL4-only and vice versa. Labs can easily maintain strains without worrying about any ill effects of the mutants, as the transgenes are ineffectual until both come together in the same individual. While initially the GAL4-UAS system was used to ectopically express genes of interest, any RNA of interest can now be expressed using the system. The tool can be used to express genes from other species, overexpress genes of interest or even silence expression of specific genes using RNAi (Dietzl et al., 2007).

While the ability to over- and under-express genes has been very fruitful, perhaps the biggest innovation for neurogenetics research was the use of the GAL4-UAS system to directly silence and activate selected neurons with high temporal precision. Thus, a fruit fly can develop and behave normally, while specific neurons can be activated or silenced in a context of interest. By expressing a temperature-sensitive variant of the GTPase *shibire* (shi^{ts1}) under UAS control, it is possible to inhibit neurotransmitter release. At temperatures above 29 °C, shi^{ts1} inhibits recycling of synaptic vesicles and causes a cessation of neurotransmission from neurons expressing GAL4 (Kitamoto, 2001). Because of the temperature-sensitive nature of this allele, GAL4-expressing neurons function normally at other temperatures, but neurotransmitter release can be stopped by increasing the temperature. It is also possible to directly activate neurons. One of the earliest techniques used for this purpose was Channelrhodopsin-2 (ChR2) in combination with GAL4-UAS (Schroll et al., 2006). ChR2 is a light-gated cation-selective membrane channel (Nagel et al., 2003). When cells expressing the ChR2 protein are illuminated, it causes a rapid depolarisation of any cell where it is expressed. One drawback of ChR2 is that the cell of interest needs to be illuminated directly. This works well with translucent larvae, but in order to use this method in adults, it is necessary to open the pigmented head capsule to expose neurons in the brain to light, which limits the range of behavioural experiments a where this technique can be used. It is now also possible to use a thermogenetic approach similar to shi^{ts1} to activate neurons. Using TrpA1 (a temperature-sensitive cation channel) in the same manner as ChR2, it is possible to rapidly depolarise neurons by exposing fruit flies to temperatures above 27 °C (Hamada et al., 2008). Reaching the required temperature can be achieved very rapidly by heating fruit flies directly with lasers, allowing for sub-second timing of neuronal activity (Bath et al., 2014). These techniques have made it possible to examine function and circuitry of specific neuron populations in fruit flies with unprecedented precision.

3.1.3 Honey bee experiments: Electrophysiology and pharmacology

While very recent advances have been made to make it possible to generate transgenic honey bees (Ben-Shahar, 2014; Schulte et al., 2014), nothing like the tools seen in *Drosophila* are available. Instead, the majority of honey bee studies have employed behavioural, pharmacological or electrophysiological techniques. These different approaches allow for very different sets of experiments to be performed. Despite this, until very recently there was broad agreement about the neurobiology of the reward systems in honey bees and flies, which led to the assumption that reward systems were broadly conserved across insect orders. Somewhat confusingly, the most recent studies of neurotransmitter systems in the *Drosophila* reward system seem to conflict with classic honey bee electrophysiological and pharmacological studies.

Early electrophysiological analyses by Hammer and Menzel of the neuron VUMmx1 proved to have enormous influence on the comprehension of the honey bee reward system (Hammer, 1993; Hammer and Menzel, 1995). As we discuss in Section 3.2 below, stimulation of VUMmx1 was able to substitute sucrose reward in the proboscis extension paradigm. Evidence suggests VUMmx1 is octopaminergic (Hammer and Menzel, 1998), and studies of the reward system in honey bees since then have focussed on both the role of OA and locations where VUMmx1 projects.

Most studies in honey bees, however, have used neuropharmacological methods to explore circuit functions (Scheiner et al., 2002). The pharmacological methods applied in honey bees cannot match the resolution and precision of the neurogenetic methods available for fruit fly research. Many behavioural studies have applied pharmacological agents to the whole brain or systemically (Barron et al., 2007b; Farooqui et al., 2004), and even brain microinjection targets a region rather than a specific circuit. It is also difficult to completely differentiate between octopaminergic and dopaminergic circuits with pharmacology. A range of compounds have been identified that differentially bind to OA or DA receptors, but there remains some overlap (Beggs et al., 2011; Mustard et al., 2005). This is especially true of the AmDOP2 and AmOA1 receptors (Beggs et al., 2011). Sadly, no single compound has affinity completely limited to a single biogenic amine receptor (Mustard et al., 2005); consequently, it is vital in honey bee pharmacological studies to use a range of agonists and antagonists in behavioural analyses to tease apart the most likely contributions of different receptor systems to any given behaviour.

3.2 Neuroanatomy of insect reward systems

Hymenoptera and Diptera diverged in the Carboniferous (more than 300 million years ago) concordant with the divergence of seed plants (Grimaldi and Engel, 2005). Despite this ancient divergence, many key aspects of brain morphology and function are apparently conserved between these two groups, as we describe below. The honey bee brain is an order of magnitude larger than the *Drosophila* brain (Fig. 1). The honey bee brain possesses approximately 1 million neurons compared to approximately

100,000 neurons in the fruit fly (Shimada et al., 2005; Witthöft, 1967). All the major areas of neuropil recognised in the fruit fly brain are present in the honey bee brain (Fig. 2). Below, we briefly lay out which neuropils are thought to be involved in reward processing, in honey bees and flies, and discuss their supposed functions.

Most studies examining reward in honey bees and fruit flies have done so in the context of olfactory learning, and therefore, most of our inferences about the reward system are derived from experiments using an olfactory stimulus during conditioning. Typically, in an experiment, the animal is given a reward, usually in the form of sucrose, while being simultaneously exposed to an odour. While the sucrose is detected by gustatory receptors on the proboscis or tarsi (Wang et al., 2004), odours are detected via olfactory receptor neurons (ORNs) in the antenna or the maxillary palps and a signal is sent to glomeruli in the antennal lobes (ALs). A single ORN projects to one

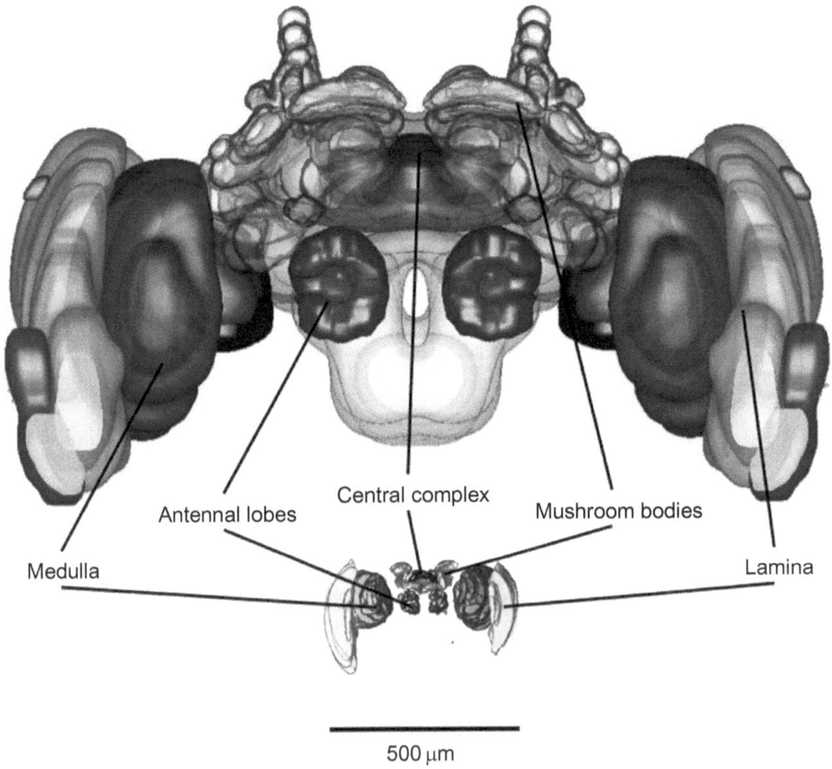

Figure 1 3D models of the honey bee brain and *Drosophila* brain to scale. *Figures provided by Paulk, and adapted from Paulk et al. (2014) with permission.*

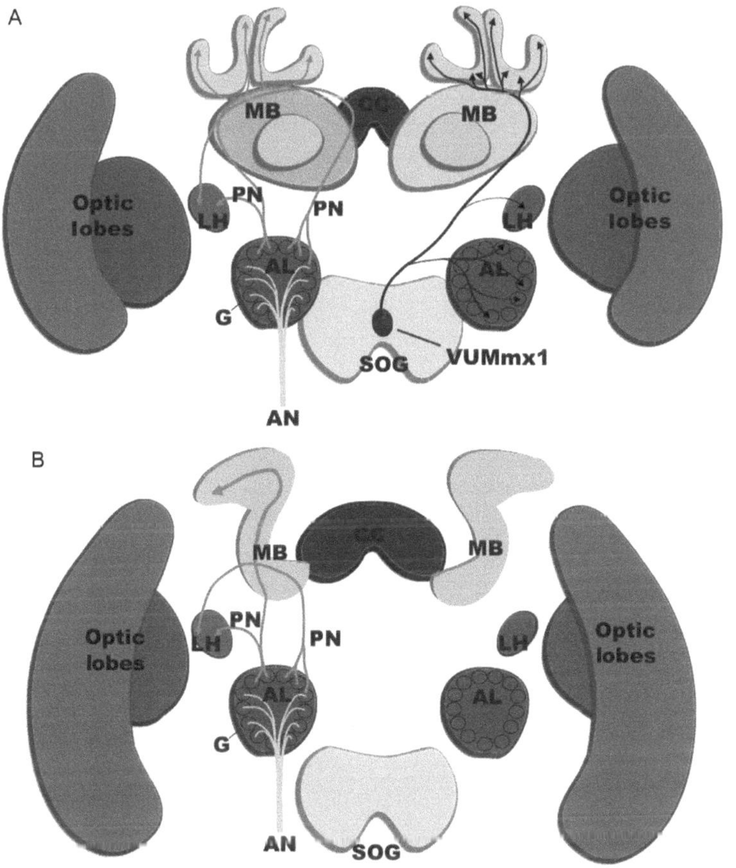

Figure 2 The reward pathways (olfactory and gustatory circuits) of the honey bee brain and fruit fly brain. Schematic frontal views of a cross-section of the central honey bee (A) and fruit fly (B) brain (head capsule and eyes removed). The olfactory (CS) pathway is depicted in light blue (light grey in the print version). Olfactory neurons send information to the brain via the antennal nerve. These neurons form synapses within the glomeruli of the antennal lobes (ALs) onto local interneurons (not shown) and projection neurons conveying olfactory information to the lateral horn (LH) and the mushroom bodies (MBs). In the honey bee brain (A), the gustatory (US) pathway involves the VUMmx1 neuron (brown, light grey in the print version), which projects from the gnathal ganglion to the LH, AL and MB. VUMmx1 is bilaterally symmetrical, but in this figure only the right side is shown. Abbreviations: CS, conditioned stimulus; US, unconditioned stimulus; VUMmx1, ventral unpaired medial neuron 1 of the maxillary neuromere. *Figure adapted from Perry and Barron (2013) with permission.*

glomerulus only. The signal from multiple ORNs is processed within a glomerulus before the output is transmitted to projection neurons (PNs). The PNs project to the mushroom bodies (MBs) and the lateral horn (LH).

The MBs have long been of interest to researchers because of their potential role in sensory integration and their location as a junction between sensory processing centres and premotor centres (Kenyon, 1896). This is also the structure where honey bees and fruit flies show the most anatomical difference. *Drosophila* MBs are rather modest occupying less than 4% of the brain volume. Each contains approximately 2500 Kenyon cells, with a rather simple calyx (the region of dendritic input to the MB) composed of two fused caps that resembles a simple button. By contrast, honey bees MBs are exuberant, filling more than 20% of the brain volume and each containing more than 170,000 Kenyon cells with complex folded double calyces (Giurfa, 2013). In the fly, the MB only receives olfactory input, whereas in honey bees the MB receives olfactory, visual, mechanosensory and gustatory inputs to different regions of the calyx (Ehmer and Gronenberg, 2002; Giurfa, 2013; Mobbs, 1982; Strausfeld, 2002). The potential for multimodal integration in the honey bee MB may be necessary to support the learning of multimodal and abstract stimulus properties associated with rewards (Avarguès-Weber et al., 2010; Perry and Barron, 2013). The MBs have been shown to be indispensable for olfactory memory formation in fruit flies (Heisenberg et al., 1985) and honey bees (Menzel et al., 1974). While there has been some dispute about the particular function of the MB, Heisenberg (2003) has argued that outputs from the calyces of the MB are necessary for olfactory reward learning (Heisenberg, 2003). In mutant flies with malformed MBs, olfactory learning was affected but visual, tactile and motor learning were not affected (Wolf et al., 1998). Although there is no direct input from the optical lobes to the MB in flies, a small subset of MB intrinsic neurons are involved in visual reward learning in flies (Vogt et al., 2014). It is currently unclear if the MB conveys valence to rewarding stimuli, or if it functions more generally as a classifier of incoming stimuli.

Recent papers with *Drosophila* have argued that valence is assigned in the LH (Galizia, 2014; Parnas et al., 2013). As mentioned above, this neuropil also receives input from the PNs in the AL. Perhaps the most convincing evidence for the LH as a valence assignor comes from the discovery of the gustatory receptor Gr43a in a subset of LH neurons. This receptor detects fructose in response to changes in haemolymph level in response to feeding (Miyamoto et al., 2012). Further, the spatial arrangement of neurons projecting to this region from the AL has been shown to be indicative of whether or not they drive approach or avoidance (Knaden et al., 2012). When multiple competing odours are being detected, the combined valence is computed in the LH (Parnas et al., 2013). The LH is ideally situated to be a valence assignor, as it is a premotor area and therefore activation of neurons in this area could activate motor systems for either approach or avoidance.

Another potentially important neuropil for reward processing is the gnathal ganglion (GNG). This ganglion receives input from chemosensory receptor neurons on the proboscis and tarsi (Stocker, 1994). Large widely branching octopaminergic neurons with their soma in the GNG appear to be a conserved feature of insect brain anatomy across diverse insect orders. In honey bees, *Drosophila*, *Locusta migratoria* and *Periplaneta americana*, the GNG contains a very small number of octopaminergic neurons (Bräunig, 1991; Bräunig and Burrows, 2004; Sinakevitch and Strausfeld, 2006; Sinakevitch et al., 2005). These neurons, 26 in *Drosophila* (Sinakevitch and Strausfeld, 2006) and 12 in the honey bee (Sinakevitch et al., 2005), are located in small and sparsely distributed clusters (Bräunig, 1991; Bräunig and Burrows, 2004; Sinakevitch and Strausfeld, 2006; Sinakevitch et al., 2005). Many of these neurons have very extensive innervations so that most, if not all, regions of neuropil have octopaminergic input (Busch and Tanimoto, 2010; Busch et al., 2009; Sinakevitch and Strausfeld, 2006; Sinakevitch et al., 2005).

In what has become a landmark study, Hammer (1993) recorded from VUMmx1 while honey bees were trained in the restrained proboscis extension paradigm (Bitterman et al., 1983; Kuwabara, 1957) to learn the association between a novel odour and sucrose reward. VUMmx1 responded strongly to sucrose with prolonged firing (Hammer, 1993). This finding was replicated by Schröter et al. (2007), although the duration of the response to sucrose was less. In a PER paradigm (Bitterman et al., 1983), honey bees were conditioned such that a novel odour immediately preceded sucrose presentation, VUMmx1 then responded to the conditioned odour as well as sucrose reward (Hammer, 1993). In this paradigm, the muscle controlling proboscis extension (M17) could be conditioned to respond to a novel odour that was forward-paired with sucrose delivery to the antennae and mouthparts after just a single learning trial (Hammer, 1993). The M17 muscle could also be conditioned if the sucrose unconditioned stimulus during training was replaced by a transient supra-threshold depolarization of VUMmx1 (Hammer, 1993). VUMmx1 did not directly activate muscle M17 or proboscis extension, but the neuron responded to both primary sucrose rewards and odour stimuli associated with reward, and the neuron was part of the mechanism by which proboscis extension could be conditioned to respond to stimuli predictive of reward. Menzel (2001) drew a parallel between these properties of VUMmx1 and the properties of the dopaminergic neurons in the mammalian midbrain which also respond to primary and conditioned rewards, and are organisational of reward-seeking behaviour (Schultz, 1997, 2002, 2007).

In *Drosophila*, a single large neuron OA-VUMa2 that shows octopamine-like immunoreactivity and expresses tyramine decarboxylase (the enzyme converting tyramine to octopamine) has a large cell body in the GNG and projects to AL, MB calyces and LH (Busch et al., 2009). This is significant because the morphology of this neuron is strikingly similar to that of a VUMmx1 (Hammer, 1993; Hammer and Menzel, 1995). However, as of yet, it is not clear if any of the OA-VUM neurons in flies perform the same function as VUMmx1 in bees (Burke et al., 2012).

While it is often assumed that memories are stored and formed in areas where input from disparate regions converges (e.g. after several steps of information processing), information does not flow in one direction in the *Drosophila* or honey bee nervous systems. In fact, neurons from the MB also give feedback to target PNs in the AL (Hu et al., 2010). This means that there is the potential for reward learning to occur within the AL. So far, it has been demonstrated in flies that plasticity in odour coding in the AL is sufficient to result in change of behaviour independent of activity in the MB (Thum et al., 2007). However, only appetitive memory storage within the AL has been demonstrated, while aversive conditioning failed to produce a similar result (Thum et al., 2007). Based on these findings, it is important that we remain cognisant of the possible involvement of feedback to areas of sensory input if we want to understand how insect reward systems work.

3.2.1 Neuroanatomy of larval reward pathways

So far, we have only considered the structure of the nervous systems of adult insects, but all holometabolous insects also have a larval form with its own structurally distinct nervous system. The adult and larval forms represent two distinct organisms with very different lifestyles and ecologies (Truman and Riddiford, 1999), and their nervous systems reflect this reality. There are some structural overlaps between the two nervous systems; both larvae and adult fruit flies have MBs, for example, but during metamorphosis the majority of MB-projecting neurons are pruned away and new ones are formed (Technau and Heisenberg, 1982). This is necessary as the sensory input neurons for the adult form are not the same as for the larvae (Chapman, 2013). This dramatic neural transformation may contribute to the memory discontinuity that seems to exist between larval and adult insects. It is commonly reported that appetitive memories formed in larvae are abolished during metamorphosis (Barron and Corbet, 1999). Therefore, when discussing *Drosophila* reward pathways we are potentially talking about at least two distinct reward systems: the larva and adult.

While honey bees have free-living larvae, they are not "free living" in the same way one would describe *Drosophila* larvae. Rather than foraging and fending for themselves like *Drosophila* larvae, honey bee larvae are very fragile and remain mostly immobile in a wax cell where they are fed by adult bees. Currently, we know nothing about the potential for honey bee larvae to associate rewards with stimuli, or how they might otherwise react to rewards.

3.3 Octopamine signals reward in insects

3.3.1 Octopamine signals reward in honey bees

Early studies in honey bees implicated OA in reward processing. Application of OA to the brain increased bees' sensitivity to sugars (Mercer and Menzel, 1982) and enhanced storage acquisition and retrieval of reward associations (Erber et al., 1993; Menzel et al., 1999). These studies directed attention to what possible octopaminergic neurons might be involved. Shortly thereafter, an immunohistochemical study provided some evidence that the neuron identified by Hammer (1993), VUMmx1, was octopaminergic (Kreissl et al., 1994). Concluding that VUMmx1 was octopaminergic, Hammer and Menzel (1998) investigated the consequences of OA microinjection into the major zones of projection of VUMmx1: MB, AL and LH. Both the MB and AL had already been shown to be important foci for learning and memory (Menzel et al., 1974). Microinjection of OA into either the AL or MB, but not the LH, paired with presentation of an odour was sufficient to condition proboscis extension to the odour. This suggested that microinjection of OA could substitute for sucrose in reward learning (Hammer and Menzel, 1998). Repeated presentation of odour forward-paired with microinjection of OA into the AL resulted in a smooth progressive acquisition of a learned PER to the odour (Hammer and Menzel, 1998). In the MB calyx, however, repeated training with odour forward-paired with microinjection of OA resulted in a learned PER only 20 min post-training as if a consequence of a consolidation process rather than an acquisition process. OA microinjection into the LH paired with odour did not result in any learned proboscis extension to the odour.

Hammer and Menzel's (1998) study proved to be extremely influential; it was taken as strong evidence that VUMmx1 is octopaminergic. It also positioned octopaminergic circuits within the MB and AL as key elements of the honey bee reward system. A model of insect learning developed in which simultaneous presentation of a conditioned odour stimulus with an unconditioned rewarding stimulus such as sugar will activate both the olfactory stimulus pathway and reward pathways in the brain (Menzel, 2001). For the olfactory stimulus, this will include a specific pattern of neural activity in the glomeruli of the AL, PNs from AL, and MB for the CS (Fig. 3). For the US, this will include motor pathways controlling proboscis extension, and also VUMmx1. VUMmx1 releases OA into the AL and MB calyces where the neuromodulator acts to change the likelihood that future odour presentation will be sufficient to activate proboscis extension. This model focused attention on the role of OA in the MB and AL for reward learning, and

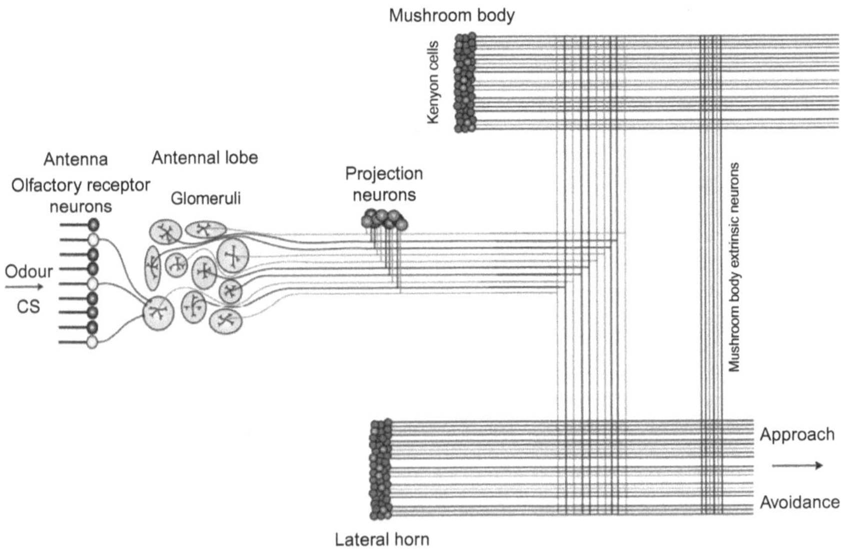

Figure 3 Circuit model of odour memory. Odour stimuli activate different sets of glomeruli within the antennal lobe. Odour information is conveyed to the MB by projection neurons that synapse with the MB calyx and separately within the lateral horn. Odours are represented in the MBs as distinct patterns of activity across sets of Kenyon cells. Extrinsic MB output neurons are connected to the Kenyon cells by latent synapses. *Figure adapted from Perry and Barron (2013) with permission.*

perhaps unfortunately it led to a perception that the LH was not involved in reward learning. Models of the insect reward system, and the possible roles of the LH and VUMmx1 within it, have continued to evolve as we discuss below.

In the AL, modulatory actions of OA released by VUMmx1 very likely contribute to the changes in odour coding across glomeruli that occur in this region when odours are paired with sucrose reward (Denker et al., 2010; Faber et al., 1999; Locatelli et al., 2013; Rath et al., 2011). Differential conditioning with one rewarded and one unrewarded odour causes the patterns of glomerular activation for the two odours to diverge, making the patterns for these odours more distinct after training (Faber et al., 1999; Fernandez et al., 2009; Galizia and Menzel, 2001; Locatelli et al., 2013). Additionally, overall activity for the reward-associated odour increases in both the ALs (Denker et al., 2010; Faber et al., 1999) resulting presumably in increased discriminability, changing the valence of the odours and increasing their salience (Fernandez et al., 2009). OA modulates network activity within the AL (Rein et al., 2013), and OA signalling via the OA receptor AmOAR in the AL is necessary for the acquisition and recall of odour–sucrose associations (Farooqui et al., 2003).

For odour learning, the cross-glomeruli pattern of activity representing a specific odour activates a specific pattern of PNs, which project to the LH and MB. In the MB, PNs synapse with Kenyon cells within the lip of the calyx. The primary role of the MB is odour identification (Galizia, 2014). Each identifiable odour activates a distinct pattern of Kenyon cells, with each odour sparsely coded as a cross fibre pattern across the very large number of Kenyon cells. Like the AL, odour coding within the MB is also plastic and serves to increase the valence and discriminability of rewarded odours (Szyszka et al., 2005, 2008). Odour pairing with sucrose results in a stabilisation and strengthening of the activity pattern within the Kenyon cells, whereas non-rewarded odour presentations lead to a weakened odour response in the Kenyon cell population (Szyszka et al., 2008). Kenyon cells output to extrinsic neurons which project to the LH, and within this connection matrix there is an additional layer of learning-related plasticity (Strube-Bloss et al., 2011). Reward learning increases the number of extrinsic neurons responding to an odour and causes the activity patterns arising from rewarded and non-rewarded odours to diverge (Strube-Bloss et al., 2011). One identified neuron, PE1, which is inhibitory of the downstream LH, showed reduced firing to odours that have been paired with sugar reward, but not to unrewarded odours (Menzel and Manz, 2005; Okada et al., 2007).

There is abundant evidence showing OA signalling in the MB is involved in reward learning in the bee (Hammer and Menzel, 1998), but here the precise role of VUMmx1 is not clear. Heisenberg (2003) argued convincingly that (in *Drosophila* at least) learning required cAMP signalling within Kenyon cells and that the necessary locus of synaptic activity for reward learning was the outputs of the Kenyon cells in the lobes of the MB where they synapse with the extrinsic neurons (Fig. 3). VUMmx1, however, arbourizes within the calyx of the MB rather than the lobes, and within the calyx arbours are rather sparse and may be parasynaptic with Kenyon cell fibres rather than making one-on-one connections (Sinakevitch et al., 2005). OA released by VUMmx1 could generate cAMP signals within Kenyon cells in the region of the calyx, but it seemed unfeasible for a cAMP signal to diffuse from there to the entire length of the axon of the Kenyon cells to affect pre-synapses at the axon terminals in the lobes (Heisenberg, 2003). Even so, OA released by VUMmx1 within the calyx could be imagined to influence the activity of Kenyon cells by altering weights of synaptic connection within the calyx. The calyx is a region of synaptic contact between the Kenyon cells and the GABA-immunoreactive inhibitory neurons of the protocerebral tract which provides inhibitory feedback from the MB lobes to the calyces (Bicker et al., 1985; Grünewald, 1999a,b). Further, in *Drosophila* Kenyon cells in the calyx are not exclusively post-synaptic (Christiansen et al., 2011), and Kenyon cell dendrites contain both pre- and post-synaptic terminals. This raises the possibility of feedback connections between Kenyon cells or with PNs. OA released into the calyx could modulate synaptic connections there to alter the strengths of feedback signals and thereby alter activity within the Kenyon cells (Strube-Bloss et al., 2011).

3.3.2 Octopamine in fruit flies

After the experiment demonstrating that the octopaminergic VUMmx1 neuron could replace positive appetitive stimuli in a learning experiment (Hammer, 1993), it was widely assumed that reward learning involved octopaminergic signalling in insects (Barron et al., 2010; Hammer and Menzel, 1998; Mizunami et al., 2009; Perry and Barron, 2013; Schwaerzel et al., 2003; Unoki et al., 2005, 2006). This was surprising because in mammals it was assumed that the main neurochemical signal for reward is DA (Schultz, 2002; Schultz et al., 1997). Early studies with *Drosophila* supported the emerging hypothesis that in insects, reward was signalled by OA (Honjo and Furukubo-Tokunaga, 2009; Schwaerzel et al., 2003). Schwaerzel et al. (2003) found that mutants lacking tyrosine-beta-hydroxylase (a synthesis enzyme necessary for production of octopamine) were unable to form appetitive memories of sucrose paired with an odour, indicating that OA was necessary for the formation of appetitive memories. Further, the behavioural phenotype of these mutants could be rescued by either feeding fruit flies OA or by expressing tyramine-beta-hydroxylase thermogenetically. Perception of sucrose was not affected in either of these mutants, suggesting that the role of OA is not directly related to sensing sucrose. Schroll et al. (2006) did a series of experiments in *Drosophila* larvae using ChR2 in combination with tyramine-beta-hydroxyase under GAL4 control to show that activating octopaminergic neurons alone is sufficient to generate positive valence for an odour stimulus. Using the shi^{ts1} method, Honjo and Furukubo-Tokunaga (2009) found that when OA release was inhibited in larvae, formation of appetitive memories was blocked. Taken together, these findings added strong support to the hypothesis that in insects OA neurons signal reward.

However, unlike in bees, the origin of the key octopaminergic signal has remained elusive. Several octopaminergic neurons project to the MB and LH as in honey bees (Busch et al., 2009; Sinakevitch and Strausfeld, 2006), and the MB in particular contains many octopaminergic receptors (Han et al., 1998). Interestingly, activating individual GNG OA neurons that innervate the MB tightly temporally paired with an odour (including OA-VUMa2) did not establish appetitive memories (Burke et al., 2012). Blocking these neurons during training did not inhibit memory formation either (Burke et al., 2012). It would seem that the ventral paired and unpaired neurons in *Drosophila* have a very different function than VUMmx1. The research of Schwaerzel et al. (2003) suggests the *Drosophila* reward system involves OA signalling to the MB, but the specific cell populations generating that signal are currently unknown. Individual VUM neurons are not sufficient in fruit flies for reward learning. Currently, we cannot say whether this means they are not involved in reward signalling, or whether the fruit fly octopaminergic system fails to function correctly when dissected to individual components.

New research with *Drosophila* suggests that the evaluation of odour valence occurs within the LH (Galizia, 2014). Galizia (2014) proposes the MB can be considered loci for odour identification whereas the LH is the locus for determining whether the odour is rewarding or punishing, and whether the animal is in a motivational state to respond (Galizia, 2014). This new perspective places the LH as key to a functional reward system. If this view is correct, it presents a difficulty in reconciling Hammer and Menzel's (1998) classic studies with microinjection of OA into honey bee brain neuropils. Given the well-established roles for OA in arousal, food seeking and food learning in bees, why did microinjection of OA into the LH not establish reward learning? The answer may be that the principal known input to the LH from the MB is inhibitory (Okada et al., 2007; Rybak and Menzel, 1998). The inhibitory output from the MB in response to an odour decreases after that odour has been paired with reward (Okada et al., 2007; Rybak and Menzel, 1998). This would release the LH from inhibition and enable motor pathways to be activated. In bees, depolarisation of VUMmx1 could potentially release OA at three points through the honey bee reward system causing coordinated changes in the circuit to alter the valence of a stimulus. A microinjection of OA to the LH alone, however, would not release the LH from inhibition by the MB and would not be sufficient for a learned PER to a stimulus to be expressed.

3.4 Dopamine in insect reward

3.4.1 Dopamine signals punishment in insects

The focus on OA as a neuromodulator of reward learning marked a deviation from mammalian research in which DA was clearly the principal neurochemical system of reward signalling, and OA is only present in trace amounts. However, the distinction between insect and mammals became even sharper after early fruit fly studies started finding that DA signalled punishment instead of reward. DA was first implicated in aversive learning when Tempel et al. (1984) demonstrated that mutants lacking dopa-decarboxylase (synthesis enzyme for dopamine) were impaired in an aversive learning paradigm utilising electric shocks. This finding was corroborated by experiments showing that aversive learning was impaired when dopaminergic neurons were silenced with shi^{ts1} (Schwaerzel et al., 2003). Inhibiting DA release with shi^{ts1} impaired formation of aversive memories in larvae as well (Honjo and Furukubo-Tokunaga, 2009). Further, activating dopaminergic neurons with ChR2 in larvae was sufficient to replace aversive stimuli (Schroll et al., 2006). Since these experiments, several studies have narrowed down the anatomical sites necessary for aversive memory formation (Aso et al., 2010, 2012; Claridge-Chang et al., 2009). Based on these studies, it became clear that in fruit flies, specific DA neuron populations signal punishment.

Pharmacological studies in honey bees have shown that DA antagonists inhibit aversive learning in honey bees (Vergoz et al., 2007; Wright et al., 2010). This was first shown with electric shock (Vergoz et al., 2007) and later with distasteful food (Wright et al., 2010). In these experiments, treatment with antagonists of OA receptors impaired reward learning but left punishment learning intact, whereas treatment with DA receptor antagonists impaired punishment learning but left reward learning intact. Consistent with this story, pharmacological studies in crickets (*Gryllus bimaculatus*) (Mizunami et al., 2009; Unoki et al., 2005, 2006) suggest that OA signals reward, while DA signals punishment (but see Agarwal et al., 2011). However, as we discussed in Section 3.1.3, we caution that the pharmacological agents applied are not entirely specific to either OA or DA receptors (Beggs et al., 2011). Careful studies applied multiple OA receptor and DA receptor antagonists in order to differentiate the function of the two systems (Vergoz et al., 2007), but even this refined approach would not entirely separate the functions of different individual neural circuits or receptors. This has only become possible thanks to new neurogenetic tools in *Drosophila* that have allowed exploration of the functions of specific dopaminergic signals. These new studies have shown that some dopaminergic neurons are involved in aversive learning, and others in reward learning.

3.4.2 Dopamine signals some aspects of reward

New evidence from fruit flies has shown that DA is involved in some aspects of reward learning. Two mutant genotypes defective in DopR (a D1-like DA receptor, also known as dDA1) in the MB and central complex demonstrated impaired abilities to form appetitive memories (Kim et al., 2007). This defect could be rescued by ectopic expression of DopR in the MB. The same mutants displayed similar behavioural defects in larvae (Selcho et al., 2009). Based on these results, it was clear that dopaminergic signals in the MB via the DopR receptor played a role in appetitive learning.

In *Drosophila*, it now seems that DA signals nutritive value as reward while OA signals sweetness as reward. Feeding fruit flies sucrose transiently activates neurons in the protocerebral anterior medial (PAM) cluster within the inferior neuropils (Liu et al., 2012). This is a bundle of approximately 100 dopaminergic neurons that primarily project to the medial lobes of the MB (Liu et al., 2012). Activating these neurons with TrpA1 could replace sucrose during appetitive conditioning (Liu et al., 2012), suggesting that these neurons are involved in signalling some aspect of food reward. Further, conditioning with TrpA1 activation of PAM was not impaired in fruit flies lacking tyramine-beta-hydroxylase (and therefore lacking OA), suggesting that these dopaminergic neurons are functionally downstream from octopaminergic reward signalling. Burke et al. (2012) further dissected the role of OA and DA in reward with a series of ingenious experiments with nutritive and non-nutritive sugars; both sweetness and caloric content are sufficient to induce appetitive memories in fruit flies (Burke and Waddell, 2011; Fujita and Tanimura, 2011). By training fruit flies with OA neurons silenced with shi^{ts1} using arabinose (which is only sweet) and sucrose (which is both sweet and nutritious), they found that the role of OA was to signal sweetness, not nutritive content. Based on this, they concluded that the role of OA in the fruit fly reward system is to signal sweetness, while caloric value is signalled by other means. The neurons in the PAM cluster respond to OA via the OA receptor OAMB (α-adrenergic-like receptor) and are thus functionally downstream from the OA signal of sweetness (Burke et al., 2012).

Further, octopaminergic input is necessary to inhibit aversive DA signals. A dopaminergic neuron in the protocerebral posterior lateral (PPL1) cluster known as MB-MP1 (mushroom body pedunculus-medial lobe arbourizing neuron 1) gives aversive input into the medial lobe of the MB (Aso et al., 2010). OA-dependent reward signalling depends on OCTβR2 (β-adrenergic-like OA receptor) to inhibit the activity of MB-MP1 (Burke et al., 2012). This demonstrates that octopaminergic input is necessary to two separate sets of dopaminergic neurons (PAM and PPL1) in order for appetitive conditioning to take place. In fruit flies, a picture has now emerged of DA as the principle signal of nutritive reward, and OA circuits activated by sweet taste and operating upstream of DA signals. The emphasis on DA in reward in fruit flies stands in contrast to the long-standing emphasis on OA in reward in bees. While the neurogenetic studies in *Drosophila* have identified specific DA signals for reward in insects, so far nothing is really known about the role of DA in reward signalling in honey bees. While a recent study has shown that there is at least the capacity for interaction between OA and DA systems in the honey bee MB (McQuillan et al., 2012), there is as of yet no evidence to show that it plays a role in reward signalling. This marks a presently unresolved divergence in our understanding of the reward systems of honey bees and flies.

3.4.3 Dopamine signals presence of amino acids

All experiments discussed thus far have used carbohydrate rewards for sweetness. It is, therefore, not clear how these findings might translate to other food items, such as proteins. Adult *Drosophila* carefully regulates both their protein and carbohydrate intake (Lee et al., 2008), so it would be strange if only carbohydrates, but not protein, were rewarding to these animals. Initial experiments have found that, at least in larvae, amino acid intake is regulated by three dopaminergic neurons (Bjordal et al., 2014). When *Drosophila* larvae were presented with a diet deficient in lysine and tryptophan, it caused the activation of three dopaminergic neurons that are both necessary and sufficient for the larvae to reject their current food source and start looking for a new food source, which could be considered reward-seeking behaviour (Bjordal et al., 2014). It remains to be seen if this is generalizable to adults, and if these neurons feed into the same pathways as carbohydrate rewards. Regardless, these findings are highly suggestive of DA as a signal for nutritive value in fruit flies regardless of nutrient group.

3.5 Modulation of reward seeking

In *Drosophila*, two homologues of the mammalian neuropeptide Y are transcribed in response to hunger and affect fruit fly behaviour. One of the functions of these neuropeptides is to motivate fruit flies to seek out appetitive stimuli when they are hungry. The first of these two homologues is neuropeptide F (dNPF), which is primarily expressed in the brain and midgut of larva and adult alike (Brown et al., 1999). Satiated fruit flies respond less to food-associated odours than hungry flies. This is due to tonic signalling from inhibitory dopaminergic neuron MB-MP1. However, when fruit flies are hungry, dNPF is expressed, and this signal prevents tonic DA release from MB-MP1 thus making the fruit fly more responsive to food-associated cues (Krashes et al., 2009). Thus, NPF regulates a dopaminergic circuit that prevents satiated fruit flies from seeking out food. Further, dNPF neurons are activated by sex pheromones (Gendron et al., 2014) and modulated neural and behavioural responses of male flies to copulation as a reward (Shohat-Ophir et al., 2012) indicating that in the context of sexual reward NPF modulates reward responses also. The second homologue of neuropeptide Y, the short neuropeptide F (sNPF), has been shown to be necessary for formation and recall of appetitive memory (Knapek et al., 2013). Knocking down sNPF in the MB Kenyon cells or in the neurons of α/β- and γ-lobes with RNAi caused significant impairment of the ability to form appetitive memories. However, knocking down the sNPF receptor in Kenyon cells did not affect olfactory memory, suggesting that the target is elsewhere. Sucrose preference was not affected in these mutants. sNPF also affects neuronal response directly in the AL (Root et al., 2011). Thus, it appears in fruit flies that homologues of neuropeptide Y function as a context signal of hunger state, rather than a direct signal of reward. It can, therefore, modulate the valence assigned to a food reward by a fruit fly given the present context the fruit fly operates in (Beshel and Zhong, 2013).

Honey bees have homologues of both dNPF and sNPF that are expressed in relation to food. However, they have not yet been shown to mediate food intake (Ament et al., 2011). At this stage, it would be reasonable to assume that they perform similar functions in the honey bee as in the flies.

3.6 Non-food rewards

While non-food rewards have received a great deal of attention in mammalian studies (O'Connell and Hofmann, 2011), very little is known of how non-food rewards are signalled in the insect brain. This may be an important oversight. When fruit flies are trained to an odour that signals the cessation of an electric shock, they attribute a positive valence to this odour (Tanimoto et al., 2004). Interestingly, neither dopaminergic nor octopaminergic neurons affect this form of learning (Yarali and Gerber, 2010). This suggests that olfactory relief learning probably occurs via another, as of yet, undiscovered pathway. A recent study demonstrated that OA might be involved in aversive behaviours in a modified conditioned place preference paradigm (Agarwal et al., 2011). In their experiment, honey bees did less well at learning to avoid an area paired with electric shock when treated with OA antagonists. While not directly examining reward, it suggests that the view that OA only signals rewards in insects may not be correct. In flies, responses to sexual reward are affected by NPF (Gendron et al., 2014; Shohat-Ophir et al., 2012), which suggests that examination of different forms of reward may reveal new modulatory pathways altering reward responses.

3.7 Summary of findings from honey bees and fruit flies

The reward systems of honey bees and fruit flies have both been very well studied. They are similar, but not the same. Evidence gathered to date suggests DA appears to be important for reward signalling in flies, but no evidence has yet linked DA to reward signalling in bees. Considering the conserved nature of dopaminergic reward signalling (Barron et al., 2010), is this a genuine biological difference? The findings in the fruit fly have been generated using fine-grained neurogenetic tools designed to alter the function of very specific circuits. The pharmacological tools employed in honey bees thus far to probe the OA and DA systems lack the anatomical and neurochemical specificity to adequately distinguish between the two (Beggs et al., 2011; Mustard et al., 2005). It is, therefore, possible that specific DA circuits could also play important roles in reward in the bee. As a second point of confusion, while the GNG neuron VUMmx1 appears to be sufficient for reward learning in bees, comparable (and likely homologous) neurons in the fruit fly do not appear to play an important role in learning of food reward (Burke et al., 2012). Potentially, this could reveal a significant difference in the circuitry of honey bee and fruit fly reward systems and poses some urgent unanswered questions. The structural similarity of the *Drosophila* octopaminergic OA-VUMa2 neuron with the honey bee VUMmx1 is so striking it seems perplexing that the two neurons do not have similar functions. We will, however, urge caution in the interpretation of the significance of this difference until the responses have been studied with similar behavioural assays. Currently, the VUMmx1 honey bee study measured successful learning based on the activation of the proboscis extension muscle (M17), while in the fruit flies this was measured as an approach by the whole animal. It could be that the octopaminergic GNG neurons are only sufficient to elicit proboscis extension as a response in reward learning, while additional reward signalling neurons are necessary for approach conditioning.

4. REWARD SYSTEMS AND SOCIAL EVOLUTION

The reward system is a major organiser and motivator of behaviour, and it is, therefore, not surprising that the reward system has been implicated in many different forms of social behaviour in both mammals and insects. Examples of social behaviours that are linked to reward processing include parental care, juvenile social play, adolescent social interactions, sexual activities, affiliative behaviours and aggressive territorial behaviours (O'Connell and Hofmann, 2011). Neurobiological analyses suggest that new forms of social behaviour can evolve by changes in the nature of stimuli or actions that activate a pre-existing reward system. This could either involve evolutionary changes such that the reward system becomes activated by social stimuli to promote new forms of social affiliation, or alternatively that existing reward-directed behaviour is modified to enable new forms of social behaviour. In mammals, there is evidence of the former, whereas in social insects currently there is more evidence for the latter.

For mammals, the most pertinent example comes from the neurobiological analyses of social pair bonding in voles (genus *Microtus*). Prairie voles *Microtus ochrogaster* are socially monogamous and form enduring pair bonds between males and females, whereas the closely related meadow and montane voles *Microtus pennsylvanicus* and *Microtus montanus* form weak pair bonds and consequently are more solitary and promiscuous (Williams et al., 1992). Differences in the strength of the pair bond have been related to differences between the species in the density and distribution of vasopressin (males) or oxytocin (female) receptors in specific brain regions. In prairie voles, males express vasopressin receptors at high densities in regions of the brain known to be involved in hedonic reactions and pleasure (Donaldson and Young, 2008). Similarly, females express oxytocin receptors at high density in the nucleus accumbens. Vasopressin and oxytocin are involved in social recognition in mammals (Ferguson et al., 2000, 2001; Young and Wang, 2004) and are released in response to the specific odour signatures of familiar individuals. Young and colleagues argue that the neuroanatomical basis of the strong pair bond in prairie voles is the strong anatomical coupling of the mesolimbic dopaminergic reward system and the peptidergic social recognition systems (Young and Wang, 2004; Young et al., 2005). Sex with a known partner would activate both systems, resulting in the formation of a conditioned preference for the partner. In the non-monogamous species, the reward system and social recognition systems are both active, but they are only weakly coupled; the inference being that the specific social recognition cues of a sexual partner are less rewarding in these species (Donaldson and Young, 2008; Young and Wang, 2004; Young et al., 2005).

This classic mammalian example highlights how new forms of social interaction can evolve by the linking of social recognition cues to the reward system to alter the valence of recognition cues and thereby promote social affiliation. There is currently no evidence that a similar mechanism has contributed to the evolution of the insect societies, but this may be because thus far we have not looked. Nestmate recognition systems are well developed across social insects (Breed, 1983; Breed et al., 1984, 1988, 1994; Page and Breed, 1987), and in one species of paper wasp (*Polistes fuscatus*) individual recognition has been demonstrated based on individually distinctive facial markings (Tibbetts, 2002). In many species of social insect, nestmates instinctively cluster and readily exchange food or groom each other demonstrating a very high level of innate social affiliation and cooperation (Seeley, 1989, 1995), but currently it is not clear if the insect reward system is activated by social affiliation or whether social recognition cues interact with the reward system for any insect species.

In honey bees, there is evidence that new forms of social behaviour have evolved by modification of existing reward-directed behaviour. Specialised behaviour supporting social foraging is regulated by octopaminergic systems that feature in solitary foraging roles in other insects. As we have discussed above in both honey bees and flies, OA circuits mediate the perception of sucrose reward, learning of sucrose and hunger (Hammer, 1993; Long and Murdock, 1983; Perisse et al., 2013; Scheiner et al., 2006, 2014). In honey bees, OA also modulates social foraging. OA treatment decreases the mean age at which honey bees begin foraging (Barron et al., 2002; Schulz et al., 2002) and changes how honey bees react to social stimuli within the hive to promote foraging (Barron and Robinson, 2005; Barron et al., 2002). Returning foragers sometimes dance as a social signal to communicate to nestmates the location and value of resources needed by the hive. The dance is heralded as the only known example of symbolic communication in insects and represents a remarkable new form of social communication that only occurs in the genus *Apis*. Dance behaviour of returning nectar and pollen foragers was modulated by OA in a manner that was dose dependent and could be blocked by the OA receptor antagonist mianserin (Barron et al., 2007a). Dance parameters reporting the value of located food sources (dance circuit vigour and dance circuit number) were by far the most sensitive to OA treatment, while positional information represented in dances was largely unchanged, suggesting that OA altered the communication of food value in dances rather specifically (Barron et al., 2007a). Similar effects were seen following cocaine treatment (Barron et al., 2009), which alters biogenic amine reuptake in the insect brain (Borue et al., 2010; Søvik, 2013; Søvik and Barron, 2013; Vickrey et al., 2009). Since the dance signals a workers' assessment of reward quality, this may be the closest measure we have in insects to assess an affective state: the subjective assessment of the quality of a reward.

When honey bees forage for the colony, they do not consume the resources they have gathered; rather, they immediately surrender them to their nest mates in the hive or deposit the floral resources in cells in the colony. This is especially true for pollen foragers. Forager honey bees do not consume pollen, and the gathered pollen is collected in corbiculae on the hind legs before being deposited in storage cells in the colony (Seeley, 1995; Winston, 1987). Pollen foragers even leave the colony satiated with a crop full of honey from the hive to fuel their pollen-collecting trip (A.B.

Barron, unpublished). Dance is also purely of social benefit: the signal serves to recruit nestmates to any profitable resource discovered by a forager. Both of these forms of social behaviour appear linked to the honey bee reward system.

How these new forms of social foraging behaviour may have evolved from pre-existing foraging behaviour of the imagined solitary or sub-social ancestor of the honey bee is a matter for speculation. In his considerations of animal motivation and drive, Tinbergen (1951) imagined that reward-seeking behaviour was generated by internal motivational energy that built up in an animal over time and directed behaviour towards the attainment of specific resources or circumstances. This internal energy was only released by the execution of a specific consummatory act that marked the successful attainment of the sought-for outcome (Tinbergen, 1951). One interpretation of the evolution of social foraging could be that the pre-existing reward-seeking system has been exapted by the evolutionary process to respond to the collection of food as a consummatory act more than the consumption of food. Tinbergen's terminology has fallen out of usage as neurobiological analyses of reward seeking have developed; however, it has become clear from current comparative neurobiological analyses that different elements of the reward system are activated by different stimuli, which to use Tinbergen's terminology could be related to different consummatory acts. As we have discussed, new findings from *Drosophila* have shown that sweet taste and nutrition activate different elements of the reward system: sweet taste is signalled by specific (but presently unidentified) OA circuits, and the nutritive value of reward is signalled by identified dopaminergic circuits (Burke et al., 2012). We can speculate that increasing the valence of sweet taste as a reward might yield an animal that is more motivated to gather food than to consume food. A full phylogenetically informed comparative analyses of the relative roles of octopaminergic and dopaminergic elements of the reward system in the rewarding nature of sweet taste, nutrition and social foraging across social and solitary species may provide a way to test this hypothesis.

The examples available across insects and mammals indicate that evolutionary modification of reward-directed behaviour has been important in the evolution of new forms of social behaviour (O'Connell and Hofmann, 2011). The adaptations have involved changes in stimuli or behaviour that are activational of the reward system more than structural changes in the circuitry of the reward system itself. In voles, for example, the essential mesolimbic dopaminergic reward system is essentially the same in non-monogamous and monogamous species: the key difference is in the extent to which the reward system is activated by social recognition cues (Donaldson and Young, 2008; Young and Wang, 2004). As has been demonstrated by the comparison of honey bees and fruit flies, the essential reward system may be quite similar across divergent insect lineages, but the basis of behavioural differences between lineages may lie in the behavioural modules that are linked to the reward system. The extensive changes in gene regulation that have been linked to the evolution of eusociality (Simola et al., 2013) may have been part of the mechanism by which an ancestral reward system was "remodelled" in the process of social evolution.

5. CONCLUDING REMARKS

The honey bee and *Drosophila* are the two canonical model systems for insect neuroscience research, but in many ways these two insects could not be more different. They occur in divergent and derived lineages, and it is important to consider to what extent the findings from one system can generalise to the other. The reward systems of the two species provide an informative point of comparison. Reward is involved in many different aspects of behaviour, but despite marked differences in behaviour, ecology and evolutionary history between honey bees and fruit flies the reward systems are quite similar. Both involve networks of aminergic circuits to process different elements of reward. The honey bee literature has emphasised the role of OA in reward, whereas the fruit fly literature is increasingly emphasising specific dopaminergic circuits interacting with octopaminergic and peptidergic circuits in reward processing. At the time of writing, this apparent difference is more likely due to the different resolution of analysis that has been possible in the two systems rather than a difference in biology. Neurogenetic tools available for *Drosophila* have allowed extremely focused and precise investigation of specific circuits. By comparison, the pharmacological methods that have predominated the honey bee literature are coarser and may struggle to differentiate the roles of OA and DA and/or specific circuits. We propose an analysis of the possible role of DA in reward in honey bees as a priority for research.

There remain some significant points of contrast between the fruit fly and honey bee literature: most notably the findings that direct stimulation of the octopaminergic neuron VUMmx1 is sufficient for reward learning in bees, but direct stimulation of the extremely similar (and likely homologous) neuron in flies, OA-VUMa2 did not have the same behavioural outcome.

Whether this difference is due to a difference in the learning assays used across the two systems or a genuine difference in the weighting of OA signals between the two species is currently unclear.

A consideration of the role of the reward system in the evolution of new forms of social behaviour would suggest that this can be achieved by evolutionary modification of the stimuli or actions that activate the reward system, rather than by direct modification of the reward system itself. Consequently, we might expect the reward system to remain stable and reasonably conserved across insect lineages, and to be a useful point of reference for comparative neuroscience research.

ACKNOWLEDGEMENTS

We would like to thank Angelique Paulk for contributing images for Fig. 1 and Scott Waddell for helpful discussions.

REFERENCES

Agarwal, M., Giannoni Guzmán, M., Morales-Matos, C., Del Valle Díaz, R.A., Abramson, C.I., Giray, T., 2011. Dopamine and octopamine influence avoidance learning of honey bees in a place preference assay. PLoS One 6, e25371.

Ament, S.A., Velarde, R.A., Kolodkin, M.H., Moyse, D., Robinson, G.E., 2011. Neuropeptide Y-like signalling and nutritionally mediated gene expression and behaviour in the honey bee. Insect Mol. Biol. 20, 335–345.

Aso, Y., Siwanowicz, I., Bräcker, L., Ito, K., Kitamoto, T., Tanimoto, H., 2010. Specific dopaminergic neurons for the formation of labile aversive memory. Curr. Biol. 20, 1445–1451.

Aso, Y., Herb, A., Ogueta, M., Siwanowicz, I., Templier, T., Friedrich, A.B., Ito, K., Scholz, H., Tanimoto, H., 2012. Three dopamine pathways induce aversive odor memories with different stability. PLoS Genet. 8, e1002768.

Avarguès-Weber, A., Portelli, G., Benard, J., Dyer, A., Giurfa, M., 2010. Configural processing enables discrimination and categorization of face-like stimuli in honeybees. J. Exp. Biol. 213, 593–601.

Bailey, C., 1926. Epicurus: The Extant Remains. Clarendon Press, Oxford, UK.

Barron, A., Corbet, S., 1999. Preimaginal conditioning in *Drosophila* revisited. Anim. Behav. 58, 621–628.

Barron, A.B., Robinson, G.E., 2005. Selective modulation of task performance by octopamine in honey bee (*Apis mellifera*) division of labour. J. Comp. Physiol. A Neuroethol. Sens. Neural Behav. Physiol. 191, 659–668.

Barron, A.B., Schulz, D.J., Robinson, G.E., 2002. Octopamine modulates responsiveness to foraging-related stimuli in honey bees (*Apis mellifera*). J. Comp. Physiol. A Neuroethol. Sens. Neural Behav. Physiol. 188, 603–610.

Barron, A.B., Maleszka, R., Vander Meer, R.K., Robinson, G.E., 2007a. Octopamine modulates honey bee dance behavior. Proc. Natl. Acad. Sci. U.S.A. 104, 1703–1707.

Barron, A.B., Vander Meer, R.K., Maleszka, J., Robinson, G.E., Maleszka, R., 2007b. Comparing injection, feeding and topical application methods for treatment of honeybees with octopamine. J. Insect Physiol. 53, 187–194.

Barron, A.B., Maleszka, R., Helliwell, P.G., Robinson, G.E., 2009. Effects of cocaine on honey bee dance behaviour. J. Exp. Biol. 212, 163–168.

Barron, A.B., Søvik, E., Cornish, J.L., 2010. The roles of dopamine and related compounds in reward-seeking behavior across animal phyla. Front. Behav. Neurosci. 4, 163.
Bath, D.E., Stowers, J.R., Hörmann, D., Poehlmann, A., Dickson, B.J., Straw, A.D., 2014. FlyMAD: rapid thermogenetic control of neuronal activity in freely walking *Drosophila*. Nat. Methods 11, 756–762.
Beggs, K.T., Tyndall, J.D.A., Mercer, A.R., 2011. Honey bee dopamine and octopamine receptors linked to intracellular calcium signaling have a close phylogenetic and pharmacological relationship. PLoS One 6, e26809.
Ben-Shahar, Y., 2014. A piggyBac route to transgenic honeybees. Proc. Natl. Acad. Sci. U.S.A. 111, 8708–8709.
Berridge, K.C., Kringelbach, M.L., 2008. Affective neuroscience of pleasure: reward in humans and animals. Psychopharmacology (Berl) 199, 457–480.
Berridge, K.C., Valenstein, E.S., 1991. What psychological process mediates feeding evoked by electrical stimulation of the lateral hypothalamus? Behav. Neurosci. 105, 3–14.
Berridge, K.C., Robinson, T.E., Aldridge, J.W., 2009. Dissecting components of reward: "liking", "wanting", and learning. Curr. Opin. Pharmacol. 9, 65–73.
Beshel, J., Zhong, Y., 2013. Graded encoding of food odor value in the *Drosophila* brain. J. Neurosci. 33, 15693–15704.
Bicker, G., Schäfer, S., Kingan, T.G., 1985. Mushroom body feedback interneurones in the honeybee show GABA-like immunoreactivity. Brain Res. 360, 394–397.
Bitterman, M.E., Menzel, R., Fietz, A., Schäfer, S., 1983. Classical conditioning of proboscis extension in honeybees (*Apis mellifera*). J. Comp. Psychol. 97, 107–119.
Bjordal, M., Arquier, N., Kniazeff, J., Pin, J.P., Léopold, P., 2014. Sensing of amino acids in a dopaminergic circuitry promotes rejection of an incomplete diet in *Drosophila*. Cell 156, 510–521.
Borue, X., Condron, B., Venton, B.J., 2010. Both synthesis and reuptake are critical for replenishing the releasable serotonin pool in *Drosophila*. J. Neurochem. 113, 188–199.
Brand, A.H., Perrimon, N., 1993. Targeted gene expression as a means of altering cell fates and generating dominant phenotypes. Development 118, 401–415.
Bräunig, P., 1991. Suboesophageal DUM neurons innervate the principal neuropiles of the locust brain. Philos. Trans. R. Soc. B Biol. Sci. 332, 221–240.
Bräunig, P., Burrows, M., 2004. Projection patterns of posterior dorsal unpaired median neurons of the locust subesophageal ganglion. J. Comp. Neurol. 478, 164–175.
Breed, M., 1983. Nestmate recognition in honey bees. Anim. Behav. 31, 86–91.
Breed, M.D., Velthuis, H.H.W., Robinson, G.E., 1984. Do worker honey bees discriminate among unrelated and related larval phenotypes. Ann. Entomol. Soc. Am. 77, 737–739.
Breed, M.D., Stiller, T.M., Moor, M.J., 1988. The ontogeny of kin discrimination cues in the honey bee, Apis mellifera. Behav. Genet. 18, 439–448.
Breed, M., Welch, C., Cruz, R., 1994. Kin discrimination within honey bee (*Apis mellifera*) colonies: an analysis of the evidence. Behav. Processes 33, 25–39.
Brown, M.R., Crim, J.W., Arata, R.C., Cai, H.N., Chun, C., Shen, P., 1999. Identification of a *Drosophila* brain-gut peptide related to the neuropeptide Y family. Peptides 20, 1035–1042.
Burke, C.J., Waddell, S., 2011. Remembering nutrient quality of sugar in *Drosophila*. Curr. Biol. 21, 746–750.
Burke, C.J., Huetteroth, W., Owald, D., Perisse, E., Krashes, M.J., Das, G., Gohl, D., Silies, M., Certel, S., Waddell, S., 2012. Layered reward signalling through octopamine and dopamine in *Drosophila*. Nature 492, 433–437.
Busch, S., Tanimoto, H., 2010. Cellular configuration of single octopamine neurons in *Drosophila*. J. Comp. Neurol. 518, 2355–2364.

Busch, S., Selcho, M., Ito, K., Tanimoto, H., 2009. A map of octopaminergic neurons in the *Drosophila* brain. J. Comp. Neurol. 513, 643–667.
Carr, D.B., Sesack, S.R., 2000. GABA-containing neurons in the rat ventral tegmental area project to the prefrontal cortex. Synapse 38, 114–123.
Chapman, R.F., 2013. The Insects: Structure and Function. Cambridge University Press, Cambridge, UK.
Christiansen, F., Zube, C., Andlauer, T.F.M., Wichmann, C., Fouquet, W., Owald, D., Mertel, S., Leiss, F., Tavosanis, G., Luna, A.J.F., et al., 2011. Presynapses in Kenyon cell dendrites in the mushroom body calyx of *Drosophila*. J. Neurosci. 31, 9696–9707.
Claridge-Chang, A., Roorda, R.D., Vrontou, E., Sjulson, L., Li, H., Hirsh, J., Miesenböck, G., 2009. Writing memories with light-addressable reinforcement circuitry. Cell 139, 405–415.
Davis, R.L., 2005. Olfactory memory formation in *Drosophila*: from molecular to systems neuroscience. Annu. Rev. Neurosci. 28, 275–304.
Denker, M., Finke, R., Schaupp, F., Grün, S., Menzel, R., 2010. Neural correlates of odor learning in the honeybee antennal lobe. Eur. J. Neurosci. 31, 119–133.
Dietzl, G., Chen, D., Schnorrer, F., Su, K.-C., Barinova, Y., Fellner, M., Gasser, B., Kinsey, K., Oppel, S., Scheiblauer, S., et al., 2007. A genome-wide transgenic RNAi library for conditional gene inactivation in *Drosophila*. Nature 448, 151–156.
Donaldson, Z.R., Young, L.J., 2008. Oxytocin, vasopressin, and the neurogenetics of sociality. Science 322, 900–904.
Dreher, J.C., Tremblay, L., 2009. Handbook of Reward and Decision Making. Academic Press, New York.
Ehmer, B., Gronenberg, W., 2002. Segregation of visual input to the mushroom bodies in the honeybee (*Apis mellifera*). J. Comp. Neurol. 451, 362–373.
Erber, J., Kloppenburg, P., Scheidler, A., 1993. Neuromodulation by serotonin and octopamine in the honeybee: behaviour, neuroanatomy and electrophysiology. Experientia 49, 1073–1083.
Faber, T., Joerges, J., Menzel, R., 1999. Associative learning modifies neural representations of odors in the insect brain. Nat. Neurosci. 2, 74–78.
Farooqui, T., Robinson, K., Vaessin, H., Smith, B.H., 2003. Modulation of early olfactory processing by an octopaminergic reinforcement pathway in the honeybee. J. Neurosci. 23, 5370–5380.
Farooqui, T., Vaessin, H., Smith, B.H., 2004. Octopamine receptors in the honeybee (*Apis mellifera*) brain and their disruption by RNA-mediated interference. J. Insect Physiol. 50, 701–713.
Ferguson, J.N., Young, L.J., Hearn, E.F., Matzuk, M.M., Insel, T.R., Winslow, J.T., 2000. Social amnesia in mice lacking the oxytocin gene. Nat. Genet. 25, 284–288.
Ferguson, J.N., Aldag, J.M., Insel, T.R., Young, L.J., 2001. Oxytocin in the medial amygdala is essential for social recognition in the mouse. J. Neurosci. 21, 8278–8285.
Fernandez, P.C., Locatelli, F.F., Person-Rennell, N., Deleo, G., Smith, B.H., 2009. Associative conditioning tunes transient dynamics of early olfactory processing. J. Neurosci. 29, 10191–10202.
Fujita, M., Tanimura, T., 2011. Drosophila evaluates and learns the nutritional value of sugars. Curr. Biol. 21, 751–755.
Galizia, C.G., 2014. Olfactory coding in the insect brain: data and conjectures. Eur. J. Neurosci. 39, 1784–1795.
Galizia, C., Menzel, R., 2001. The role of glomeruli in the neural representation of odours: results from optical recording studies. J. Insect Physiol. 47, 115–130.
Gardner, E.L., 2005. Endocannabinoid signaling system and brain reward: emphasis on dopamine. Pharmacol. Biochem. Behav. 81, 263–284.

Gendron, C.M., Kuo, T.-H., Harvanek, Z.M., Chung, B.Y., Yew, J.Y., Dierick, H.A., Pletcher, S.D., 2014. *Drosophila* life span and physiology are modulated by sexual perception and reward. Science 343, 544–548.

Giurfa, M., 2013. Cognition with few neurons: higher-order learning in insects. Trends Neurosci. 36, 285–294.

Grimaldi, D., Engel, M.S., 2005. Evolution of the Insects. Cambridge University Press, Cambridge, UK.

Grünewald, B., 1999a. Physiological properties and response modulations of mushroom body feedback neurons during olfactory learning in the honeybee, *Apis mellifera*. J. Comp. Physiol. A 185, 565–576.

Grünewald, B., 1999b. Morphology of feedback neurons in the mushroom body of the honeybee, *Apis mellifera*. J. Comp. Neurol. 404, 114–126.

Hamada, F.N., Rosenzweig, M., Kang, K., Pulver, S.R., Ghezzi, A., Jegla, T.J., Garrity, P.a., 2008. An internal thermal sensor controlling temperature preference in *Drosophila*. Nature 454, 217–220.

Hammer, M., 1993. An identified neuron mediates the unconditioned stimulus in associative olfactory learning in honeybees. Nature 366, 59–63.

Hammer, M., Menzel, R., 1995. Learning and memory in the honeybee. J. Neurosci. 15, 1617–1630.

Hammer, M., Menzel, R., 1998. Multiple sites of associative odor learning as revealed by local brain microinjections of octopamine in honeybees. Learn. Mem. 5, 146–156.

Han, K.A., Millar, N.S., Davis, R.L., 1998. A novel octopamine receptor with preferential expression in *Drosophila* mushroom bodies. J. Neurosci. 18, 3650–3658.

Heisenberg, M., 2003. Mushroom body memoir: from maps to models. Nat. Rev. Neurosci. 4, 266–275.

Heisenberg, M., Borst, A., Wagner, S., Byers, D., 1985. *Drosophila* mushroom body mutants are deficient in olfactory learning. J. Neurogenet. 2, 1–30.

Honjo, K., Furukubo-Tokunaga, K., 2009. Distinctive neuronal networks and biochemical pathways for appetitive and aversive memory in *Drosophila* larvae. J. Neurosci. 29, 852–862.

Hu, A., Zhang, W., Wang, Z., 2010. Functional feedback from mushroom bodies to antennal lobes in the *Drosophila* olfactory pathway. Proc. Natl. Acad. Sci. U.S.A. 107, 10262–10267.

Ikemoto, S., 2010. Brain reward circuitry beyond the mesolimbic dopamine system: a neurobiological theory. Neurosci. Biobehav. Rev. 35, 129–150.

Kenyon, F., 1896. The meaning and structure of the so-called "mushroom bodies" of the hexapod brain. Am. Nat. 30, 643–650.

Kim, Y.-C., Lee, H.-G., Han, K.-A., 2007. D1 dopamine receptor dDA1 is required in the mushroom body neurons for aversive and appetitive learning in *Drosophila*. J. Neurosci. 27, 7640–7647.

Kitamoto, T., 2001. Conditional modification of behavior in *Drosophila* by targeted expression of a temperature-sensitive *shibire* allele in defined neurons. J. Neurobiol. 47, 81–92.

Knaden, M., Strutz, A., Ahsan, J., Sachse, S., Hansson, B.S., 2012. Spatial representation of odorant valence in an insect brain. Cell Rep. 1, 392–399.

Knapek, S., Kahsai, L., Winther, A.M.E., Tanimoto, H., Nässel, D.R., 2013. Short neuropeptide F acts as a functional neuromodulator for olfactory memory in Kenyon cells of *Drosophila* mushroom bodies. J. Neurosci. 33, 5340–5345.

Kranz, G.S., Kasper, S., Lanzenberger, R., 2010. Reward and the serotonergic system. Neuroscience 166, 1023–1035.

Krashes, M.J., DasGupta, S., Vreede, A., White, B., Armstrong, J.D., Waddell, S., 2009. A neural circuit mechanism integrating motivational state with memory expression in *Drosophila*. Cell 139, 416–427.

Kreissl, S., Eichmüller, S., Bicker, G., Rapus, J., Eckert, M., Eichmüller, S., 1994. Octopamine-like immunoreactivity in the brain and subesophageal ganglion of the honeybee. J. Comp. Neurol. 348, 583–595.

Kuwabara, M., 1957. Bildung des bedingten Reflexes von Pavlovs Typus bei der Honigbiene, *Apis mellifica*. J. Fac. Sci. Hokaido Univ. 8, 458–464.

Lee, K.P., Simpson, S.J., Clissold, F.J., Brooks, R., Ballard, J.W.O., Taylor, P.W., Soran, N., Raubenheimer, D., 2008. Lifespan and reproduction in *Drosophila*: new insights from nutritional geometry. Proc. Natl. Acad. Sci. U.S.A. 105, 2498–2503.

Leknes, S., Tracey, I., 2008. A common neurobiology for pain and pleasure. Nat. Rev. Neurosci. 9, 314–320.

Liu, C., Plaçais, P.-Y., Yamagata, N., Pfeiffer, B.D., Aso, Y., Friedrich, A.B., Siwanowicz, I., Rubin, G.M., Preat, T., Tanimoto, H., 2012. A subset of dopamine neurons signals reward for odour memory in *Drosophila*. Nature 488, 512–516.

Locatelli, F.F., Fernandez, P.C., Villareal, F., Muezzinoglu, K., Huerta, R., Galizia, C.G., Smith, B.H., 2013. Nonassociative plasticity alters competitive interactions among mixture components in early olfactory processing. Eur. J. Neurosci. 37, 63–79.

Loeb, J., 1918. Forced Movements, Tropisms, and Animal Conduct. Dover, New York.

Long, T.F., Murdock, L.L., 1983. Stimulation of blowfly feeding behavior by octopaminergic drugs. Proc. Natl. Acad. Sci. U.S.A. 80, 4159–4163.

Mahler, S.V., Smith, K.S., Berridge, K.C., 2007. Endocannabinoid hedonic hotspot for sensory pleasure: anandamide in nucleus accumbens shell enhances "liking" of a sweet reward. Neuropsychopharmacology 32, 2267–2278.

McQuillan, H.J., Nakagawa, S., Mercer, A.R., 2012. Mushroom bodies of the honeybee brain show cell population-specific plasticity in expression of amine-receptor genes. Learn. Mem. 19, 151–158.

Menzel, R., 2001. Searching for the memory trace in a mini-brain, the honeybee. Learn. Mem. 8, 53–62.

Menzel, R., Manz, G., 2005. Neural plasticity of mushroom body-extrinsic neurons in the honeybee brain. J. Exp. Biol. 208, 4317–4332.

Menzel, R., Erber, J., Masuhr, T., 1974. Learning and memory in the honeybee. In: Browne, L.B. (Ed.), Experimental Analysis of Insect Behaviour. Springer-Verlag, Berlin, pp. 195–217.

Menzel, R., Heyne, A., Kinzel, C., Gerber, B., Fiala, A., 1999. Pharmacological dissociation between the reinforcing, sensitizing, and response-releasing functions of reward in honeybee classical conditioning. Behav. Neurosci. 113, 744–754.

Mercer, A., Menzel, R., 1982. The effects of biogenic amines on conditioned and unconditioned responses to olfactory stimuli in the honey bee *Apis mellifera*. J. Comp. Physiol. A Neuroethol. Sens. Neural Behav. Physiol. 145, 363–368.

Miyamoto, T., Slone, J., Song, X., Amrein, H., 2012. A fructose receptor functions as a nutrient sensor in the Drosophila brain. Cell 151, 1113–1125.

Mizunami, M., Unoki, S., Mori, Y., Hirashima, D., Hatano, A., Matsumoto, Y., 2009. Roles of octopaminergic and dopaminergic neurons in appetitive and aversive memory recall in an insect. BMC Biol. 7, 46.

Mobbs, P., 1982. The brain of the honey bee *Apis mellifera*. I. The connections and spatial organization of the mushroom bodies. Philos. Trans. R. Soc. B Biol. Sci. 298, 309–354.

Mustard, J.A., Beggs, K.T., Mercer, A.R., 2005. Molecular biology of the invertebrate dopamine receptors. Arch. Insect Biochem. Physiol. 59, 103–117.

Nagel, G., Szellas, T., Huhn, W., Kateriya, S., Adeishvili, N., Berthold, P., Ollig, D., Hegemann, P., Bamberg, E., 2003. Channelrhodopsin-2, a directly light-gated cation-selective membrane channel. Proc. Natl. Acad. Sci. U.S.A. 100, 13940–13945.

O'Connell, L.A., Hofmann, H.A., 2011. The vertebrate mesolimbic reward system and social behavior network: a comparative synthesis. J. Comp. Neurol. 519, 3599–3639.

Okada, R., Rybak, J., Manz, G., Menzel, R., 2007. Learning-related plasticity in PE1 and other mushroom body-extrinsic neurons in the honeybee brain. J. Neurosci. 27, 11736–11747.
Omelchenko, N., Sesack, S.R., 2009. Ultrastructural analysis of local collaterals of rat ventral tegmental area neurons: GABA phenotype and synapses onto dopamine and GABA cells. Synapse 63, 895–906.
Page Jr., R.E., Breed, M., 1987. Kin recognition in social bees. Trends Ecol. Evol. 2, 272–275.
Parnas, M., Lin, A.C., Huetteroth, W., Miesenböck, G., 2013. Odor discrimination in *Drosophila*: from neural population codes to behavior. Neuron 79, 932–944.
Paulk, A.C., Stacey, J.A., Pearson, T.W.J., Taylor, G.J., Moore, R.J.D., Srinivasan, M.V., van Swinderen, B., 2014. Selective attention in the honeybee optic lobes precedes behavioral choices. Proc. Natl. Acad. Sci. U.S.A. 111, 5006–5011.
Pavlov, I.P., 1927. Conditioned Reflexes. Oxford University Press, London, UK.
Peciña, S., Berridge, K.C., 2005. Hedonic hot spot in nucleus accumbens shell: where do mu-opioids cause increased hedonic impact of sweetness? J. Neurosci. 25, 11777–11786.
Perisse, E., Burke, C., Huetteroth, W., Waddell, S., 2013. Shocking revelations and saccharin sweetness in the study of *Drosophila* olfactory memory. Curr. Biol. 23, R752–R763.
Perry, C.J., Barron, A.B., 2013. Neural mechanisms of reward in insects. Annu. Rev. Entomol. 58, 543–562.
Quinn, W., Harris, W., Benzer, S., 1974. Conditioned behavior in *Drosophila melanogaster*. Proc. Natl. Acad. Sci. U.S.A. 71, 708–712.
Rath, L., Giovanni Galizia, C., Szyszka, P., 2011. Multiple memory traces after associative learning in the honey bee antennal lobe. Eur. J. Neurosci. 34, 352–360.
Rein, J., Mustard, J.A., Strauch, M., Smith, B.H., Galizia, C.G., 2013. Octopamine modulates activity of neural networks in the honey bee antennal lobe. J. Comp. Physiol. A Neuroethol. Sens. Neural Behav. Physiol. 199, 947–962.
Root, C.M., Ko, K.I., Jafari, A., Wang, J.W., 2011. Presynaptic facilitation by neuropeptide signaling mediates odor-driven food search. Cell 145, 133–144.
Rybak, J., Menzel, R., 1998. Integrative properties of the Pe1 neuron, a unique mushroom body output neuron. Learn. Mem. 5, 133–145.
Scheiner, R., Plückhahn, S., Oney, B., Blenau, W., Erber, J., 2002. Behavioural pharmacology of octopamine, tyramine and dopamine in honey bees. Behav. Brain Res. 136, 545–553.
Scheiner, R., Baumann, A., Blenau, W., 2006. Aminergic control and modulation of honeybee behaviour. Curr. Neuropharmacol. 4, 259–276.
Scheiner, R., Steinbach, A., Claßen, G., Strudthoff, N., Scholz, H., 2014. Octopamine indirectly affects proboscis extension response habituation in *Drosophila melanogaster* by controlling sucrose responsiveness. J. Insect Physiol. 69C, 107–117.
Schroll, C., Riemensperger, T., Bucher, D., Ehmer, J., Vo, T., Erbguth, K., Gerber, B., Hendel, T., Nagel, G., Buchner, E., et al., 2006. Report light-induced activation of distinct modulatory neurons triggers appetitive or aversive learning in *Drosophila* larvae. Curr. Biol. 16, 1741–1747.
Schröter, U., Malun, D., Menzel, R., 2007. Innervation pattern of suboesophageal ventral unpaired median neurones in the honeybee brain. Cell Tissue Res. 327, 617–667.
Schulte, C., Theilenberg, E., Müller-Borg, M., Gempe, T., Beye, M., 2014. Highly efficient integration and expression of piggyBac-derived cassettes in the honeybee (*Apis mellifera*). Proc. Natl. Acad. Sci. U.S.A. 111, 9003–9008.
Schultz, W., 1997. Dopamine neurons and their role in reward mechanisms. Curr. Opin. Neurobiol. 7, 191–197.
Schultz, W., 2002. Getting formal with dopamine and reward. Neuron 36, 241–263.
Schultz, W., 2007. Behavioral dopamine signals. Trends Neurosci. 30, 203–210.

Schultz, W., 2010. Dopamine signals for reward value and risk: basic and recent data. Behav. Brain Funct. 6, 24.
Schultz, W., Dayan, P., Montague, P.R., 1997. A neural substrate of prediction and reward. Science 275, 1593–1599.
Schulz, D.J., Barron, A.B., Robinson, G.E., 2002. A role for octopamine in honey bee division of labor. Brain Behav. Evol. 60, 350–359.
Schwaerzel, M., Monastirioti, M., Scholz, H., Friggi-Grelin, F., Birman, S., Heisenberg, M., 2003. Dopamine and octopamine differentiate between aversive and appetitive olfactory memories in *Drosophila*. J. Neurosci. 23, 10495–10502.
Seeley, T.D., 1989. The bee colony as a superorganism. Am. Sci. 77, 546–553.
Seeley, T.D., 1995. The Wisdom of the Hive. Harvard University Press, Cambridge, MA.
Selcho, M., Pauls, D., Han, K.-A., Stocker, R.F., Thum, A.S., 2009. The role of dopamine in Drosophila larval classical olfactory conditioning. PLoS One 4, e5897.
Shimada, T., Kato, K., Kamikouchi, A., Ito, K., 2005. Analysis of the distribution of the brain cells of the fruit fly by an automatic cell counting algorithm. Phys. A Stat. Mech. Appl. 350, 144–149.
Shohat-Ophir, G., Kaun, K.R., Azanchi, R., Heberlein, U., 2012. Sexual deprivation increases ethanol intake in *Drosophila*. Science 335, 1351–1355.
Simola, D.F., Wissler, L., Donahue, G., Waterhouse, R.M., Helmkampf, M., Roux, J., Nygaard, S., Glastad, K.M., Hagen, D.E., Viljakainen, L., et al., 2013. Social insect genomes exhibit dramatic evolution in gene composition and regulation while preserving regulatory features linked to sociality. Genome Res. 23, 1235–1247.
Sinakevitch, I., Strausfeld, N.J., 2006. Comparison of octopamine-like immunoreactivity in the brains of the fruit fly and blow fly. J. Comp. Neurol. 494, 460–475.
Sinakevitch, I., Niwa, M., Strausfeld, N.J., 2005. Octopamine-like immunoreactivity in the honey bee and cockroach: comparable organization in the brain and subesophageal ganglion. J. Comp. Neurol. 488, 233–254.
Smith, K.S., Berridge, K.C., 2005. The ventral pallidum and hedonic reward: neurochemical maps of sucrose "liking" and food intake. J. Neurosci. 25, 8637–8649.
Smith, K.S., Mahler, S.V., Peciña, S., Berridge, K.C., 2007. Hedonic hotspots: generating sensory pleasure in the brain. In: Kringelbach, M.L., Berridge, K.C. (Eds.), Pleasures of the Brain. Oxford University Press, New York, pp. 27–49.
Smith, K.S., Tindell, A.J., Aldridge, J.W., Berridge, K.C., 2009. Ventral pallidum roles in reward and motivation. Behav. Brain Res. 196, 155–167.
Søvik, E., 2013. Reward Processing and Responses to Drugs of Abuse in the Honey Bee, *Apis mellifera*. Macquarie University, Australia.
Søvik, E., Barron, A.B., 2013. Invertebrate models in addiction research. Brain Behav. Evol. 82, 153–165.
Stocker, R.F., 1994. The organization of the chemosensory system in *Drosophila melanogaster*: a review. Cell Tissue Res. 275, 3–26.
Strausfeld, N.J., 2002. Organization of the honey bee mushroom body: representation of the calyx within the vertical and gamma lobes. J. Comp. Neurol. 450, 4–33.
Strube-Bloss, M.F., Nawrot, M.P., Menzel, R., 2011. Mushroom body output neurons encode odor-reward associations. J. Neurosci. 31, 3129–3140.
Szyszka, P., Ditzen, M., Galkin, A., Galizia, C.G., Menzel, R., 2005. Sparsening and temporal sharpening of olfactory representations in the honeybee mushroom bodies. J. Neurophysiol. 94, 3303–3313.
Szyszka, P., Galkin, A., Menzel, R., 2008. Associative and non-associative plasticity in Kenyon cells of the honeybee mushroom body. Front. Syst. Neurosci. 2, 3.
Tanimoto, H., Heisenberg, M., Gerber, B., 2004. Event timing turns punishment to reward. Nature 430, 983.

Technau, G., Heisenberg, M., 1982. Neural reorganization during metamorphosis of the corpora pedunculata in *Drosophila melanogaster*. Nature 295, 405–407.
Tempel, B.L., Bonini, N., Dawson, D.R., Quinn, W.G., 1983. Reward learning in normal and mutant *Drosophila*. Proc. Natl. Acad. Sci. U.S.A. 80, 1482–1486.
Tempel, B.L., Livingstone, M.S., Quinn, W.G., 1984. Mutations in the dopa decarboxylase gene affect learning in *Drosophila*. Proc. Natl. Acad. Sci. U.S.A. 81, 3577–3581.
Thum, A.S., Jenett, A., Ito, K., Heisenberg, M., Tanimoto, H., 2007. Multiple memory traces for olfactory reward learning in *Drosophila*. J. Neurosci. 27, 11132–11138.
Tibbetts, E.A., 2002. Visual signals of individual identity in the wasp *Polistes fuscatus*. Proc. R. Soc. B Biol. Sci. 269, 1423–1428.
Tinbergen, N., 1951. The Study of Instinct. Oxford University Press, New York.
Truman, J.W., Riddiford, L.M., 1999. The origins of insect metamorphosis. Nature 401, 447–452.
Unoki, S., Matsumoto, Y., Mizunami, M., 2005. Participation of octopaminergic reward system and dopaminergic punishment system in insect olfactory learning revealed by pharmacological study. Eur. J. Neurosci. 22, 1409–1416.
Unoki, S., Matsumoto, Y., Mizunami, M., 2006. Roles of octopaminergic and dopaminergic neurons in mediating reward and punishment signals in insect visual learning. Eur. J. Neurosci. 24, 2031–2038.
Van Bockstaele, E.J., Pickel, V.M., 1995. GABA-containing neurons in the ventral tegmental area project to the nucleus accumbens in rat brain. Brain Res. 682, 215–221.
Van Ree, J.M., Niesink, R.J., Van Wolfswinkel, L., Ramsey, N.F., Kornet, M.M., Van Furth, W.R., Vanderschuren, L.J., Gerrits, M.A., Van den Berg, C.L., 2000. Endogenous opioids and reward. Eur. J. Pharmacol. 405, 89–101.
Vergoz, V., Roussel, E., Sandoz, J.-C., Giurfa, M., 2007. Aversive learning in honeybees revealed by the olfactory conditioning of the sting extension reflex. PLoS One 2, e288.
Vickrey, T.L., Condron, B., Venton, B.J., 2009. Detection of endogenous dopamine changes in *Drosophila melanogaster* using fast-scan cyclic voltammetry. Anal. Chem. 81, 9306–9313.
Vlachou, S., Markou, A., 2010. $GABA_B$ receptors in reward processes. Adv. Pharmacol. 58, 315–371.
Vogt, K., Schnaitmann, C., Dylla, K.V., Knapek, S., Aso, Y., Rubin, G.M., Tanimoto, H., 2014. Shared mushroom body circuits operate visual and olfactory memories in *Drosophila*. eLife 3, e02395.
von Frisch, K., 1914. Der farbensinn und Formensinn der Biene. Zool. Jahrb. Abt. Allg. Zool. Physiol. Tiere 35, 1–188.
Wang, Z., Singhvi, A., Kong, P., Scott, K., 2004. Taste representations in the *Drosophila* brain. Cell 117, 981–991.
Weiner, J., 1999. Time, Love, Memory: A Great Biologist and His Quest for the Origins of Behavior. Knopf, New York.
Williams, J.R., Catania, K.C., Carter, C.S., 1992. Development of partner preferences in female prairie voles (*Microtus ochrogaster*): the role of social and sexual experience. Horm. Behav. 26, 339–349.
Winston, M.L., 1987. The Biology of the Honey Bee. Harvard University Press, Cambridge, MA.
Wise, R.A., Rompre, P.P., 1989. Brain dopamine and reward. Annu. Rev. Psychol. 40, 191–225.
Witthöft, W., 1967. Absolute Anzahl und Verteilung der Zellen im Hirn der Honigbiene. Zoomorphology 61, 160–184.

Wolf, R., Wittig, T., Liu, L., Wustmann, G., Eyding, D., Heisenberg, M., 1998. *Drosophila* mushroom bodies are dispensable for visual, tactile, and motor learning. Learn. Mem. 5, 166–178.
Wright, G.A., Mustard, J.A., Simcock, N.K., Ross-Taylor, A.A.R., McNicholas, L.D., Popescu, A., Marion-Poll, F., 2010. Parallel reinforcement pathways for conditioned food aversions in the honeybee. Curr. Biol. 20, 2234–2240.
Yarali, A., Gerber, B., 2010. A neurogenetic dissociation between punishment-, reward-, and relief-learning in *Drosophila*. Front. Behav. Neurosci. 4, 189.
Young, L.J., Wang, Z., 2004. The neurobiology of pair bonding. Nat. Neurosci. 7, 1048–1054.
Young, L.J., Murphy Young, A.Z., Hammock, E.A.D., 2005. Anatomy and neurochemistry of the pair bond. J. Comp. Neurol. 493, 51–57.

CHAPTER SEVEN

Epigenetics in Social Insects

Karl M. Glastad, Linh M. Chau, Michael A.D. Goodisman[1]
School of Biology, Georgia Institute of Technology, Atlanta, Georgia, USA
[1]Corresponding author: e-mail address: michael.goodisman@biology.gatech.edu

Contents

Abstract

Social insects enjoy great ecological success. The success of social insects stems from the remarkable levels of phenotypic plasticity displayed by colony members. This developmental plasticity is mediated by epigenetic inheritance. This review explores epigenetic inheritance and considers the function of epigenetic information systems in social insect taxa. Epigenetic information in eukaryotic organisms is passed across cell divisions through three major mechanisms: the methylation of DNA, the modification of histone proteins, and the activity of non-coding RNAs. Several studies have found that patterns of DNA methylation in social insects are widely conserved and that there are strong associations between levels of DNA methylation and gene function. Studies of histone proteins demonstrate the presence of conserved modifications associated with patterns of gene expression. Finally, preliminary research suggests that non-coding RNAs

Advances in Insect Physiology, Volume 48
ISSN 0065-2806
http://dx.doi.org/10.1016/bs.aiip.2014.12.007

function within social insect genomes and can influence social insect phenotype. Overall, much remains to be learned about epigenetic inheritance in social insects. However, initial studies suggest that epigenetic mechanisms play a large role in the development of social insects and may help explain the success of insect societies.

1. SOCIAL INSECTS AND DEVELOPMENTAL PLASTICITY

Social insects rank 'among the greatest achievements of organic evolution' because of the extraordinary levels of cooperation displayed by group members (Oster and Wilson, 1978; Wilson, 1971). Insect societies function efficiently because they are composed of phenotypically distinct castes, which specialize in performing particular tasks (Oster and Wilson, 1978; Wilson, 1971). For instance, in advanced social insect species, queens and kings specialize in reproduction and dispersal, whereas workers and soldiers build the nest, collect food, rear the young, and defend the colony.

The differences in behaviours undertaken by the distinct castes are often accompanied by extreme differences in morphology and physiology. Despite these impressive differences, castes generally do not differ genetically (Nijhout, 2003; Wheeler, 1986; but see Schwander et al., 2010). Rather, social insect castes represent an instance of phenotypic plasticity, whereby environmental cues trigger changes in gene expression, which lead to the production of distinct phenotypes (Nijhout, 2003). Virtually, all eukaryotic organisms display some form of phenotypic plasticity (Pfennig et al., 2010). Such plasticity is generally beneficial and allows maturing individuals to develop phenotypes well suited to the environment (Dewitt and Scheiner, 2004; Pigliucci, 2001; West-Eberhard, 2003).

The process by which social insects and other organisms produce different phenotypes through plastic development is roughly analogous to the mechanism by which different cell types within multicellular organisms differentiate in distinct tissues (Maynard Smith and Szathmary, 1998). That is, distinct castes within insect societies develop through the process of differential gene expression triggered by environmental cues received by the developing larva. In much the same way, distinct tissues within multicellular organisms develop through the process of differential gene expression triggered by environmental cues received by the differentiating cell. Consequently, differential gene expression associated with development represents a fundamentally important molecular mechanism affecting phenotype at all levels of biological organization (Maynard Smith and Szathmary, 1998).

2. EPIGENETIC INHERITANCE

2.1 The scope of epigenetic inheritance

The term epigenetics was originally used to describe the 'whole complex of developmental processes' (Waddington, 1942). However, the interpretation and scope of epigenetics have changed (Haig, 2004a), and it is, therefore, worth reviewing the meaning of epigenetics in contemporary scientific study. Waddington originally envisioned epigenetics as the study of the mechanisms that led to the development of a particular phenotype from a given genotype (Jablonka and Lamm, 2012; Waddington, 1942). However, this definition was modified by Nanney (1958), who noted that cellular phenotype was 'not determined entirely by the DNA present in the cell—that other devices, epigenetic systems, regulate the expression of the genetically determined potentialities'. The development of the field of molecular biology led to a better understanding of molecular processes that allowed cells of the same genotype to produce different phenotypes. Consequently, Nanney's definition was further refined by Riggs et al. (1996) who fashioned the modern definition of epigenetics as the 'study of mitotically and/or meiotically heritable changes in gene function that cannot be explained by changes in DNA sequence'.

Epigenetic inheritance, as defined, includes the inheritance of information through both mitotic cell division within generations and meiotic cell division between generations. However, the different types of epigenetic inheritance have very different implications for development and evolution. Thus, it is worth considering the consequences of epigenetic inheritance in these different contexts.

Intragenerational epigenetic inheritance occurs when epigenetic information is transmitted through mitotic cell divisions within an individual. In this case, a particular cell receives some type of environmental stimulus, which leads to generation of epigenetic information which is then inherited through future mitotic cell divisions. The use of the word 'inheritance' in the context of mitotic cell divisions may require further explanation, because 'inheritance' is if often implicitly thought to denote transmission through meiotic cell division only (i.e. transmission from parent to offspring) (Bossdorf et al., 2008). However, the epigenetics literature considers transmission of information through mitosis and meiosis, and we will follow this convention. Thus, intragenerational epigenetics typically focuses on the molecular mechanisms that allow the single set of DNA instructions found within a somatic cell to develop into different cell and tissue types through mitotic cell divisions (Lim and Brunet, 2013).

Intragenerational epigenetic inheritance can be contrasted with epigenetic processes that occur across generations and through meiotic cell division (Bossdorf et al., 2008; Danchin et al., 2011; Heard and Martienssen, 2014). For example, *intergenerational* epigenetic inheritance (Heard and Martienssen, 2014) occurs if a focal individual experiences an environmental stimulus that leads to the transmission of epigenetic information to the focal individual's offspring. In this case, the environment of the parent affects information carried in its germ cells, leading to one generation of epigenetic information transfer.

Intergenerational epigenetic inheritance is sometimes distinguished from *transgenerational* epigenetic inheritance (Heard and Martienssen, 2014; Lim and Brunet, 2013). Transgenerational epigenetic inheritance occurs when epigenetic effects persist beyond the offspring of the focal, affected individual. Transgenerational inheritance is generally viewed as rare, particularly in animals, because the germline is sequestered from the soma early in development and undergoes epigenetic reprogramming (Bonduriansky and Day, 2009). However, transgenerational epigenetic inheritance is of great theoretical interest, at least, because it changes the process of biological evolution (Danchin et al., 2011; Heard and Martienssen, 2014; Jablonka and Raz, 2009).

In summary, epigenetic inheritance can occur both within (intragenerational) and between (intergenerational and transgenerational) generations (Bossdorf et al., 2008; Richards, 2006). However, this review will be largely focused on intragenerational epigenetic inheritance because the majority of research on epigenetics in social insects has been conducted at this scale. Nevertheless, there is a great deal of interest in epigenetic inheritance that may occur across generations in social systems (Drewell et al., 2012; Queller, 2003), and intergenerational or transgenerational epigenetics is likely to be an important area of future research.

2.2 Epigenetic information systems

Epigenetic states are initiated by some type of environmental stimulus (Bonasio et al., 2010a). This stimulus leads to a change in an epigenetic information system, which will ultimately be transmitted through cell division. One of the most well-studied epigenetic information systems is the methylation of DNA (Cedar and Bergman, 2012; Jones, 2012). DNA methylation can affect gene function in a variety of ways and has been the focus of the most extensive investigations of epigenetics in social insects. Epigenetic information can also be passed down through modifications to DNA-associated proteins (Kharchenko et al., 2011; Suganuma and Workman, 2011). Specifically, histone proteins, which are closely associated with DNA in eukaryotic organisms, may carry epigenetic signals (Felsenfeld, 2014). A third major type of epigenetic inheritance system is concerned with the function of non-coding RNAs (ncRNAs) (Lim and Brunet, 2013). Certain classes of ncRNAs interact with DNA or with other cellular products to modify how genes are expressed. Importantly, there is a great degree of interplay among different epigenetic information systems (Bonasio et al., 2010a). One system may regulate another or multiple information systems may interact to affect gene function. Consequently, the division of epigenetic inheritance systems into DNA methylation, histone modification, and ncRNAs represents a convenient simplification (Sharma et al., 2010).

This review focuses on epigenetics in social insects. Social insects represent promising systems in which to study epigenetics for several reasons (Bonasio, 2012, 2014; Yan et al., 2014): first, social insect castes are a prime example of a polyphenism that arises through the function of epigenetic factors, with discrete (as opposed to continuous) phenotypes arising from the same genotype (Miura, 2004; Nijhout, 2003). Second, highly developed caste systems have arisen multiple independent times in social insects (Wilson, 1971). This allows for a comparative approach to studying epigenetics. Third, the caste system of social insects represents an important evolutionary innovation (Maynard Smith and Szathmary, 1998; Wilson, 1971). Knowledge of the molecular basis underlying caste differences thus promotes our understanding of how epigenetic information can lead to organismal success. Fourth, social insects show a diversity of behaviours that are hypothesized to have been influenced by epigenetic processes (Ledon-Rettig et al., 2013; Queller, 2003). And finally, some epigenetic systems in insects (i.e. DNA methylation) are relatively simple compared to those found in other eukaryotic organisms. Therefore, study of epigenetic inheritance in insects may be more tractable than in other organisms.

3. DNA METHYLATION

3.1 The mark and the machinery

Some form of DNA methylation is found in all three domains of life, suggesting that DNA methylation has an ancient and important role in biological systems (Klose and Bird, 2006; Suzuki and Bird, 2008; Zemach et al.,

2010). DNA methylation in animals usually occurs on cytosine bases found next to guanine bases (i.e. 5-methylcytosines in CpG dinucleotides) (Suzuki and Bird, 2008). Cytosine bases in mammals are methylated by DNA methyltransferases (DNMTs). DNMT3 proteins are known as the *de novo* methyltransferases (Lan et al., 2010; Law and Jacobsen, 2010), as they are responsible for establishing new genomic methylation patterns in mammals (Cedar and Bergman, 2012). Once patterns of DNA methylation are established, they are sustained by maintenance methyltransferases in the *DNMT1* gene family (Klose and Bird, 2006)—so-called due to their role in preserving established methylation patterns across cell divisions. DNMT1 is believed to use the methylation information from the methylated strand of DNA to properly methylate the nascent strand of DNA during DNA replication. In this way, epigenetic information encoded in chromatin can be passed on through cell division (Cedar and Bergman, 2012).

It has been assumed that insect DNMTs operate in the same way as their putative, mammalian orthologs (Wang et al., 2006). Therefore, the presence of one or more copies of *DNMT1* and *DNMT3* orthologs in the genome of a social insect has been considered as evidence that a focal insect species possesses a functional DNA methylation system (Yi and Goodisman, 2009). However, two insects that possess DNA methylation, the silkworm *Bombyx mori* and the locus *Schistocerca gregaria*, apparently lack a *DNMT3* gene (Falckenhayn et al., 2013; Xiang et al., 2010). Moreover, recent investigations have found very low levels of DNA methylation in *Drosophila melanogaster* (Takayama et al., 2014) even though its genome contains neither *DNMT1* nor *DNMT3*. These results suggest that our understanding of the process of DNA methylation in insects is incomplete and that other uncharacterized genes may be involved in the process of DNA methylation (Lan et al., 2010; Yi, 2012).

DNA methylation can both be transmitted through cell division and removed at certain times of organismal development (Chen and Riggs, 2011; Wu and Zhang, 2010). In particular, DNA methylation patterns in mammals are reprogrammed each generation. Genome-wide demethylation occurs in germ cells and in early embryos. Then, after implantation, methylation patterns are re-established.

Once DNA has been methylated, it can be recognized by many proteins including those in the methyl-CpG binding protein family (Clouaire and Stancheva, 2008; Lan et al., 2010). Methyl-CpG binding proteins interact with other enzymes and with chromatin, and can subsequently affect patterns of gene expression (Bogdanovic and Veenstra, 2009; Hendrich and

Tweedie, 2003). In addition, DNA methylation can directly or indirectly affect the behaviour of enzymes that interact with DNA. For instance, DNA methylation is known to affect the rate of RNA polymerase transit, which ultimately affects gene function (Lorincz et al., 2004; Zilberman et al., 2007).

3.2 Patterns of DNA methylation in insects

Levels of DNA methylation vary greatly among animal species (Feng et al., 2010; Zemach et al., 2010). For example, mammalian genomes tend to show very high levels of CpG methylation throughout the genome, with 60–90% of all CpG dinucleotides subject to methylation (Li et al., 2010; Lister et al., 2009). In contrast, invertebrates show lower levels of DNA methylation. For example, ~30% of CpG dinucleotides are methylated in the invertebrate tunicate *Ciona intestinalis* (Feng et al., 2010).

Levels of DNA methylation in insects are even lower than in other investigated invertebrates (Sarda et al., 2012). Only ~8% of CpGs are methylated in the hemimetabolous locust, *Locusta migratoria* (Robinson et al., 2011; Wang et al., 2014). DNA methylation levels appear to be lower still in holometabolous insects (Beeler et al., 2014; Kronforst et al., 2008; Simola et al., 2013a). Less than 1% of CpG nucleotides are methylated in the honeybee, *Apis mellifera* (Lyko et al., 2010), the silkworm, *B. mori* (Xiang et al., 2010), and the jewel wasp, *Nasonia vitripennis* (Beeler et al., 2014). Moreover, both the fruit fly, *D. melanogaster*, and red flour beetle, *Tribolium castaneum*, lack substantial levels of DNA methylation in their genomes (Rae and Steele, 1979; Tribolium Genome Sequencing Consortium, 2008; Urieli-Shoval et al., 1982), suggesting that many *Diptera* (Hung et al., 1999; Marhold et al., 2004; Tweedie et al., 1999) and *Coleoptera* lack DNA methylation at meaningful levels (but see Snell-Rood et al., 2013; Takayama et al., 2014). Regardless, taken together, these results suggest that DNA methylation is ancestral to insects, although it has been reduced or lost in some lineages (Field et al., 2004; Mandrioli, 2007; Sarda et al., 2012; Tweedie et al., 1997).

DNA methylation in insects tends to be found primarily in gene bodies (introns + exons) (Feng et al., 2010; Hunt et al., 2013a; Wang et al., 2013; Zemach et al., 2010). Gene body methylation is taxonomically widespread and found in other animals and plants (Feng et al., 2010; Ndlovu et al., 2011; Zemach et al., 2010). This taxonomic conservation suggests that gene body methylation has an important function.

There are subtle, but important, differences in the patterns of gene body methylation in different insect taxa. In holometabolous insects, DNA methylation levels are higher in exons than introns (Bonasio et al., 2012; Feng et al., 2010; Hunt et al., 2013b; Lyko et al., 2010). Moreover, the rest of the genome shows little evidence of DNA methylation. In contrast, DNA methylation levels appear to be as high or higher in introns than exons in hemimetabolous insects. Repetitive DNA may also be methylated in the Hemimetabola (Glastad et al., 2013; Wang et al., 2014; Fig. 1). Interestingly, although DNA methylation is generally targeted to gene bodies in insects, not all genes are methylated. In particular, insect methylation levels often show a bimodal distribution, with some genes being methylated and others not (Elango et al., 2009; Glastad et al., 2013; Hunt et al., 2010; Lyko et al., 2010; Walsh et al., 2010; Wang and Leung, 2009; Fig. 2).

Methylated and unmethylated genes show different patterns of molecular evolution. Specifically, genes with low levels of methylation are less likely to maintain their methylation status over evolutionary time, whereas heavily

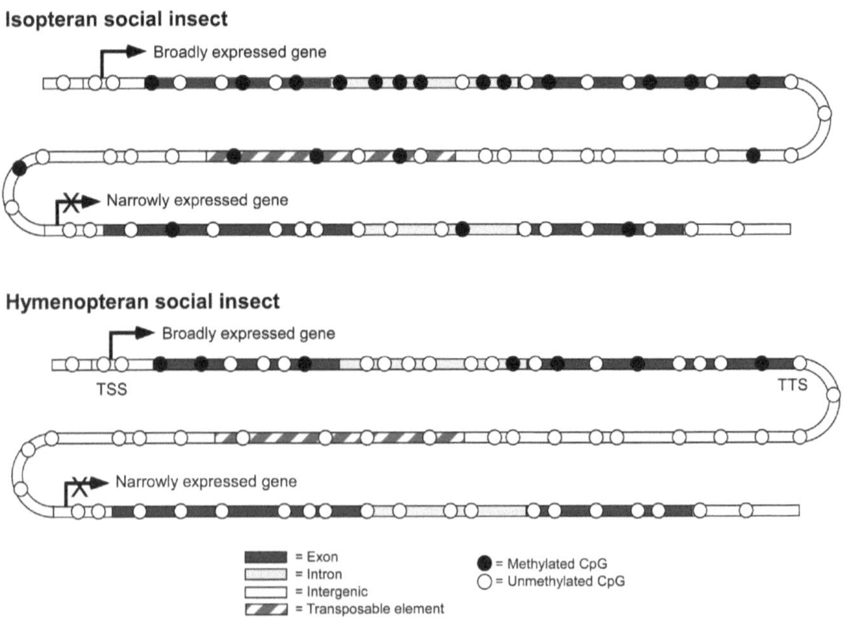

Figure 1 Putative patterns of DNA methylation in isopteran (termite) and hymenopteran (ants, bees, and wasps) social insects. Patterns in Isoptera are distinguished from patterns in Hymenoptera in that isopteran taxa have higher levels of DNA methylation, show greater methylation in introns than exons, and may show methylation in nongenic portions of the genome. *Modified from Glastad et al. (2014a).*

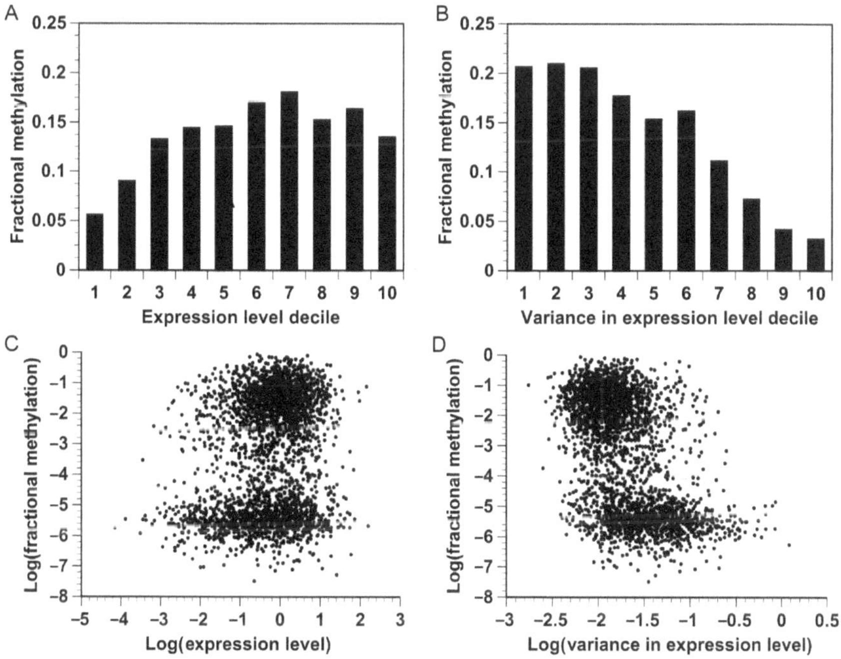

Figure 2 Relationship between gene methylation and gene expression in the fire ant *Solenopsis invicta*. (A) Moderately to highly expressed genes tend to be more highly methylated (1, low expression; 10, high expression). (B) Genes showing high variance in expression tend to show low levels of DNA methylation (1, low variance; 10, high variance). Log transformations of raw data yielding relationship between log(fractional methylation) (Lfm) for (C) gene expression level and (D) variance in gene expression illustrate the bimodal nature of DNA methylation in insects. Genes that display ~−1 to −3 Lfm are methylated whereas those that display ~−4 to −7 Lfm are essentially unmethylated. *Modified from Hunt et al. (2013b).*

methylated genes are more likely to remain heavily methylated (Hunt et al., 2010). Moreover, methylated genes tend to be involved in housekeeping functions and are usually more highly conserved at the sequence level than unmethylated genes (Glastad et al., 2011, 2013; Hunt et al., 2010, 2011a; Lyko et al., 2010; Park et al., 2011; Suzuki et al., 2007).

Finally, largely overlapping sets of orthologs are targeted by DNA methylation in insects (Hunt et al., 2013a,b; Sarda et al., 2012). That is, insects with functional DNA methylation systems exhibit substantial conservation in the patterns of genomic methylation in the sense that the same genes are targeted for methylation across taxa (Bonasio et al., 2012; Glastad et al., 2011; Hunt et al., 2013b; Lyko et al., 2010; Sarda et al., 2012; Xiang et al., 2010).

3.3 DNA methylation in social insects

Social insects have been primary model systems used for understanding the importance of insect DNA methylation (Bonasio, 2014; Glastad et al., 2011; Hunt et al., 2013a; Lyko and Maleszka, 2011; Maleszka, 2008; Schaefer and Lyko, 2007; Weiner and Toth, 2012). The interest in understanding the function of DNA methylation in social insects lies primarily in the expectation that DNA methylation may act as an epigenetic modifier affecting the development of castes and help explain the elaboration of complex societies in insects (Bonasio et al., 2012; Flores and Amdam, 2011; Foret et al., 2012; Kucharski et al., 2008; Ledon-Rettig et al., 2013; Lyko et al., 2010; Moczek and Snell-Rood, 2008; Patalano et al., 2012; Roberts and Gavery, 2012; Simola et al., 2013b; Yan et al., 2014).

DNA methylation was first noted in a social insect in the honeybee, *A. mellifera* (Wang et al., 2006). Shortly thereafter, DNA methylation was found in many of the social Hymenoptera (Bonasio et al., 2010b; Cingolani et al., 2013; Gadau et al., 2012; Glastad et al., 2011; Kocher et al., 2013; Kronforst et al., 2008; Oxley et al., 2014; Smith et al., 2011a,b; Suen et al., 2011; Weiner et al., 2013; Wurm et al., 2011; Zeng and Yi, 2010). DNA methylation has now also been found in termites (Glastad et al., 2013; Lo et al., 2012; Simola et al., 2013a; Terrapon et al., 2014). In fact, all social insects studied so far appear to possess functional systems of DNA methylation (Table 1).

A central focus of social insect epigenetics is the study of DNA methylation differences between social insect castes. For example, several studies have identified differences in methylation between *A. mellifera* queen and worker larvae (Cameron et al., 2013; Foret et al., 2012; Ikeda et al., 2011; Shi et al., 2011, 2013) and adults (Foret et al., 2012; Lyko et al., 2010; although other studies have failed to find substantial differences in methylation patterns between queens and workers; Herb et al., 2012). Queen–worker methylation differences have also been identified in the ants *Camponotus floridanus*, *Harpegnathos saltator*, *Solenopsis invicta*, and *Pogonomyrmex* sp. (Bonasio et al., 2012; Glastad et al., 2014b; Hunt et al., 2013b; Smith et al., 2012). These studies demonstrate that the reproductive and sterile female castes do show methylation differences that may have functional effects.

Differences in methylation patterns between the sexes have also been investigated in a few social insect taxa. For example, females and males show distinct methylation patterns in *S. invicta*, *C. floridanus*, and *H. saltator*

Table 1 Social insects whose genomes have been surveyed for evidence of DNA methylation

Group	Species	Evidence	References
Termite	*Coptotermes formasanus*	C, D	Glastad et al. (2013)
	Coptotermes lacteus	M	Lo et al. (2012)
	Reticulitermes flavipes	C	Glastad et al. (2013)
	Zootermopsis nevadensis	C, D	Terrapon et al. (2014)
Bee	*Apis mellifera*	B, C, D	Wang et al. (2006)
	Bombus impatiens	M	Weiner et al. (2013)
	Bombus terrestris	M	Bigot et al. (2011)
	Melipona bicolor	M	Kronforst et al. (2008)
	Trigona spinipes	M	Kronforst et al. (2008)
	Lasioglossum albipes	C, D	Kocher et al. (2013)
Ant	*Acromyrmex echinatior*	C	Simola et al. (2013a)
	Aphaenogaster albisetosa	M	Kronforst et al. (2008)
	Atta cephalotes	C, D	Suen et al. (2011)
	Camponotus festinatus	M	Kronforst et al. (2008)
	Camponotus floridanus	B, C, D	Bonasio et al. (2012)
	Cerapachys biroi	D	Oxley et al. (2014)
	Harpegnathos saltator	B, C, D	Bonasio et al. (2012)
	Linepithema humile	C, D	Smith et al. (2011a)
	Messor pergandei	M	Kronforst et al. (2008)
	Pheidole obtusospinosa	M	Kronforst et al. (2008)
	Pogonomyrmex barbatus	C, D	Smith et al. (2011b)
	Solenopsis invicta	B, C, D	Wurm et al. (2011)
Wasp	*Apoica flavissima*	M	Weiner et al. (2013)
	Dolichovespula maculata	M	Weiner et al. (2013)
	Liostenogaster flavolineata	M	Kronforst et al. (2008)
	Metapolybia cingulata	M	Kronforst et al. (2008)
	Metapolybia docilis	M	Weiner et al. (2013)

Continued

Table 1 Social insects whose genomes have been surveyed for evidence of DNA methylation—cont'd

Group	Species	Evidence	References
	Polistes dominula	M	Kronforst et al. (2008)
	Polybia occidentalis	M	Weiner et al. (2013)
	Polybia paulista	M	Weiner et al. (2013)
	Polybia sericea	M	Kronforst et al. (2008)
	Vespula pensylvanica	M	Kronforst et al. (2008)

Evidence: B, bisulfite sequencing; C, computational analyses; D, identification of *DNMT* genes; M, methylation-sensitive AFLPs. All social insects investigated thus far possess functional systems of DNA methylation.

(Bonasio et al., 2012; Glastad et al., 2014b; Hunt et al., 2013b). Variation in methylation patterns between the sexes has also been investigated in bumblebees (Bigot et al., 2011). Moreover, DNA methylation patterns have been found to differ among *A. mellifera* male developmental stages and tissues (Drewell et al., 2014).

A. mellifera has, once again, acted as an important model in understanding if worker subcastes show variation in patterns of methylation (Herb et al., 2012; Lockett et al., 2012). One of the most compelling studies demonstrated that methylation patterns can be reversed if a worker is induced to revert to a different behavioural subcaste (Herb et al., 2012). Other studies outside of honeybees have found evidence for methylation differences between worker subcastes in ants (Bonasio et al., 2012), termites (Glastad et al., 2013), and bumblebees (Amarasinghe et al., 2014). In addition, substantial differences in levels of methylation between workers of European and African subspecies of *A. mellifera* have been identified (Guzman-Novoa et al., 2005).

Importantly, despite the focus on dissimilarities in methylation profiles between social insect phenotypes, it should be noted that methylation profiles within (and even between) social insect species show far more similarities than differences. This conservation of patterns of methylation is, perhaps, more striking than the differences (Hunt et al., 2013b).

3.4 The function of DNA methylation in social insects

Gene expression: Methylated genes in insects tend to be ubiquitously expressed among cell types or phenotypes. In contrast, unmethylated genes tend to show tissue- or phenotype-specific expression patterns. This association between DNA methylation and gene expression has been discovered in the social insects *A. mellifera*, *S. invicta*, *C. floridanus*, *H. saltator*, *Reticulitermes flavipes*, *Coptotermes formosanus*, and *Zootermopsis nevadensis* (Bonasio et al., 2012; Elango et al., 2009; Foret et al., 2009; Glastad et al., 2011, 2013, 2014a; Hunt et al., 2010, 2013b; Lyko et al., 2010; Sarda et al., 2012; Terrapon et al., 2014; Fig. 2). In addition, DNA methylation was recently found to be associated with ploidy in *S. invicta*. In this case, haploid males showed higher levels of methylation than their diploid counterparts (Glastad et al., 2014b), consistent with the general trend of an association between DNA methylation and gene expression.

Some evidence exists that gene body DNA methylation can play a role in affecting transcription. For example, DNA methylation is known to affect the position and nature of histone proteins (Cedar and Bergman, 2009; Chodavarapu et al., 2010), which may affect the binding of complexes that affect transcription (Coleman-Derr and Zilberman, 2012; Henikoff, 2008; Zilberman et al., 2008). However, despite these results, the mechanistic link between gene body methylation and gene expression is still somewhat lacking, particularly in insects (Shenker and Flanagan, 2012; Takuno and Gaut, 2012). Indeed, it has been suggested that the positive correlation between gene body methylation and gene expression might merely be a by-product of the fact that highly expressed genes are more accessible to DNA methylation machinery (Takuno and Gaut, 2012). Instead, some studies suggest that DNA methylation may act to inhibit spurious intragenic transcription (Cheong et al., 2006; Huh et al., 2013; Maunakea et al., 2010) or play other regulatory roles in the bodies of actively expressed genes.

It is notable that the generally positive correlation between gene body methylation and gene expression is opposite to that typically found between *promoter* methylation and gene expression. That is, the methylation of promoter regions has been directly linked to transcriptional repression in vertebrates (Bernstein et al., 2007; Weber et al., 2007). Furthermore, DNA methylation may help repress transposable element (TE) activity in vertebrates (O'Neill et al., 1998; Yoder et al., 1997). Regardless, the contrasting correlations between DNA methylation and gene expression for gene bodies and for promoters indicate that DNA methylation can function in multiple ways depending on its genomic context.

Alternative splicing: Patterns of alternative messenger RNA (mRNA) splicing vary throughout development (Barberan-Soler and Zahler, 2008) and play a role in generating phenotypic variation (Ast, 2004). DNA methylation may directly alter the binding and function of enzymes that affect splicing leading to splice-form variation (Gelfman et al., 2013; Jones, 2012; Shukla et al., 2011). DNA methylation has been found to influence alternative splicing by either preventing or permitting the inclusion of exons (Maunakea et al., 2010; Shukla et al., 2011). DNA methylation can also affect the speed and elongation efficiency of RNA Pol II (Lorincz et al., 2004; Zilberman et al., 2007), which strongly influences splice site selection during transcription and can lead to changes in alternative splicing patterns (Kornblihtt et al., 2013; Luco et al., 2011). Finally, DNA methylation has been linked to the positioning of nucleosomes (Chodavarapu et al., 2010), which have in turn been implicated in modulating intragenic RNA Pol II kinetics (Bintu et al., 2012; Luco et al., 2011; Schwartz et al., 2009) which subsequently can affect patterns of gene splicing (de la Mata et al., 2003; Kornblihtt et al., 2013; Luco et al., 2011). Therefore, DNA methylation may influence alternative splicing through a variety of molecular mechanisms.

A link between alternative splicing and DNA methylation has been repeatedly identified in *A. mellifera* (Cingolani et al., 2013; Flores et al., 2012; Foret et al., 2012; Herb et al., 2012; Li-Byarlay et al., 2013; Lockett et al., 2012; Lyko et al., 2010). DNA methylation has also been associated with alternative splicing in the ants *C. floridanus* and *H. saltator* (Bonasio et al., 2012), and the termite *Z. nevadensis* (Terrapon et al., 2014). However, it should be noted that not all studies in insects have detected a relationship between DNA methylation and splicing (Wang et al., 2013). Moreover, the link between levels of DNA methylation and splicing in social insects is not always very strong (Flores et al., 2012; Herb et al., 2012; Li-Byarlay et al., 2013; Lyko et al., 2010). This suggests that our understanding of the interaction between alternative splicing and DNA methylation in social insects remains incomplete.

Experimental manipulation of DNA methylation: Some of the most promising research on understanding the function of DNA methylation in social insects has come from studies where levels of DNA methylation have been experimentally altered (Kucharski et al., 2008; Li-Byarlay et al., 2013; Lockett et al., 2010). Indeed, much of the interest in studying DNA methylation in social insects was initiated by Kucharski et al. (2008), who demonstrated that a knockdown of DNMT3 in developing *A. mellifera* larvae resulted in a shift in caste phenotype. Similarly, Li-Byarlay et al. (2013) used RNA interference (RNAi) to decrease DNMT3 levels in *A. mellifera* and found that the resulting decreased levels of DNA methylation affected alternative splicing. Most recently, DNA methylation levels were inhibited pharmacologically in *Bombus terrestris* (Amarasinghe et al., 2014). In this case, experimental changes in DNA methylation were associated with observed changes in aggression in workers.

DNA methylation and imprinting: One of the most exciting potential roles of DNA methylation in social insects may be that of imprinting. Imprinting is the process by which the expression of an allele depends on the parent from which that allele originates (Ferguson-Smith, 2011). Several hypotheses have been proposed to explain the evolution of imprinting (Patten et al., 2014; Spencer and Clark, 2014). However, perhaps the most attractive explanation is that imprinting arises because of conflicts between relatives (Haig, 2004b; Queller, 2003). The type of conflict that generates imprinting under the kinship theory is thought to play a particularly important role in social insects (Dobata and Tsuji, 2012; Haig, 2000; Kronauer, 2008; Queller, 2003).

A small number of genes (~100) appear to be imprinted in plants and mammals (Kohler et al., 2012; Li and Sasaki, 2011). However, imprinting seems to be rare or absent in many other taxa (Coolon et al., 2012; Moore and Mills, 2008). Regardless, DNA methylation appears to be the main method of gene imprinting, when it occurs (Kohler et al., 2012; Li and Sasaki, 2011). Thus, a great deal of interest is focused on determining if methylation-based imprinting takes place in social insects (Drewell et al., 2012; Strassmann and Queller, 2007).

4. HISTONE MODIFICATIONS AND VARIANTS

4.1 The eukaryotic nucleosome

The fundamental, repeating unit of metazoan chromatin is the nucleosome. The nucleosome is composed of approximately 146 bp of DNA wrapped around an 8-unit complex composed of histone proteins (Fig. 3). Each nucleosome is composed of two tetramers of histone proteins, each containing histones H2A, H2B, H3, and H4. The majority of metazoan nuclear DNA is incorporated into nucleosomes (Badeaux and Shi, 2013; Bell et al., 2011; Margueron and Reinberg, 2010; Zhou et al., 2011). As such, nucleosomes represent highly conserved fundamental units of chromatin.

The modification or replacement of nucleosomes has been linked to diverse regulatory processes in eukaryotes (Badeaux and Shi, 2013; Margueron and Reinberg, 2010). For example, transcription factors usually bind only to sites depleted of nucleosomes (Bell et al., 2011; Henikoff, 2008). Furthermore, many DNA-binding and chromatin-associated

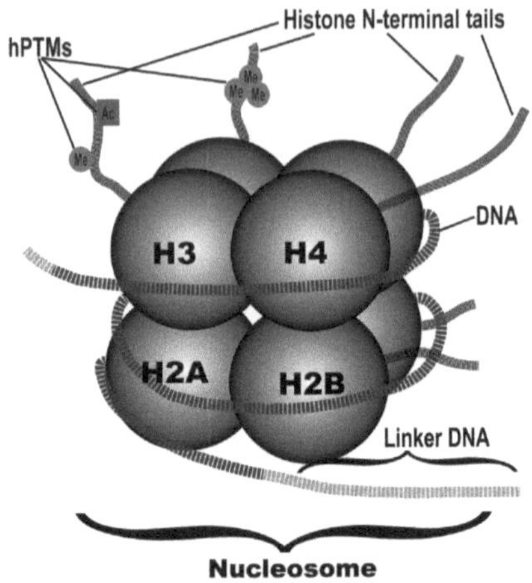

Figure 3 Diagram of a typical eukaryotic nucleosome illustrating the four canonical histone proteins (H2A, H2B, H3, and H4) present in paired histone tetramers, encircling and linking DNA, and histone tails which are modified by the addition of example methyl or acetyl groups (marked with circles and squares, respectively).

proteins bind to histone proteins, making nucleosomes important to the function of higher gene regulatory processes.

Notably, there are at least two fundamentally distinct ways in which histone proteins can be altered to affect gene regulation. First, the histone protein may be modified after it has been translated (histone post-translational modification, hPTM). Second, the histone protein may be replaced with an alternative sequence variant that shows different biochemical characteristics than the canonical histone (histone replacement). Both types of nucleosome alterations have been shown to affect gene function and be stably maintained across cell divisions. Therefore, both hPTMs and histone replacement represent important epigenetic modifications.

4.2 Histone modifications

Post-translational modifications to histone proteins represent a diverse set of epigenetic signals (Zhou et al., 2011). Common hPTMs include the addition of methyl, phosphorus, or acetyl groups. Modifications typically occur on histone protein's N-terminal tails, and the most commonly modified residues of these histone tails are lysine and arginine (Zhou et al., 2011; Fig. 3). Specific hPTMs are commonly referenced by first listing the histone protein that is being modified (e.g. H3: histone H3), then the specific residue by amino acid and position (e.g. K4: lysine 4), and finally, the specific modification made (e.g. me3: trimethylation). The full name, for example, of an hPTM would be listed as H3K4me3 (trimethylation of the lysine in the 4th amino acid position on histone H3).

hPTMs can result in an alteration of transcription in several ways. The addition of a chemical group to a histone protein can directly impact the strength of the association between the target histone and underlying DNA. This can increase or decrease the ability of proteins to bind to the DNA, depending upon how histone–DNA binding is impacted. Secondly, specific hPTMs can act as distinct epigenetic signals that are read by other proteins. Such effector proteins can translate the modification status of histone residues into a specific functional outcome (Badeaux and Shi, 2013; Greer and Shi, 2012).

The mechanisms by which hPTMs are maintained across cell division remain poorly understood (Bannister and Kouzarides, 2011; Margueron and Reinberg, 2010). However, cells may use hPTM information on existing histone proteins or integrate *trans*-acting factors to help perpetuate parent hPTM patterns during or after DNA replication (Margueron and Reinberg, 2010). Therefore, information stored in nucleosomes can apparently be transmitted across cell division.

The addition and subsequent removal of hPTMs on histones are accomplished by a suite of proteins (Bannister and Kouzarides, 2011; Margueron and Reinberg, 2010). Usually, distinct residues require distinct transferases for their modification. Histone post-translational-modifying proteins can be localized to their target regions through sequence-specific binding, by binding to other proteins, or recognizing specific chromatin signatures (Badeaux and Shi, 2013; Greer and Shi, 2012).

The behaviour of hPTMs can be well illustrated by the hPTMs H3K4me3 and H3K36me3. Both of these modifications are associated with actively expressed genes. However, H3K4me3 is largely confined to the promoter region, while H3K36me3 is targeted to regions downstream of H3K4me3 enrichment within gene bodies (Bannister and Kouzarides, 2011). These alternate patterns are established because the histone methyltransferases responsible for generating H3K4me3 and H3K36me3 marks each preferentially bind to one of the two primary forms of RNA Pol II (Bannister and Kouzarides, 2011). Through this differential targeting, regions marked by H3K4me3 and H3K36me3 often correspond with distinct transcriptional states across the bodies of actively transcribed genes.

Many different chemical groups can be applied to histone residues (Greer and Shi, 2012; Rice et al., 2003). However, perhaps the best-studied modification is that of lysine acetylation. Within histone protein sequences, positively charged residues such as lysine are responsible for facilitating the strong binding of DNA to the underlying histone proteins. Such residues are important to dictating the strength of histone–DNA interactions in the nucleosome.

The addition of acetyl groups to lysines of the histone tail strongly reduces the electrostatic attraction between the histone and negatively charged DNA, leading to a less-tightly associated DNA–histone complex. This increases the accessibility of the underlying DNA to DNA-binding elements (Zentner and Henikoff, 2013). Consequently, lysine acetylation is associated with actively expressed gene promoters and distal enhancer elements (de Ruijter et al., 2003; Suganuma and Workman, 2011). Evidence suggests that the number of acetylated residues, rather than which specific residues are acetylated, actually dictates the level of charge neutralization achieved by acetylation (Martin et al., 2004). In this way, the addition (or removal) of multiple acetyl groups to histone tails can result in increased (or decreased) accessibility of the underlying DNA (de Ruijter et al., 2003; Zentner and Henikoff, 2013; Zhou et al., 2011).

The acetylation of histone residues represents an example of how hPTMs may directly impact the strength of the nucleosome's histone–DNA association. In contrast, many hPTMs are instead recognized by specific proteins which then subsequently alter transcription. One prominently studied example of this latter class of hPTM is histone methylation.

The methylation of histone lysines (and other residues) has far less of a pronounced effect on the DNA–histone interaction than acetylation and is thus not thought to directly facilitate the physical opening of chromatin (Greer and Shi, 2012; Zentner and Henikoff, 2013). Instead, histone methylation information is often interpreted by reader proteins (Greer and Shi, 2012). In this way, different histone methylations can have differing outcomes depending upon the specific hPTMs and the proteins reading them (Rice et al., 2003). For example, H3K9me3 is associated with heterochromatin repression. In contrast, H3K4me3 is associated with activated transcription (Zhou et al., 2011).

4.3 Histone variants

Another important way in which the nucleosome can be altered to induce epigenetic change is through the replacement of one or more of the core histone proteins with sequence variants of that protein (Talbert and Henikoff, 2010). Replacement of canonical histone proteins with variants can dramatically alter gene expression, because nucleosome–DNA interactions fundamentally determine the accessibility of DNA. Furthermore, variants of a canonical histone protein may also contain binding targets for specific enzymes or modifications not present on the canonical version (Baldi and Becker, 2013).

For example, the histone variant H2A.Z is a variant of the canonical histone H2A (Baldi and Becker, 2013; Talbert and Henikoff, 2010). H2A.Z is structurally quite similar to H2A. However, there are several key differences between the two proteins that lead to changes in histone–histone binding within the nucleosome and terminal tail function (Baldi and Becker, 2013). Specifically, H2A.Z increases the mobility of nucleosomes or destabilizes the nucleosomes in which it is found (Talbert and Henikoff, 2010).

The common histone variant H3.3 differs from its parent H3 by only a few amino acids. However, H3.3 is typically found in the bodies of active genes. Interestingly, while most canonical histones are incorporated into DNA during cell division, H3.3 incorporation is replication independent and is likely induced upon nucleosome eviction from chromatin. H3.3's preferential incorporation into actively expressed genes likely arises because actively transcribed genes are often associated with transcription-associated nucleosome eviction (Kulaeva et al., 2010). H3.3 has been linked both to actively expressed gene bodies as well as to specific regions of heterochromatin. Thus, the direct functional outcome of nucleosome incorporation of H3.3 is currently unclear (Mito et al., 2005; Talbert and Henikoff, 2010).

4.4 Effects of hPTMs and histone variants on gene function

Gene expression: Promoters and surrounding regions of active and inactive genes often carry distinct chromatin signatures. The promoters of actively expressed genes often possess hPTMs that result in increased accessibility to the underlying DNA. Important hPTMs associated with active promoters include H3K4me3 and various lysine acetylations (e.g. H3K9 and H3K27). In addition, the promoters of actively expressed genes are often enriched with the histone variant H2A.z (Kharchenko et al., 2011). The impact of these modifications on the local chromatin environment results in an increase in the accessibility of the DNA. This, in turn, allows for the binding of transcription factors and RNA polymerase II (Henikoff, 2008).

Conversely, the promoters of repressed genes often show enrichment of hPTMs associated with a closed chromatin configuration (Hublitz et al., 2009). The two most well-known repressive hPTMs are the methylation of H3K9 and H3K27. For example, H3K9me3 is found in constitutive heterochromatin. The formation of heterochromatin is thought to occur because heterochromatin protein 1, which instigates heterochromatin formation, binds preferentially to methylated H3K9 (Hublitz et al., 2009). The second prominent form of repression is mediated by the polycomb group of proteins. This group of proteins is responsible for the inducible silencing of genes and is associated with the formation of facultative heterochromatin. Regions of polycomb repression are frequently induced by the establishment of H3K27me3 and subsequent binding of polycomb group proteins that then remodel the local chromatin to a less accessible state (Zhou et al., 2011).

In some instances, the functional outcome associated with a given histone modification may depend upon combinatorial action (Tsai et al., 2010). For example, H3K27me3 is generally associated with repressed gene promoters (Hublitz et al., 2009; Zhou et al., 2011). However, when H3K27me3 is found in conjunction with the classically activating hPTM H3K4me3, then H3K27me3 marks the so-called bivalent genes. Bivalent genes are maintained in a poised state, ready to rapidly transition either to active expression or to long-term repression, and are thus often targeted to genes of developmentally important function (Zhou et al., 2011).

Alternative splicing: The 3′-proximal regions of active gene bodies are often marked by a distinct set of hPTMs. The most well studied of these hPTMs is the trimethylation of H3K36 (Bannister and Kouzarides, 2011; Zhou et al., 2011). H3K36me3 has been implicated in altering the rate of turnover of nucleosomes and is associated with regions of elongating RNA Pol II within genes. Notably, at least one recent report has implicated increases or decreases in H3K36me3 levels as leading to inclusion or exclusion of alternatively spliced exons, respectively (Luco et al., 2011).

More generally, the efficiency of splicing is related to the dynamics of RNA Pol II. Specifically, changes in the speed of RNA Pol II transit have been shown to relate to exon splicing efficiency, wherein reductions in RNA Pol II speed lead to increases in exon inclusion (de la Mata et al., 2003; Dujardin et al., 2014). Consequently, it is likely that hPTMs play multiple other important roles in mediating alternative splicing, because nucleosome positioning and stability directly impact RNA Pol II speeds (Luco et al., 2011).

4.5 Histone modifications and variants in social insects

The first investigations to evaluate the nature of hPTMs in social insects were based upon comparative analyses using hPTM data from

D. melanogaster orthologs as proxies for hPTM enrichment in social insects. These studies suggested that levels of DNA methylation in social insects were strongly associated with several important hPTMs (Hunt et al., 2013b; Nanty et al., 2011). Furthermore, another early investigation probed the existence of multiple N-terminal tail modifications to honey bee histones H3 and H4 using mass spectrometry, confirming the conservation of many important hPTMs (Dickman et al., 2013).

Notably, as of the time of this writing, only one investigation has directly examined hPTMs as they relate to the production of social insect castes. Simola et al. (2013b) found that many hPTMs differed substantially between *C. floridanus* males and workers in accordance with gene expression differences. These findings support the general conclusion that hPTMs and associated transcription factors are as important in the differentiation of phenotype as seen in other model organisms. Indeed, given information from studies of hPTMs in model organisms, as well as emerging research interrogating these epigenetic signals in social insects, it is highly likely that chromatin epigenetics will become an important area of social insect research in the future.

5. NONCODING RNAs

It had long been thought that most of the transcriptome was composed of mRNAs slated to be translated into proteins (Gustincich et al., 2006). However, in recent years, it has been revealed that the transcriptome is more complex. Studies now suggest that most of the metazoan genome is transcribed, despite the fact that only a small percent of the genome consists of protein-coding genes (Berretta and Morillon, 2009).

ncRNAs are RNAs that are not translated into proteins (Mattick and Makunin, 2006). While some of these RNA products may have no specific cellular function, some classes of ncRNAs are functional and play a role in regulating cellular processes (Berretta and Morillon, 2009). For example, two of the most well-known classes of ncRNAs are ribosomal RNA and transfer RNA (Mattick and Makunin, 2006).

In addition to these well-known ncRNAs, many other classes of functional ncRNAs have been discovered. Some of these ncRNAs silence expression of specific gene targets through the RNAi pathway, whereas others act as guides for chromatin modifying enzymes. The function of many ncRNAs often lies in their ability to show sequence complementary to target nucleic acids. Sequence complementary allows ncRNAs to fulfill multiple roles in almost every step of eukaryotic gene expression and act as potential facilitators of gene regulation.

ncRNAs are generally divided into two classes based on their size (Morris and Mattick, 2014). Short ncRNAs range in size from 21 to 29 bp. These ncRNAs generally function by interacting with cellular machinery to repress gene expression. In contrast, long ncRNAs are greater than 200 bp in length (Mercer et al 2009). The mechanisms associated with long ncRNA biogenesis and function are less well defined.

A number of ncRNAs can mediate intragenerational epigenetic inheritance. For instance, some ncRNAs can mediate the repression of targeted genes. This repressed state can be transmitted across mitotic cell division and, consequently, represents epigenetic information (Mercer et al., 2009). Increasing evidence suggests that ncRNAs can also mediate intergenerational epigenetic inheritance (Lim and Brunet, 2013). For example, different types of ncRNAs can be transferred between parents and offspring (Lim and Brunet, 2013; Rassoulzadegan et al., 2006). This transfer represents the transmission of epigenetic information across organismal generations.

5.1 Short non-coding RNAs

There are three major classes of short ncRNAs: microRNAs (miRNAs), short interfering RNAs (siRNAs), and piwi-interacting RNAs (piRNAs) (Carthew and Sontheimer, 2009). In general, these ncRNAs interact with protein complexes to regulate the expression of targeted mRNAs. Short ncRNAs are also involved in other cellular processes such as chromatin remodelling and transgenerational epigenetic inheritance (Aravin et al., 2007; Carthew and Sontheimer, 2009).

microRNAs: miRNAs are short regulatory ncRNAs that are approximately 21 bp in length. miRNAs were first discovered in the 1990s and have since been found to be vital in many cellular processes like cell proliferation and embryonic pluripotency (Mattick and Makunin, 2006). The importance of miRNAs to cellular processes stems from the variety of roles that miRNAs play in gene regulation.

miRNAs are produced through a series of processing events that take place inside the cell. miRNA-encoding loci are first transcribed to produce long pri-miRNAs that form hairpin structures. These pri-miRNAs are further processed to generate pre-miRNAs that are exported out of the nucleus. Pre-miRNAs are then processed by the Dicer enzyme into mature double-stranded miRNAs. One strand of these double-stranded RNA products functions as the mature miRNA while the other is degraded.

The RNAi pathway is initiated when the mature miRNA is loaded into the RNA-induced silencing complex (RISC), a ribonucleoprotein complex with endonucleolytic activity. The RISC uses the base pairing capabilities of the miRNA and binds to the 3′-UTR or the open reading frame of the target mRNA. Successful binding of the miRNA initiates either mRNA degradation or translational repression (Filipowicz et al., 2008). mRNA degradation through endonucleolytic cleavage occurs when there is perfect complementary between the miRNA and the target mRNA. Mismatches in miRNA-mRNA binding prevent cleavage and promote translational repression (Brennecke et al., 2005; Carthew and Sontheimer, 2009).

In addition to transcriptional and translational regulation, miRNAs are also involved in epigenetic regulation. Some miRNAs can indirectly control chromatin structure by regulating the expression of genes responsible for heterochromatin formation (Carthew and Sontheimer, 2009). miRNAs can also work as mediators of intergenerational epigenetic inheritance. For example, gene regulatory states facilitated by miRNAs can be inherited from parent to offspring through the transfer of miRNAs through the gametes (Lim and Brunet, 2013; Wagner et al., 2008).

Short interfering RNAs: siRNAs are small ncRNAs that associate with RISC to regulate gene expression. siRNAs are very similar to miRNAs in that they interact with proteins from the same families and that they participate in the RNAi pathway. The main distinction between siRNAs and miRNAs is their biogenesis.

siRNAs are derived from long double-stranded RNA precursors, in contrast to miRNAs which originate from stem-loop RNA precursors (Carthew and Sontheimer, 2009; Rana, 2007). Several types of dsRNAs, such as viral RNA, repeat associated transcripts, and transgenes, can be processed into siRNAs (Carthew and Sontheimer, 2009). Mature siRNAs are then assembled into the RISC where they bind to target mRNAs and regulate gene expression through the RNAi pathway. Perfect sequence complementarity promotes mRNA cleavage, whereas mismatches induce silencing through translational repression or exonucleolytic degradation (Carthew and Sontheimer, 2009). However, the frequency at which endogenous siRNAs silence imperfectly matched targets remains unknown (Carthew and Sontheimer, 2009).

The function of siRNAs is not just restricted to transcriptional regulation. Several studies suggest that siRNAs are involved in regulating chromatin states (Carthew and Sontheimer, 2009). In this case, siRNAs are loaded into the RNA-induced transcriptional silencing complex (RITSC). Using the base pairing abilities of the siRNA, the RITSC binds to the specific target and facilitates heterochromatin formation by recruiting chromatin modifying enzymes to the target sequence (Carthew and Sontheimer, 2009). Such local changes in chromatin structure can also affect alternative splicing when siRNAs are targeted toward sequences near an alternative exon (Allo et al., 2009).

Piwi-interacting RNAs: piRNAs are short ncRNAs (~29 bp) that interact with Piwi proteins. Piwi proteins are members of the Argonaute protein family that, in general, use the base pairing abilities of small ncRNAs to bind to specific targets (Luteijn and Ketting, 2013). piRNAs bind to Piwi proteins to regulate the expression of TEs (Sienski et al., 2012). The Piwi protein–RNA complex binds to TE loci, initiating the deposition of heterochromatic marks and a silent chromatin state (Sienski et al., 2012).

piRNAs can be transferred from parent to offspring, suggesting that piRNA-mediated chromatin states can be stably inherited across generations (Brennecke et al., 2008; Luteijn and Ketting, 2013). For instance, piRNAs can be maternally inherited between generations through maternal cytoplasm (Brennecke et al., 2008). The information stored by this initial cache of piRNAs would allow for the re-establishment of chromatin states after meiosis (Luteijn and Ketting, 2013).

5.2 Long non-coding RNAs

Long ncRNAs represent a very broad category of regulatory molecules that regulate genes in both *cis* and *trans* (Mercer et al., 2009; Morris and Mattick, 2014; Ponting et al., 2009). Genes close to the long ncRNAs can be *cis*-regulated through transcriptional interference, which occurs when a long ncRNA overlaps with the promoter region of another gene. Transcription of the long ncRNA blocks the binding of regulatory factors needed for the transcription of the other target gene (Martens et al., 2004). More distant genes are *trans*-regulated when long ncRNAs bind to DNA regulatory regions to either block or activate transcription (Ponting et al., 2009). Some long ncRNAs also act as inhibitors by directly binding RNA Pol II (Ponting et al., 2009).

In addition to transcriptional regulation, some long ncRNAs may act as facilitators of epigenetic regulation. Many long ncRNAs can bind to specific targets and recruit chromatin modifying enzymes, initiating the formation of a silent chromatin state (Morris and Mattick, 2014; Ponting et al., 2009). This capability has made long ncRNAs important to the regulation of epigenetic processes such as X-chromosome dosage compensation and parental imprinting in mammals (Ponting et al., 2009). Long ncRNA-induced heterochromatin formation can also indirectly control alternative splicing because chromatin structure affects which exons are included in transcripts (Morris and Mattick, 2014).

5.3 Noncoding RNAs in social insects

Of all the ncRNA classes, miRNAs have been the most well studied in social insects. When the *A. mellifera* genome was first sequenced, miRNAs were discovered using computational methods based on sequence homology to known miRNAs in *D. melanogaster* (Weaver et al., 2007). Several subsequent investigations have examined the expression of miRNAs in different social insect phenotypes and found that miRNAs are sometimes differentially expressed between *A. mellifera* castes (Behura and Whitfield, 2010; Guo et al., 2013; Liu et al., 2012; Weaver et al., 2007; Zondag et al., 2012) and *C. floridanus* and *H. saltator* ant castes (Bonasio et al., 2010b). Such expression specificity suggests that miRNAs may play regulatory roles in the development of different phenotypes.

Studies of other ncRNAs in social insects have focused on establishing the existence of different ncRNA classes. For instance, only two piRNAs have been discovered in *A. mellifera* and both are highly expressed in drones when compared to queens and workers (Liao et al., 2010). Like piRNAs, the functions of long ncRNAs in social insects remain unclear. Several long ncRNAs have been discovered in *A. mellifera* (Humann et al., 2013; Kiya et al., 2012; Sawata et al., 2002, 2004). Furthermore, some of these long ncRNAs are differentially expressed between castes or tissues. Such differential expression suggests that these ncRNAs may play a role in phenotypic differentiation (Humann et al., 2013; Sawata et al., 2004).

We are just beginning to understand the properties of ncRNAs in 'wild-type' social insects. However, ncRNAs have successfully been used to study gene function in several social insect species. Specifically, RNAi has become an important tool to study the operation of genes. This technique can be used to knock down expression levels of specific genes by exploiting the endogenous RNAi pathway associated with miRNAs and siRNAs.

So far, RNAi has been successful in ants, bees, wasps, and termites (Table 2). Many investigations using RNAi have targeted genes associated

Table 2 RNAi studies conducted to silence expression of genes in social insects

Group	Species	Target gene	References
Termite	*Cryptotermes secundus*	*Neofem2*	Korb et al. (2009)
	Reticulitermes flavipes	*Hexamerin*	Nambu et al. (2010), Zhou et al. (2006), and Zhou et al. (2008)
	Reticulitermes flavipes	*Deviate*	Schwinghammer et al. (2011)
	Reticulitermes flavipes	*Termicin*	Hamilton and Bulmer (2012)
	Reticulitermes flavipes	*Bacteria binding proteins*	Hamilton and Bulmer (2012)
	Hodotermopsis sjostedti	*Insulin receptor*	Hattori et al. (2013)
Bee	*Apis cerana*	*Glutathione S-transferase*	Meng et al. (2014)
	Apis mellifera	*Chemosensory Protein 5*	Maleszka et al. (2007)
	Apis mellifera	*Complementary sex determiner*	Gempe et al. (2009) and Hasselmann et al. (2008)
	Apis mellifera	*Deformed wing virus*	Desai et al. (2012)
	Apis mellifera	*DNA methyltransferase 3*	Kucharski et al. (2008) and Li-Byarlay et al. (2013)
	Apis mellifera	*Dpp*	Wilson et al. (2014)
	Apis mellifera	*Epidermal growth factor receptor*	Formesyn et al. (2014) and Kamakura (2011)
	Apis mellifera	*Feminizer*	Gempe et al. (2009) and Hasselmann et al. (2008)
	Apis mellifera	*Gemini transcription factor*	Jarosch et al. (2011)
	Apis mellifera	*Glutamate-gated chloride channel*	El Hassani et al. (2012)
	Apis mellifera	*Glycerol-3-phosphate dehydrogenase*	Jarosch and Moritz (2011)
	Apis mellifera	*Hypopharyngeal amylase*	Gatehouse et al. (2004)
	Apis mellifera	*Insulin receptor substrate*	Mutti et al. (2011), Wang et al. (2010), and Wolschin et al. (2011)

Table 2 RNAi studies conducted to silence expression of genes in social insects—cont'd

Group	Species	Target gene	References
	Apis mellifera	*Israeli Acute Paralysis Virus*	Hunter et al. (2010) and Maori et al. (2009)
	Apis mellifera	*Juvenile hormone esterase*	Mackert et al. (2008)
	Apis mellifera	*Long-wavelength opsin 1*	Leboulle et al. (2013)
	Apis mellifera	*Methyl farnesoate epoxidase-encoding gene*	Bomtorin et al. (2014)
	Apis mellifera	*Notch*	Wilson et al. (2010)
	Apis mellifera	*Octopamine receptor*	Farooqui et al. (2004)
	Apis mellifera	*Relish*	Schluns and Crozier (2007)
	Apis mellifera	*Toll*	Wilson et al. (2014)
	Apis mellifera	*Tor*	Mutti et al. (2011) and Patel et al. (2007)
	Apis mellifera	*Ultrabithorax*	Medved et al. (2014)
	Apis mellifera	*Ultraspriracle*	Ament et al. (2012) and Wang et al. (2012)
	Apis mellifera	*Vitellogenin*	Amdam et al. (2003), Amdam et al. (2006), Amdam et al. (2007), Antonio et al. (2008), Ihle et al. (2010), Nelson et al. (2007), Nilsen et al. (2011), Nunes and Simoes (2009), Nunes et al. (2013), Wang et al. (2012), and Wheeler et al. (2013)
	Bombus ignitus	*Transferrin*	Kim et al. (2009)
	Bombus ignitus	*Ferritin*	Kim et al. (2009)
	Bombus ignitus	*Peroxiredoxins*	Hu et al. (2010)
	Bombus terrestris	*Defensin*, *Abaecin*, and *Nautilus*	Deshwal and Mallon (2014)
Ant	*Camponotus floridanus*	*Peptidoglycan recognition proteins*	Ratzka et al. (2013)
	Solenopsis invicta	*Pheromone biosynthesis activating neuropeptide*	Choi and Vander Meer (2012) and Choi et al. (2012)
	Solenopsis invicta	*Vitellogenin receptor*	Lu et al. (2009)
Wasp	*Polistes metricus*	*Hexamerin*	Hunt et al. (2011b)

with caste and phenotype differentiation (Amdam et al., 2003, 2006, 2007; Hunt et al., 2011b; Ihle et al., 2010; Lu et al., 2009; Nelson et al., 2007; Zhou et al., 2006, 2008). RNAi has also been used to treat and prevent viral infections (Desai et al., 2012; Deshwal and Mallon, 2014; Hunter et al., 2010; Maori et al., 2009). These studies demonstrate the potential use of RNAi for disease control in bees and other social insects.

6. FUTURE DIRECTIONS

The study of epigenetic inheritance in social insects holds great promise. However, there is a great deal of work to be done. For example, although patterns of DNA methylation have been the focus of considerable research in hymenopteran social insects, very little is known about the distribution and effects of histone posttranslational modifications or ncRNAs. Therefore, we know very little about the genomic distribution of two of the major epigenetic inheritance systems, or about how different epigenetic information systems interact in social insect species.

Moreover, the majority of what is known about the function of epigenetic systems in insects is inferred from other eukaryotic systems. The social insect community would be well served by more directed biochemical and genetic studies on the function of genes involved in epigenetic processes in social insects. In addition, more experimental manipulations of epigenetic systems through techniques such as RNAi would be welcome. Such studies are difficult, but may be particularly helpful in elucidating the function of epigenetic information.

Overall, social insects will continue to serve as fascinating and important models for understanding epigenetic inheritance. The continued research on social insect epigenetics will provide critical information on the processes underlying complex development. In addition, such studies will further our understanding of the factors influencing the elaboration of social systems, which define complex insect societies.

REFERENCES

Allo, M., Buggiano, V., Fededa, J.P., Petrillo, E., Schor, I., de la Mata, M., Agirre, E., Plass, M., Eyras, E., Abou Elela, S., Klinck, R., Chabot, B., Kornblihtt, A.R., 2009. Control of alternative splicing through siRNA-mediated transcriptional gene silencing. Nat. Struct. Mol. Biol. 16, 717–724.

Amarasinghe, H.E., Clayton, C.I., Mallon, E.B., 2014. Methylation and worker reproduction in the bumble-bee (Bombus terrestris). Proc. R. Soc. Lond. B 281, 1780. http://dx.doi.org/10.1098/Rspb.2013.2502.

Amdam, G.V., Simoes, Z.L.P., Guidugli, K.R., Norberg, K., Omholt, S.W., 2003. Disruption of vitellogenin gene function in adult honeybees by intraabdominal injection of double-stranded RNA. BMC Biotechnol. 3, Article 1.

Amdam, G.V., Csondes, A., Fondrk, M.K., Page Jr., R.E., 2006. Complex social behaviour derived from maternal reproductive traits. Nature 439, 76–78.

Amdam, G.V., Nilsen, K.A., Norberg, K., Fondrk, M.K., Hartfelder, K., 2007. Variation in endocrine signaling underlies variation in social life history. Am. Nat. 170, 37–46.

Ament, S.A., Wang, Y., Chen, C.-C., Blatti, C.A., Hong, F., Liang, Z.S., Negre, N., White, K.P., Rodriguez-Zas, S.L., Mizzen, C.A., Sinha, S., Zhong, S., Robinson, G.E., 2012. The transcription factor *ultraspiracle* influences honey bee social behavior and behavior-related gene expression. PLoS Genet. 8, e1002596.

Antonio, D.S.M., Guidugli-Lazzarini, K.R., do Nascimento, A.M., Simoes, Z.L.P., Hartfelder, K., 2008. RNAi-mediated silencing of vitellogenin gene function turns honeybee (Apis mellifera) workers into extremely precocious foragers. Naturwissenschaften 95, 953–961.

Aravin, A.A., Hannon, G.J., Brennecke, J., 2007. The Piwi-piRNA pathway provides an adaptive defense in the transposon arms race. Science 318, 761–764.

Ast, G., 2004. How did alternative splicing evolve? Nat. Rev. Genet. 5, 773–782.

Badeaux, A.I., Shi, Y., 2013. Emerging roles for chromatin as a signal integration and storage platform. Nat. Rev. Mol. Cell Biol. 14, 211–224.

Baldi, S., Becker, P., 2013. The variant histone H2A.V of Drosophila—three roles, two guises. Chromosoma 122, 245–258.

Bannister, A.J., Kouzarides, T., 2011. Regulation of chromatin by histone modifications. Cell Res. 21, 381–395.

Barberan-Soler, S., Zahler, A.M., 2008. Alternative splicing regulation during C-elegans development: splicing factors as regulated targets. PLoS Genet. 4, ArtNo.: e1000001.

Beeler, S.M., Wong, G.T., Zheng, J.M., Bush, E.C., Remnant, E.J., Oldroyd, B.P., Drewell, R.A., 2014. Whole-genome DNA methylation profile of the jewel wasp (Nasonia vitripennis). G3 (Bethesda) 4, 383–388.

Behura, S.K., Whitfield, C.W., 2010. Correlated expression patterns of microRNA genes with age-dependent behavioural changes in honeybee. Insect Mol. Biol. 19, 431–439.

Bell, O., Tiwari, V.K., Thoma, N.H., Schubeler, D., 2011. Determinants and dynamics of genome accessibility. Nat. Rev. Genet. 12, 554–564.

Bernstein, B.E., Meissner, A., Lander, E.S., 2007. The mammalian epigenome. Cell 128, 669–681.

Berretta, J., Morillon, A., 2009. Pervasive transcription constitutes a new level of eukaryotic genome regulation. EMBO Rep. 10, 973–982.

Bigot, Y., Jegot, G., Casteret, S., Aupinel, P., Tasei, J.N., 2011. DNA modifications and genome rearrangements during the development and sex differentiation of the bumble bee Bombus terrestris. Insect Mol. Biol. 20, 165–175.

Bintu, L., Ishibashi, T., Dangkulwanich, M., Wu, Y.Y., Lubkowska, L., Kashlev, M., Bustamante, C., 2012. Nucleosomal elements that control the topography of the barrier to transcription. Cell 151, 738–749.

Bogdanovic, O., Veenstra, G.J.C., 2009. DNA methylation and methyl-CpG binding proteins: developmental requirements and function. Chromosoma 118, 549–565.

Bomtorin, A.D., Mackert, A., Rosa, G.C.C., Moda, L.M., Martins, J.R., Bitondi, M.M.G., Hartfelder, K., Simoes, Z.L.P., 2014. Juvenile hormone biosynthesis gene expression in the corpora allata of honey bee (Apis mellifera L.) female castes. PLoS One 9, e86923.

Bonasio, R., 2012. Emerging topics in epigenetics: ants, brains, and noncoding RNAs. Ann. N.Y. Acad. Sci. 1260, 14–23.

Bonasio, R., 2014. The role of chromatin and epigenetics in the polyphenisms of ant castes. Brief. Funct. Genomics 13, 235–245.

Bonasio, R., Tu, S.J., Reinberg, D., 2010a. Molecular signals of epigenetic states. Science 330, 612–616.
Bonasio, R., Zhang, G.J., Ye, C.Y., Mutti, N.S., Fang, X.D., Qin, N., Donahue, G., Yang, P.C., Li, Q.Y., Li, C., Zhang, P., Huang, Z.Y., Berger, S.L., Reinberg, D., Wang, J., Liebig, J., 2010b. Genomic comparison of the ants *Camponotus floridanus* and *Harpegnathos saltator*. Science 329, 1068–1071.
Bonasio, R., Li, Q.Y., Lian, J.M., Mutti, N.S., Jin, L.J., Zhao, H.M., Zhang, P., Wen, P., Xiang, H., Ding, Y., Jin, Z.H., Shen, S.S., Wang, Z.J., Wang, W., Wang, J., Berger, S.L., Liebig, J., Zhang, G.J., Reinberg, D., 2012. Genome-wide and caste-specific DNA methylomes of the ants *Camponotus floridanus* and *Harpegnathos saltator*. Curr. Biol. 22, 1755–1764.
Bonduriansky, R., Day, T., 2009. Nongenetic inheritance and its evolutionary implications. Annu. Rev. Ecol. Evol. Syst. 40, 103–125.
Bossdorf, O., Richards, C.L., Pigliucci, M., 2008. Epigenetics for ecologists. Ecol. Lett. 11, 106–115.
Brennecke, J., Stark, A., Russell, R.B., Cohen, S.M., 2005. Principles of microRNA-target recognition. PLoS Biol. 3, 404–418.
Brennecke, J., Malone, C.D., Aravin, A.A., Sachidanandam, R., Stark, A., Hannon, G.J., 2008. An epigenetic role for maternally inherited piRNAs in transposon silencing. Science 322, 1387–1392.
Cameron, R.C., Duncan, E.J., Dearden, P.K., 2013. Biased gene expression in early honeybee larval development. BMC Genomics 14, 903. http://dx.doi.org/10.1186/1471-2164-14-903.
Carthew, R.W., Sontheimer, E.J., 2009. Origins and mechanisms of miRNAs and siRNAs. Cell 136, 642–655.
Cedar, H., Bergman, Y., 2009. Linking DNA methylation and histone modification: patterns and paradigms. Nat. Rev. Genet. 10, 295–304.
Cedar, H., Bergman, Y., 2012. Programming of DNA methylation patterns. Annu. Rev. Biochem. 81, 97–117.
Chen, Z.X., Riggs, A.D., 2011. DNA methylation and demethylation in mammals. J. Biol. Chem. 286, 18347–18353.
Cheong, J., Yamada, Y., Yamashita, R., Irie, T., Kanai, A., Wakaguri, H., Nakai, K., Ito, T., Saito, I., Sugano, S., Suzuki, Y., 2006. Diverse DNA methylation statuses at alternative promoters of human genes in various tissues. DNA Res. 13, 155–167.
Chodavarapu, R.K., Feng, S.H., Bernatavichute, Y.V., Chen, P.Y., Stroud, H., Yu, Y.C., Hetzel, J.A., Kuo, F., Kim, J., Cokus, S.J., Casero, D., Bernal, M., Huijser, P., Clark, A.T., Kramer, U., Merchant, S.S., Zhang, X.Y., Jacobsen, S.E., Pellegrini, M., 2010. Relationship between nucleosome positioning and DNA methylation. Nature 466, 388–392.
Choi, M.-Y., Vander Meer, R.K., 2012. Ant trail pheromone biosynthesis is triggered by a neuropeptide hormone. PLoS One 7, ArtNo.: e50400.
Choi, M.Y., Meer, R.K.V., Coy, M., Scharf, M.E., 2012. Phenotypic impacts of PBAN RNA interference in an ant, *Solenopsis invicta*, and a moth, *Helicoverpa zea*. J. Insect Physiol. 58, 1159–1165.
Cingolani, P., Cao, X.Y., Khetani, R.S., Chen, C.C., Coon, M., Sammak, A., Bollig-Fischer, A., Land, S., Huang, Y., Hudson, M.E., Garfinkel, M.D., Zhong, S., Robinson, G.E., Ruden, D.M., 2013. Intronic Non-CG DNA hydroxymethylation and alternative mRNA splicing in honey bees. BMC Genomics 14, ArtNo.: 666.
Clouaire, T., Stancheva, I., 2008. Methyl-CpG binding proteins: specialized transcriptional repressors or structural components of chromatin? Cell. Mol. Life Sci. 65, 1509–1522.
Coleman-Derr, D., Zilberman, D., 2012. Deposition of histone variant H2A.Z within gene bodies regulates responsive genes. PLoS Genet. 8, ArtNo.: e1002988.

Coolon, J.D., Stevenson, K.R., McManus, C.J., Graveley, B.R., Wittkopp, P.J., 2012. Genomic imprinting absent in Drosophila melanogaster adult females. Cell Rep. 2, 69–75.

Danchin, E., Charmantier, A., Champagne, F.A., Mesoudi, A., Pujol, B., Blanchet, S., 2011. Beyond DNA: integrating inclusive inheritance into an extended theory of evolution. Nat. Rev. Genet. 12, 475–486.

de la Mata, M., Alonso, C.R., Kadener, S., Fededa, J.P., Blaustein, M., Pelisch, F., Cramer, P., Bentley, D., Kornblihtt, A.R., 2003. A slow RNA polymerase II affects alternative splicing in vivo. Mol. Cell 12, 525–532.

de Ruijter, A.J.M., van Gennip, A.H., Caron, H.N., Kemp, S., van Kuilenburg, A.B.P., 2003. Histone deacetylases (HDACs): characterization of the classical HDAC family. Biochem. J. 370, 737–749.

Desai, S.D., Eu, Y.J., Whyard, S., Currie, R.W., 2012. Reduction in deformed wing virus infection in larval and adult honey bees (Apis mellifera L.) by double-stranded RNA ingestion. Insect Mol. Biol. 21, 446–455.

Deshwal, S., Mallon, E.B., 2014. Antimicrobial peptides play a functional role in bumblebee anti-trypanosome defense. Dev. Comp. Immunol. 42, 240–243.

Dewitt, T.J., Scheiner, S.M., 2004. Phenotypic Plasticity: Functional and Conceptual Approaches. Oxford University Press, Oxford.

Dickman, M.J., Kucharski, R., Maleszka, R., Hurd, P.J., 2013. Extensive histone post-translational modification in honey bees. Insect Biochem. Mol. Biol. 43, 125–137.

Dobata, S., Tsuji, K., 2012. Intragenomic conflict over queen determination favours genomic imprinting in eusocial Hymenoptera. Proc. R. Soc. Lond. B 279, 2553–2560.

Drewell, R.A., Lo, N., Oxley, P.R., Oldroyd, B.P., 2012. Kin conflict in insect societies: a new epigenetic perspective. Trends Ecol. Evol. 27, 367–373. http://dx.doi.org/10.1016/j.tree.2012.02.005.

Drewell, R.A., Bush, E.C., Remnant, E.J., Wong, G.T., Beeler, S.M., Stringham, J.L., Lim, J., Oldroyd, B.P., 2014. The dynamic DNA methylation cycle from egg to sperm in the honey bee Apis mellifera. Development 141, 2702–2711.

Dujardin, G., Lafaille, C., de la Mata, M., Marasco, Luciano E., Muñoz, Manuel J., Le Jossic-Corcos, C., Corcos, L., Kornblihtt, Alberto R., 2014. How slow RNA polymerase II elongation favors alternative exon skipping. Mol. Cell 54, 683–690.

El Hassani, A.K., Schuster, S., Dyck, Y., Demares, F., Leboulle, G., Armengaud, C., 2012. Identification, localization and function of glutamate-gated chloride channel receptors in the honeybee brain. Eur. J. Neurosci. 36, 2409–2420.

Elango, N., Hunt, B.G., Goodisman, M.A.D., Yi, S.V., 2009. DNA methylation is widespread and associated with differential gene expression in castes of the honeybee, *Apis mellifera*. Proc. Natl. Acad. Sci. U.S.A. 106, 11206–11211.

Falckenhayn, C., Boerjan, B., Raddatz, G., Frohme, M., Schoofs, L., Lyko, F., 2013. Characterization of genome methylation patterns in the desert locust Schistocerca gregaria. J. Exp. Biol. 216, 1423–1429.

Farooqui, T., Vaessin, H., Smith, B.H., 2004. Octopamine receptors in the honeybee (Apis mellifera) brain and their disruption by RNA-mediated interference. J. Insect Physiol. 50, 701–713.

Felsenfeld, G., 2014. A brief history of epigenetics. Cold Spring Harb. Perspect. Biol. 6, ArtNo.: a018200.

Feng, S.H., Cokus, S.J., Zhang, X.Y., Chen, P.Y., Bostick, M., Goll, M.G., Hetzel, J., Jain, J., Strauss, S.H., Halpern, M.E., Ukomadu, C., Sadler, K.C., Pradhan, S., Pellegrini, M., Jacobsen, S.E., 2010. Conservation and divergence of methylation patterning in plants and animals. Proc. Natl. Acad. Sci. U.S.A. 107, 8689–8694.

Ferguson-Smith, A.C., 2011. Genomic imprinting: the emergence of an epigenetic paradigm. Nat. Rev. Genet. 12, 565–575.
Field, L.M., Lyko, F., Mandrioli, M., Prantera, G., 2004. DNA methylation in insects. Insect Mol. Biol. 13, 109–115.
Filipowicz, W., Bhattacharyya, S.N., Sonenberg, N., 2008. Mechanisms of post-transcriptional regulation by microRNAs: are the answers in sight? Nat. Rev. Genet. 9, 102–114.
Flores, K.B., Amdam, G.V., 2011. Deciphering a methylome: what can we read into patterns of DNA methylation? J. Exp. Biol. 214, 3155–3163.
Flores, K., Wolschin, F., Corneveaux, J.J., Allen, A.N., Huentelman, M.J., Amdam, G.V., 2012. Genome-wide association between DNA methylation and alternative splicing in an invertebrate. BMC Genomics 13, ArtNo.: 480.
Foret, S., Kucharski, R., Pittelkow, Y., Lockett, G.A., Maleszka, R., 2009. Epigenetic regulation of the honey bee transcriptome: unravelling the nature of methylated genes. BMC Genomics 10, ArtNo.: 472.
Foret, S., Kucharski, R., Pellegrini, M., Feng, S.H., Jacobsen, S.E., Robinson, G.E., Maleszka, R., 2012. DNA methylation dynamics, metabolic fluxes, gene splicing, and alternative phenotypes in honey bees. Proc. Natl. Acad. Sci. U.S.A. 109, 4968–4973.
Formesyn, E.M., Cardoen, D., Ernst, U.R., Danneels, E.L., Van Vaerenbergh, M., De Koker, D., Verleyen, P., Wenseleers, T., Schoofs, L., de Graaf, D.C., 2014. Reproduction of honeybee workers is regulated by epidermal growth factor receptor signaling. Gen. Comp. Endocrinol. 197, 1–4.
Gadau, J., Helmkampf, M., Nygaard, S., Roux, J., Simola, D.F., Smith, C.R., Suen, G., Wurm, Y., Smith, C.D., 2012. The genomic impact of 100 million years of social evolution in seven ant species. Trends Genet. 28, 14–21.
Gatehouse, H.S., Gatehouse, L.N., Malone, L.A., Hodges, S., Tregidga, E., Todd, J., 2004. Amylase activity in honey bee hypopharyngeal glands reduced by RNA interference. J. Apicult. Res. 43, 9–13.
Gelfman, S., Cohen, N., Yearim, A., Ast, G., 2013. DNA-methylation effect on cotranscriptional splicing is dependent on GC architecture of the exon-intron structure. Genome Res. 23, 789–799.
Gempe, T., Hasselmann, M., Schiott, M., Hause, G., Otte, M., Beye, M., 2009. Sex determination in honeybees: two separate mechanisms induce and maintain the female pathway. PLoS Biol. 7, ArtNo.: e1000222.
Glastad, K.M., Hunt, B.G., Yi, S.V., Goodisman, M.A.D., 2011. DNA methylation in insects: on the brink of the epigenomic era. Insect Mol. Biol. 20, 553–565.
Glastad, K.M., Hunt, B.G., Goodisman, M.A.D., 2013. Evidence of a conserved functional role for DNA methylation in termites. Insect Mol. Biol. 22, 143–154.
Glastad, K.M., Hunt, B.G., Goodisman, M.A.D., 2014a. Evolutionary insights into DNA methylation in insects. Curr. Opin. Insect Sci. 1, 25–30.
Glastad, K.M., Hunt, B.G., Yi, S.V., Goodisman, M.A.D., 2014b. Epigenetic inheritance and genome regulation: is DNA methylation linked to ploidy in haplodiploid insects? Proc. R. Soc. Lond. B 281, ArtNo.: 20140411.
Greer, E.L., Shi, Y., 2012. Histone methylation: a dynamic mark in health, disease and inheritance. Nat. Rev. Genet. 13, 343–357.
Guo, X.Q., Su, S.K., Skogerboe, G., Dai, S.A.J., Li, W.F., Li, Z.G., Liu, F., Ni, R.F., Guo, Y., Chen, S.L., Zhang, S.W., Chen, R.S., 2013. Recipe for a busy bee: microRNAs in honey bee caste determination. PLoS One 8, e81661. http://dx.doi.org/10.1371/journal.pone.0081661.
Gustincich, S., Sandelin, A., Plessy, C., Katayama, S., Simone, R., Lazarevic, D., Hayashizaki, Y., Carninci, P., 2006. The complexity of the mammalian transcriptome. J. Physiol. Lond. 575, 321–332.

Guzman-Novoa, E., Hunt, G.J., Page, R.E., Uribe-Rubio, J.L., Prieto-Merlos, D., Becerra-Guzman, F., 2005. Paternal effects on the defensive behavior of honeybees. J. Hered. 96, 376–380.

Haig, D., 2000. The kinship theory of genomic imprinting. Annu. Rev. Ecol. Syst. 31, 9–32.

Haig, D., 2004a. The (dual) origin of epigenetics. Cold Spring Harb. Symp. Quant. Biol. 69, 67–70.

Haig, D., 2004b. Genomic imprinting and kinship: how good is the evidence? Annu. Rev. Genet. 38, 553–585.

Hamilton, C., Bulmer, M.S., 2012. Molecular antifungal defenses in subterranean termites: RNA interference reveals in vivo roles of termicins and GNBPs against a naturally encountered pathogen. Dev. Comp. Immunol. 36, 372–377.

Hasselmann, M., Gempe, T., Schiott, M., Nunes-Silva, C.G., Otte, M., Beye, M., 2008. Evidence for the evolutionary nascence of a novel sex determination pathway in honeybees. Nature 454, 519–522.

Hattori, A., Sugime, Y., Sasa, C., Miyakawa, H., Ishikawa, Y., Miyazaki, S., Okada, Y., Cornette, R., Lavine, L.C., Emlen, D.J., Koshikawa, S., Miura, T., 2013. Soldier morphogenesis in the damp-wood termite is regulated by the insulin signaling pathway. J. Exp. Zool. B Mol. Dev. Evol. 320B, 295–306.

Heard, E., Martienssen, R.A., 2014. Transgenerational epigenetic inheritance: myths and mechanisms. Cell 157, 95–109.

Hendrich, B., Tweedie, S., 2003. The methyl-CpG binding domain and the evolving role of DNA methylation in animals. Trends Genet. 19, 269–277.

Henikoff, S., 2008. Nucleosome destabilization in the epigenetic regulation of gene expression. Nat. Rev. Genet. 9, 15–26.

Herb, B.R., Wolschin, F., Hansen, K.D., Aryee, M.J., Langmead, B., Irizarry, R., Amdam, G.V., Feinberg, A.P., 2012. Reversible switching between epigenetic states in honeybee behavioral subcastes. Nat. Neurosci. 15, 1371–1373.

Hu, Z., Lee, K.S., Choo, Y.M., Yoon, H.J., Lee, S.M., Lee, J.H., Kim, D.H., Sohn, H.D., Jin, B.R., 2010. Molecular cloning and characterization of 1-Cys and 2-Cys peroxiredoxins from the bumblebee Bombus ignitus. Comp. Biochem. Physiol. B 155, 272–280.

Hublitz, P., Albert, M., Peters, A.H.F.M., 2009. Mechanisms of transcriptional repression by histone lysine methylation. Int. J. Dev. Biol. 53, 335–354.

Huh, I., Zeng, J., Park, T., Yi, S.V., 2013. DNA methylation and transcriptional noise. Epigenet. Chromatin 6, 9. http://dx.doi.org/10.1186/1756-8935-6-9.

Humann, F.C., Tiberio, G.J., Hartfelder, K., 2013. Sequence and expression characteristics of long noncoding RNAs in honey bee caste development—potential novel regulators for transgressive ovary size. PLoS One 8, e78915. http://dx.doi.org/10.1371/journal.pone.0078915.

Hung, M.S., Karthikeyan, N., Huang, B.L., Koo, H.C., Kiger, J., Shen, C.K.J., 1999. Drosophila proteins related to vertebrate DNA (5-cytosine) methyltransferases. Proc. Natl. Acad. Sci. U.S.A. 96, 11940–11945.

Hunt, B.G., Brisson, J.A., Yi, S.V., Goodisman, M.A.D., 2010. Functional conservation of DNA methylation in the pea aphid and the honeybee. Genome Biol. Evol. 2, 719–728.

Hunt, B.G., Ometto, L., Wurm, Y., Shoemaker, D., Yi, S.V., Keller, L., Goodisman, M.A. D., 2011a. Relaxed selection is a precursor to the evolution of phenotypic plasticity. Proc. Natl. Acad. Sci. U.S.A. 108, 15936–15941.

Hunt, J.H., Mutti, N.S., Havukainen, H., Henshaw, M.T., Amdam, G.V., 2011b. Development of an RNA interference tool, characterization of its target, and an ecological test of caste differentiation in the eusocial wasp *Polistes*. PLoS One 6, ArtNo.: e26641.

Hunt, B.G., Glastad, K.M., Yi, S.V., Goodisman, M.A.D., 2013a. The function of intragenic DNA methylation: insights from insect epigenomes. Integr. Comp. Biol. 53, 319–328.

Hunt, B.G., Glastad, K.M., Yi, S.V., Goodisman, M.A.D., 2013b. Patterning and regulatory associations of DNA methylation are mirrored by histone modifications in insects. Genome Biol. Evol. 5, 591–598.

Hunter, W., Ellis, J., Vanengelsdorp, D., Hayes, J., Westervelt, D., Glick, E., Williams, M., Sela, I., Maori, E., Pettis, J., Cox-Foster, D., Paldi, N., 2010. Large-scale field application of RNAi technology reducing Israeli acute paralysis virus disease in honey bees (Apis mellifera, Hymenoptera: Apidae). PLoS Pathog. 6. http://dx.doi.org/10.1371/journal.ppat.1001160.

Ihle, K.E., Page, R.E., Frederick, K., Fondrk, M.K., Amdam, G.V., 2010. Genotype effect on regulation of behaviour by vitellogenin supports reproductive origin of honeybee foraging bias. Anim. Behav. 79, 1001–1006.

Ikeda, T., Furukawa, S., Nakamura, J., Sasaki, M., Sasaki, T., 2011. CpG methylation in the hexamerin 110 gene in the European honeybee, Apis mellifera. J. Insect. Sci. 11, ArtNo.: 74.

Jablonka, E., Lamm, E., 2012. Commentary: the epigenotype—a dynamic network view of development. Int. J. Epidemiol. 41, 16–20.

Jablonka, E., Raz, G., 2009. Transgenerational epigenetic inheritance: prevalence, mechanisms, and implications for the study of heredity and evolution. Q. Rev. Biol. 84, 131–176.

Jarosch, A., Moritz, R.F.A., 2011. Systemic RNA-interference in the honeybee Apis mellifera: tissue dependent uptake of fluorescent siRNA after intra-abdominal application observed by laser-scanning microscopy. J. Insect Physiol. 57, 851–857.

Jarosch, A., Stolle, E., Crewe, R.M., Moritz, R.F.A., 2011. Alternative splicing of a single transcription factor drives selfish reproductive behavior in honeybee workers (*Apis mellifera*). Proc. Natl. Acad. Sci. U.S.A. 108, 15282–15287.

Jones, P.A., 2012. Functions of DNA methylation: islands, start sites, gene bodies and beyond. Nat. Rev. Genet. 13, 484–492.

Kamakura, M., 2011. Royalactin induces queen differentiation in honeybees. Nature 473, 478–483.

Kharchenko, P.V., Alekseyenko, A.A., Schwartz, Y.B., Minoda, A., Riddle, N.C., Ernst, J., Sabo, P.J., Larschan, E., Gorchakov, A.A., Gu, T.T., Linder-Basso, D., Plachetka, A., Shanower, G., Tolstorukov, M.Y., Luquette, L.J., Xi, R.B., Jung, Y.L., Park, R.W., Bishop, E.P., Canfield, T.K., Sandstrom, R., Thurman, R.E., MacAlpine, D.M., Stamatoyannopoulos, J.A., Kellis, M., Elgin, S.C.R., Kuroda, M.I., Pirrotta, V., Karpen, G.H., Park, P.J., 2011. Comprehensive analysis of the chromatin landscape in *Drosophila melanogaster*. Nature 471, 480–485.

Kim, B.Y., Lee, K.S., Yoon, H.J., Kim, I., Li, J.H., Sohn, H.D., Jin, B.R., 2009. Expression profile of the iron-binding proteins transferrin and ferritin heavy chain subunit in the bumblebee *Bombus ignitus*. Comp. Biochem. Physiol. 153, 165–170.

Kiya, T., Ugajin, A., Kunieda, T., Kubo, T., 2012. Identification of kakusei, a nuclear non-coding RNA, as an immediate early gene from the honeybee, and its application for neuroethological study. Int. J. Mol. Sci. 13, 15496–15509.

Klose, R.J., Bird, A.P., 2006. Genomic DNA methylation: the mark and its mediators. Trends Biochem. Sci. 31, 89–97.

Kocher, S.D., Li, C., Yang, W., Tan, H., Yi, S.V., Yang, X.Y., Hoekstra, H.E., Zhang, G.J., Pierce, N.E., Yu, D.W., 2013. The draft genome of a socially polymorphic halictid bee, Lasioglossum albipes. Genome Biol. 14, ArtNo.: R142.

Kohler, C., Wolff, P., Spillane, C., 2012. Epigenetic mechanisms underlying genomic imprinting in plants. Annu. Rev. Plant Biol. 63, 331–352.

Korb, J., Weil, T., Hoffmann, K., Foster, K.R., Rehli, M., 2009. A gene necessary for reproductive suppression in termites. Science 324, 758.
Kornblihtt, A.R., Schor, I.E., Allo, M., Dujardin, G., Petrillo, E., Munoz, M.J., 2013. Alternative splicing: a pivotal step between eukaryotic transcription and translation. Nat. Rev. Mol. Cell Biol. 14, 153–165.
Kronauer, D.J.C., 2008. Genomic imprinting and kinship in the social Hymenoptera: what are the predictions? J. Theor. Biol. 254, 737–740.
Kronforst, M.R., Gilley, D.C., Strassmann, J.E., Queller, D.C., 2008. DNA methylation is widespread across social Hymenoptera. Curr. Biol. 18, R287–R288.
Kucharski, R., Maleszka, J., Foret, S., Maleszka, R., 2008. Nutritional control of reproductive status in honeybees via DNA methylation. Science 319, 1827–1830.
Kulaeva, O.I., Hsieh, F.K., Studitsky, V.M., 2010. RNA polymerase complexes cooperate to relieve the nucleosomal barrier and evict histones. Proc. Natl. Acad. Sci. U.S.A. 107, 11325–11330.
Lan, J., Hua, S., He, X.N., Zhang, Y., 2010. DNA methyltransferases and methyl-binding proteins of mammals. Acta Biochim. Biophys. Sin. 42, 243–252.
Law, J.A., Jacobsen, S.E., 2010. Establishing, maintaining and modifying DNA methylation patterns in plants and animals. Nat. Rev. Genet. 11, 204–220.
Leboulle, G., Niggebrugge, C., Roessler, R., Briscoe, A.D., Menzel, R., de Ibarra, N.H., 2013. Characterisation of the RNA interference response against the long-wavelength receptor of the honeybee. Insect Biochem. Mol. Biol. 43, 959–969.
Ledon-Rettig, C.C., Richards, C.L., Martin, L.B., 2013. Epigenetics for behavioral ecologists. Behav. Ecol. 24, 311–324.
Li, Y.F., Sasaki, H., 2011. Genomic imprinting in mammals: its life cycle, molecular mechanisms and reprogramming. Cell Res. 21, 466–473.
Li, Y.R., Zhu, J.D., Tian, G., Li, N., Li, Q.B., Ye, M.Z., Zheng, H.C., Yu, J.A., Wu, H.L., Sun, J.H., Zhang, H.Y., Chen, Q.A., Luo, R.B., Chen, M.F., He, Y.H., Jin, X., Zhang, Q.H., Yu, C., Zhou, G.Y., Sun, J.F., Huang, Y.B., Zheng, H.S., Cao, H.Z., Zhou, X.Y., Guo, S.C., Hu, X.D., Li, X., Kristiansen, K., Bolund, L., Xu, J.J., Wang, W., Yang, H.M., Wang, J.A., Li, R.Q., Beck, S., Wang, J., Zhang, X.Q., 2010. The DNA methylome of human peripheral blood mononuclear cells. PLoS Biol. 8, ArtNo.: e1000533.
Liao, Z., Jia, Q.D., Li, F., Han, Z.J., 2010. Identification of two piwi genes and their expression profile in honeybee, Apis mellifera. Arch. Insect Biochem. Physiol. 74, 91–102.
Li-Byarlay, H., Li, Y., Stroud, H., Feng, S.H., Newman, T.C., Kaneda, M., Hou, K.K., Worley, K.C., Elsik, C.G., Wickline, S.A., Jacobsen, S.E., Ma, J., Robinson, G.E., 2013. RNA interference knockdown of DNA methyltransferase 3 affects gene alternative splicing in the honey bee. Proc. Natl. Acad. Sci. U.S.A. 110, 12750–12755.
Lim, J.P., Brunet, A., 2013. Bridging the transgenerational gap with epigenetic memory. Trends Genet. 29, 176–186.
Lister, R., Pelizzola, M., Dowen, R.H., Hawkins, R.D., Hon, G., Tonti-Filippini, J., Nery, J.R., Lee, L., Ye, Z., Ngo, Q.M., Edsall, L., Antosiewicz-Bourget, J., Stewart, R., Ruotti, V., Millar, A.H., Thomson, J.A., Ren, B., Ecker, J.R., 2009. Human DNA methylomes at base resolution show widespread epigenomic differences. Nature 462, 315–322.
Liu, F., Peng, W., Li, Z., Li, W., Li, L., Pan, J., Zhang, S., Miao, Y., Chen, S., Su, S., 2012. Next-generation small RNA sequencing for microRNAs profiling in Apis mellifera: comparison between nurses and foragers. Insect Mol. Biol. 21, 297–303.
Lo, N., Li, B., Ujvari, B., 2012. DNA methylation in the termite Coptotermes lacteus. Insect. Soc. 59, 257–261.
Lockett, G.A., Helliwell, P., Maleszka, R., 2010. Involvement of DNA methylation in memory processing in the honey bee. Neuroreport 21, 812–816.

Lockett, G.A., Kucharski, R., Maleszka, R., 2012. DNA methylation changes elicited by social stimuli in the brains of worker honey bees. Genes Brain Behav. 11, 235–242.

Lorincz, M.C., Dickerson, D.R., Schmitt, M., Groudine, M., 2004. Intragenic DNA methylation alters chromatin structure and elongation efficiency in mammalian cells. Nat. Struct. Mol. Biol. 11, 1068–1075.

Lu, H.L., Vinson, S.B., Pietrantonio, P.V., 2009. Oocyte membrane localization of vitellogenin receptor coincides with queen flying age, and receptor silencing by RNAi disrupts egg formation in fire ant virgin queens. FEBS J. 276, 3110–3123.

Luco, R.F., Allo, M., Schor, I.E., Kornblihtt, A.R., Misteli, T., 2011. Epigenetics in alternative pre-mRNA splicing. Cell 144, 16–26.

Luteijn, M.J., Ketting, R.F., 2013. PIWI-interacting RNAs: from generation to transgenerational epigenetics. Nat. Rev. Genet. 14, 523–534.

Lyko, F., Maleszka, R., 2011. Insects as innovative models for functional studies of DNA methylation. Trends Genet. 27, 127–131.

Lyko, F., Foret, S., Kucharski, R., Wolf, S., Falckenhayn, C., Maleszka, R., 2010. The honey bee epigenomes: differential methylation of brain DNA in queens and workers. PLoS Biol. 8, ArtNo.: e1000506.

Mackert, A., do Nascimento, A.M., Bitondi, M.M.G., Hartfelder, K., Simoes, Z.L.P., 2008. Identification of a juvenile hormone esterase-like gene in the honey bee, Apis mellifera L.—expression analysis and functional assays. Comp. Biochem. Physiol. B Biochem. Mol. Biol. 150, 33–44.

Maleszka, R., 2008. Epigenetic integration of environmental and genomic signals in honey bees. Epigenetics 3, 188–192.

Maleszka, J., Foret, S., Saint, R., Maleszka, R., 2007. RNAi-induced phenotypes suggest a novel role for a chemosensory protein CSP5 in the development of embryonic integument in the honeybee (Apis mellifera). Dev. Genes Evol. 217, 189–196.

Mandrioli, M., 2007. A new synthesis in epigenetics: towards a unified function of DNA methylation from invertebrates to vertebrates. Cell. Mol. Life Sci. 64, 2522–2524.

Maori, E., Paldi, N., Shafir, S., Kalev, H., Tsur, E., Glick, E., Sela, I., 2009. IAPV, a bee-affecting virus associated with Colony Collapse Disorder can be silenced by dsRNA ingestion. Insect Mol. Biol. 18, 55–60.

Margueron, R., Reinberg, D., 2010. Chromatin structure and the inheritance of epigenetic information. Nat. Rev. Genet. 11, 285–296.

Marhold, J., Rothe, N., Pauli, A., Mund, C., Kuehle, K., Brueckner, B., Lyko, F., 2004. Conservation of DNA methylation in dipteran insects. Insect Mol. Biol. 13, 117–123.

Martens, J.A., Laprade, L., Winston, F., 2004. Intergenic transcription is required to repress the Saccharomyces cerevisiae SER3 gene. Nature 429, 571–574.

Martin, A.M., Pouchnik, D.J., Walker, J.L., Wyrick, J.J., 2004. Redundant roles for histone H3 N-terminal lysine residues in subtelomeric gene repression in Saccharomyces cerevisiae. Genetics 167, 1123–1132.

Mattick, J.S., Makunin, I.V., 2006. Non-coding RNA. Hum. Mol. Genet. 15, R17–R29.

Maunakea, A.K., Nagarajan, R.P., Bilenky, M., Ballinger, T.J., D'Souza, C., Fouse, S.D., Johnson, B.E., Hong, C.B., Nielsen, C., Zhao, Y.J., Turecki, G., Delaney, A., Varhol, R., Thiessen, N., Shchors, K., Heine, V.M., Rowitch, D.H., Xing, X.Y., Fiore, C., Schillebeeckx, M., Jones, S.J.M., Haussler, D., Marra, M.A., Hirst, M., Costello, J.F., 2010. Conserved role of intragenic DNA methylation in regulating alternative promoters. Nature 466, 253–257.

Maynard Smith, J., Szathmary, E., 1998. The Major Transitions in Evolution. Oxford University Press, Oxford.

Medved, V., Huang, Z.Y., Popadic, A., 2014. Ubx promotes corbicular development in Apis mellifera. Biol. Lett. 10. http://dx.doi.org/10.1098/rsbl.2013.1021.

Meng, F., Zhang, Y.Y., Liu, F., Guo, X.Q., Xu, B.H., 2014. Characterization and mutational analysis of omega-class GST (GSTO1) from Apis cerana cerana, a gene involved in response to oxidative stress. PLoS One 9, e93100. http://dx.doi.org/10.1371/journal.pone.0093100.

Mercer, T.R., Dinger, M.E., Mattick, J.S., 2009. Long non-coding RNAs: insights into functions. Nat. Rev. Genet. 10, 155–159.

Mito, Y., Henikoff, J.G., Henikoff, S., 2005. Genome-scale profiling of histone H3.3 replacement patterns. Nat. Genet. 37, 1090–1097.

Miura, T., 2004. Proximate mechanisms and evolution of caste polyphenism in social insects: from sociality to genes. Ecol. Res. 19, 141–148.

Moczek, A.P., Snell-Rood, E.C., 2008. The basis of bee-ing different: the role of gene silencing in plasticity. Evol. Dev. 10, 511–513.

Moore, T., Mills, W., 2008. Evolutionary theories of imprinting—enough already! Adv. Exp. Med. Biol. 626, 116–122.

Morris, K.V., Mattick, J.S., 2014. The rise of regulatory RNA. Nat. Rev. Genet. 15, 423–437.

Mutti, N.S., Dolezal, A.G., Wolschin, F., Mutti, J.S., Gill, K.S., Amdam, G.V., 2011. IRS and TOR nutrient-signaling pathways act via juvenile hormone to influence honey bee caste fate. J. Exp. Biol. 214, 3977–3984.

Nambu, Y., Tanaka, H., Enoki, A., Itakura, S., 2010. RNA interference in the termite Reticulitermes speratus: silencing of the hexamerin gene using a single 21 nucleotide small interfering RNA-promoted differentiation of nymph to nymphoid. Sociobiology 55, 527–546.

Nanney, D.L., 1958. Epigenetic control systems. Proc. Natl. Acad. Sci. U.S.A. 44, 712–717.

Nanty, L., Carbajosa, G., Heap, G.A., Ratnieks, F., van Heel, D.A., Down, T.A., Rakyan, V.K., 2011. Comparative methylomics reveals gene-body H3K36me3 in *Drosophila* predicts DNA methylation and CpG landscapes in other invertebrates. Genome Res. 21, 1841–1850.

Ndlovu, M.N., Denis, H., Fuks, F., 2011. Exposing the DNA methylome iceberg. Trends Biochem. Sci. 36, 381–387.

Nelson, C.M., Ihle, K.E., Fondrk, M.K., Page, R.E., Amdam, G.V., 2007. The gene vitellogenin has multiple coordinating effects on social organization. PLoS Biol. 5, e62. http://dx.doi.org/10.1371/journal.pbio.0050062.

Nijhout, H.F., 2003. Development and evolution of adaptive polyphenisms. Evol. Dev. 5, 9–18.

Nilsen, K.A., Ihle, K.E., Frederick, K., Fondrk, M.K., Smedal, B., Hartfelder, K., Amdam, G.-V., 2011. Insulin-like peptide genes in honey bee fat body respond differently to manipulation of social behavioral physiology. J. Exp. Biol. 214, 1488–1497.

Nunes, F.M.F., Simoes, Z.L.P., 2009. A non-invasive method for silencing gene transcription in honeybees maintained under natural conditions. Insect Biochem. Mol. Biol. 39, 157–160.

Nunes, F.M.F., Ihle, K.E., Mutti, N.S., Simoes, Z.L.P., Amdam, G.V., 2013. The gene vitellogenin affects microRNA regulation in honey bee (Apis mellifera) fat body and brain. J. Exp. Biol. 216, 3724–3732.

O'Neill, R.J.W., O'Neill, M.J., Graves, J.A.M., 1998. Undermethylation associated with retroelement activation and chromosome remodelling in an interspecific mammalian hybrid. Nature 393, 68–72.

Oster, G.F., Wilson, E.O., 1978. Caste and Ecology in the Social Insects. Princeton University Press, Princeton, NJ.

Oxley, P.R., Ji, L., Fetter-Pruneda, I., McKenzie, S.K., Li, C., Hu, H.F., Zhang, G.J., Kronauer, D.J.C., 2014. The genome of the clonal raider ant Cerapachys biroi. Curr. Biol. 24, 451–458.

Park, J., Peng, Z.G., Zeng, J., Elango, N., Park, T., Wheeler, D., Werren, J.H., Yi, S.V., 2011. Comparative analyses of DNA methylation and sequence evolution using *Nasonia* genomes. Mol. Biol. Evol. 28, 3345–3354.

Patalano, S., Hore, T.A., Reik, W., Sumner, S., 2012. Shifting behaviour: epigenetic reprogramming in eusocial insects. Curr. Opin. Cell Biol. 24, 367–373.

Patel, A., Fondrk, M.K., Kaftangolu, O., Emore, C., Hunt, G., Frederick, R., Amdam, G.V., 2007. The making of a queen: TOR pathway is a key player in diphenic caste development. PLoS One 2, e509.

Patten, M.M., Ross, L., Curley, J.P., Queller, D.C., Bonduriansky, R., Wolf, J.B., 2014. The evolution of genomic imprinting: theories, predictions and empirical tests. Heredity 113, 119–128.

Pfennig, D.W., Wund, M.A., Snell-Rood, E.C., Cruickshank, T., Schlichting, C.D., Moczek, A.P., 2010. Phenotypic plasticity's impacts on diversification and speciation. Trends Ecol. Evol. 25, 459–467.

Pigliucci, M., 2001. Phenotypic Plasticity: Beyond Nature and Nurture. Johns Hopkins University Press, Baltimore, MD.

Ponting, C.P., Oliver, P.L., Reik, W., 2009. Evolution and functions of long noncoding RNAs. Cell 136, 629–641.

Queller, D.C., 2003. Theory of genomic imprinting conflict in social insects. BMC Evol. Biol. 3, 15.

Rae, P.M.M., Steele, R.E., 1979. Absence of cytosine methylation at C-C-G-G and G-C-G-C sites in the rDNA coding regions and intervening sequences of Drosophila and the rDNA of other higher insects. Nucleic Acids Res. 6, 2987–2995.

Rana, T.M., 2007. Illuminating the silence: understanding the structure and function of small RNAs. Nat. Rev. Mol. Cell Biol. 8, 23–36.

Rassoulzadegan, M., Grandjean, V., Gounon, P., Vincent, S., Gillot, I., Cuzin, F., 2006. RNA-mediated non-Mendelian inheritance of an epigenetic change in the mouse. Nature 441, 469–474.

Ratzka, C., Gross, R., Feldhaar, H., 2013. Systemic gene knockdown in *Camponotus floridanus* workers by feeding of dsRNA. Insect. Soc. 60, 475–484.

Rice, J.C., Briggs, S.D., Ueberheide, B., Barber, C.M., Shabanowitz, J., Hunt, D.F., Shinkai, Y., Allis, C.D., 2003. Histone methyltransferases direct different degrees of methylation to define distinct chromatin domains. Mol. Cell 12, 1591–1598.

Richards, E.J., 2006. Opinion—inherited epigenetic variation—revisiting soft inheritance. Nat. Rev. Genet. 7, 395–401.

Riggs, A.D., Martienssen, R.A., Russo, V.E.A., 1996. Introduction. In: Russo, V.E.A., Martienssen, R.A., Riggs, A.D. (Eds.), Epigenetic Mechanisms of Gene Regulation. Cold Spring Harbor Laboratory Press, Woodbury, NY, pp. 1–4.

Roberts, S.B., Gavery, M.R., 2012. Is there a relationship between DNA methylation and phenotypic plasticity in invertebrates? Front. Physiol. 2, ArtNo.: 116.

Robinson, K.L., Tohidi-Esfahani, D., Lo, N., Simpson, S.J., Sword, G.A., 2011. Evidence for widespread genomic methylation in the migratory locust, Locusta migratoria (Orthoptera: Acrididae). PLoS One 6, ArtNo.: e28167.

Sarda, S., Zeng, J., Hunt, B.G., Yi, S.V., 2012. The evolution of invertebrate gene body methylation. Mol. Biol. Evol. 29, 1907–1916.

Sawata, M., Yoshino, D., Takeuchi, H., Kamikouchi, A., Ohashi, K., Kubo, T., 2002. Identification and punctate nuclear localization of a novel noncoding RNA, Ks-1, from the honeybee brain. RNA 8, 772–785.

Sawata, M., Takeuchi, H., Kubo, T., 2004. Identification and analysis of the minimal promoter activity of a novel noncoding nuclear RNA gene, AncR-1, from the honeybee (Apis mellifera L.). RNA 10, 1047–1058.

Schaefer, M., Lyko, F., 2007. DNA methylation with a sting: an active DNA methylation system in the honeybee. BioEssays 29, 208–211.

Schluns, H., Crozier, R.H., 2007. Relish regulates expression of antimicrobial peptide genes in the honeybee, Apis mellifera, shown by RNA interference. Insect Mol. Biol. 16, 753–759.
Schwander, T., Lo, N., Beekman, M., Oldroyd, B.P., Keller, L., 2010. Nature versus nurture in social insect caste differentiation. Trends Ecol. Evol. 25, 275–282.
Schwartz, S., Meshorer, E., Ast, G., 2009. Chromatin organization marks exon-intron structure. Nat. Struct. Mol. Biol. 16, 990–995.
Schwinghammer, M.A., Zhou, X.G., Kambhampati, S., Bennett, G.W., Scharf, M.E., 2011. A novel gene from the takeout family involved in termite trail-following behavior. Gene 474, 12–21.
Sharma, S., Kelly, T.K., Jones, P.A., 2010. Epigenetics in cancer. Carcinogenesis 31, 27–36.
Shenker, N., Flanagan, J.M., 2012. Intragenic DNA methylation: implications of this epigenetic mechanism for cancer research. Br. J. Cancer 106, 248–253.
Shi, Y.Y., Huang, Z.Y., Zeng, Z.J., Wang, Z.L., Wu, X.B., Yan, W.Y., 2011. Diet and cell size both affect queen-worker differentiation through DNA methylation in honey bees (Apis mellifera, Apidae). PLoS One 6, ArtNo.: e18808.
Shi, Y.Y., Yan, W.Y., Huang, Z.Y., Wang, Z.L., Wu, X.B., Zeng, Z.J., 2013. Genomewide analysis indicates that queen larvae have lower methylation levels in the honey bee (Apis mellifera). Naturwissenschaften 100, 193–197.
Shukla, S., Kavak, E., Gregory, M., Imashimizu, M., Shutinoski, B., Kashlev, M., Oberdoerffer, P., Sandberg, R., Oberdoerffer, S., 2011. CTCF-promoted RNA polymerase II pausing links DNA methylation to splicing. Nature 479, 74–79.
Sienski, G., Donertas, D., Brennecke, J., 2012. Transcriptional silencing of transposons by Piwi and maelstrom and its impact on chromatin state and gene expression. Cell 151, 964–980.
Simola, D.F., Wissler, L., Donahue, G., Waterhouse, R.M., Helmkampf, M., Roux, J., Nygaard, S., Glastad, K.M., Hagen, D.E., Viljakainen, L., Reese, J.T., Hunt, B.G., Graur, D., Elhaik, E., Kriventseva, E.V., Wen, J.Y., Parker, B.J., Cash, E., Privman, E., Childers, C.P., Munoz-Torres, M.C., Boomsma, J.J., Bornberg-Bauer, E., Currie, C.R., Elsik, C.G., Suen, G., Goodisman, M.A.D., Keller, L., Liebig, J., Rawls, A., Reinberg, D., Smith, C.D., Smith, C.R., Tsutsui, N., Wurm, Y., Zdobnov, E.M., Berger, S.L., Gadau, J., 2013a. Social insect genomes exhibit dramatic evolution in gene composition and regulation while preserving regulatory features linked to sociality. Genome Res. 23, 1235–1247.
Simola, D.F., Ye, C.Y., Mutti, N.S., Dolezal, K., Bonasio, R., Liebig, J., Reinberg, D., Berger, S.L., 2013b. A chromatin link to caste identity in the carpenter ant *Camponotus floridanus*. Genome Res. 23, 486–496.
Smith, C.D., Zimin, A., Holt, C., Abouheif, E., Benton, R., Cash, E., Croset, V., Currie, C.R., Elhaik, E., Elsik, C.G., Fave, M.J., Fernandes, V., Gadau, J., Gibson, J.D., Graur, D., Grubbs, K.J., Hagen, D.E., Helmkampf, M., Holley, J.A., Hu, H., Viniegra, A.S.I., Johnson, B.R., Johnson, R.M., Khila, A., Kim, J.W., Laird, J., Mathis, K.A., Moeller, J.A., Munoz-Torres, M.C., Murphy, M.C., Nakamura, R., Nigam, S., Overson, R.P., Placek, J.E., Rajakumar, R., Reese, J.T., Robertson, H.M., Smith, C.R., Suarez, A.V., Suen, G., Suhr, E.L., Tao, S., Torres, C.W., van Wilgenburg, E., Viljakainen, L., Walden, K.K.O., Wild, A.L., Yandell, M., Yorke, J.A., Tsutsui, N.D., 2011a. Draft genome of the globally widespread and invasive Argentine ant (*Linepithema humile*). Proc. Natl. Acad. Sci. U.S.A. 108, 5673–5678.
Smith, C.R., Smith, C.D., Robertson, H.M., Helmkampf, M., Zimin, A., Yandell, M., Holt, C., Hu, H., Abouheif, E., Benton, R., Cash, E., Croset, V., Currie, C.R., Elhaik, E., Elsik, C.G., Fave, M.J., Fernandes, V., Gibson, J.D., Graur, D., Gronenberg, W., Grubbs, K.J., Hagen, D.E., Viniegra, A.S.I., Johnson, B.R., Johnson, R.M., Khila, A., Kim, J.W., Mathis, K.A., Munoz-Torres, M.C., Murphy, M.C., Mustard, J.A., Nakamura, R., Niehuis, O., Nigam, S., Overson, R.P., Placek, J.E., Rajakumar, R., Reese, J.T., Suen, G., Tao, S., Torres, C.W., Tsutsui, N.D.,

Viljakainen, L., Wolschin, F., Gadau, J., 2011b. Draft genome of the red harvester ant *Pogonomyrmex barbatus*. Proc. Natl. Acad. Sci. U.S.A. 108, 5667–5672.

Smith, C.R., Mutti, N.S., Jasper, W.C., Naidu, A., Smith, C.D., Gadau, J., 2012. Patterns of DNA methylation in development, division of labor and hybridization in an ant with genetic caste determination. PLoS One 7, ArtNo.: e42433.

Snell-Rood, E.C., Troth, A., Moczek, A.P., 2013. DNA methylation as a mechanism of nutritional plasticity: limited support from horned beetles. J. Exp. Zool. 320B, 22–34.

Spencer, H.G., Clark, A.G., 2014. Non-conflict theories for the evolution of genomic imprinting. Heredity 113, 112–118.

Strassmann, J.E., Queller, D.C., 2007. Insect societies as divided organisms: the complexities of purpose and cross-purpose. Proc. Natl. Acad. Sci. U.S.A. 104, 8619–8626.

Suen, G., Teiling, C., Li, L., Holt, C., Abouheif, E., Bornberg-Bauer, E., Bouffard, P., Caldera, E.J., Cash, E., Cavanaugh, A., Denas, O., Elhaik, E., Fave, M.J., Gadau, J., Gibson, J.D., Graur, D., Grubbs, K.J., Hagen, D.E., Harkins, T.T., Helmkampf, M., Hu, H., Johnson, B.R., Kim, J., Marsh, S.E., Moeller, J.A., Munoz-Torres, M.C., Murphy, M.C., Naughton, M.C., Nigam, S., Overson, R., Rajakumar, R., Reese, J.T., Scott, J.J., Smith, C.R., Tao, S., Tsutsui, N.D., Viljakainen, L., Wissler, L., Yandell, M.D., Zimmer, F., Taylor, J., Slater, S.C., Clifton, S.W., Warren, W.C., Elsik, C.G., Smith, C.D., Weinstock, G.M., Gerardo, N.M., Currie, C.R., 2011. The genome sequence of the leaf-cutter ant *Atta cephalotes* reveals insights into Its obligate symbiotic lifestyle. PLoS Genet. 7, ArtNo.: e1002007.

Suganuma, T., Workman, J.L., 2011. Signals and combinatorial functions of histone modifications. Annu. Rev. Biochem. 80, 473–499.

Suzuki, M.M., Bird, A., 2008. DNA methylation landscapes: provocative insights from epigenomics. Nat. Rev. Genet. 9, 465–476.

Suzuki, M.M., Kerr, A.R.W., De Sousa, D., Bird, A., 2007. CpG methylation is targeted to transcription units in an invertebrate genome. Genome Res. 17, 625–631.

Takayama, S., Dhahbi, J., Roberts, A., Mao, G.X., Heo, S.J., Pachter, L., Martin, D.I.K., Boffelli, D., 2014. Genome methylation in D. melanogaster is found at specific short motifs and is independent of DNMT2 activity. Genome Res. 24, 821–830.

Takuno, S., Gaut, B.S., 2012. Body-methylated genes in Arabidopsis thaliana are functionally important and evolve slowly. Mol. Biol. Evol. 29, 219–227.

Talbert, P.B., Henikoff, S., 2010. Histone variants—ancient wrap artists of the epigenome. Nat. Rev. Mol. Cell Biol. 11, 264–275.

Terrapon, N., Li, C., Robertson, H.M., Ji, L., Meng, X.H., Booth, W., Chen, Z.S., Childers, C.P., Glastad, K.M., Gokhale, K., Gowin, J., Gronenberg, W., Hermansen, R.A., Hu, H.F., Hunt, B.G., Huylmans, A.K., Khalil, S.M.S., Mitchell, R.D., Munoz-Torres, M.C., Mustard, J.A., Pan, H.L., Reese, J.T., Scharf, M.E., Sun, F.M., Vogel, H., Xiao, J., Yang, W., Yang, Z.K., Yang, Z.Q., Zhou, J.J., Zhu, J.W., Brent, C.S., Elsik, C.G., Goodisman, M.A.D., Liberles, D.A., Roe, R.M., Vargo, E.L., Vilcinskas, A., Wang, J., Bornberg-Bauer, E., Korb, J., Zhang, G.J., Liebig, J., 2014. Molecular traces of alternative social organization in a termite genome. Nat. Commun. 5, ArtNo.: 3636.

Tribolium Genome Sequencing Consortium, Richards, S., Gibbs, R.A., Weinstock, G.M., Brown, S.J., Denell, R., Beeman, R.W., Bucher, G., Friedrich, M., Grimmelikhuijzen, C.J.P., Klingler, M., Lorenzen, M., Roth, S., Schroder, R., Tautz, D., Zdobnov, E.M., 2008. The genome of the model beetle and pest Tribolium castaneum. Nature 452, 949–955.

Tsai, W.-W., Wang, Z., Yiu, T.T., Akdemir, K.C., Xia, W., Winter, S., Tsai, C.-Y., Shi, X., Schwarzer, D., Plunkett, W., Aronow, B., Gozani, O., Fischle, W., Hung, M.-C., Patel, D.J., Barton, M.C., 2010. TRIM24 links a non-canonical histone signature to breast cancer. Nature 468, 927–932.

Tweedie, S., Charlton, J., Clark, V., Bird, A., 1997. Methylation of genomes and genes at the invertebrate-vertebrate boundary. Mol. Cell. Biol. 17, 1469–1475.
Tweedie, S., Ng, H.H., Barlow, A.L., Turner, B.M., Hendrich, B., Bird, A., 1999. Vestiges of a DNA methylation system in Drosophila melanogaster? Nat. Genet. 23, 389–390.
Urieli-Shoval, S., Gruenbaum, Y., Sedat, J., Razin, A., 1982. The absence of detectable methylated bases in Drosophila melanogaster DNA. FEBS Lett. 146, 148–152.
Waddington, C.H., 1942. The epigenotype. Endeavour 1, 18–20, Reprinted in International J. Epidemiology 2012. 41:10–13.
Wagner, K.D., Wagner, N., Ghanbarian, H., Grandjean, V., Gounon, P., Cuzin, F., Rassoulzadegan, M., 2008. RNA induction and inheritance of epigenetic cardiac hypertrophy in the mouse. Dev. Cell 14, 962–969.
Walsh, T.K., Brisson, J.A., Robertson, H.M., Gordon, K., Jaubert-Possamai, S., Tagu, D., Edwards, O.R., 2010. A functional DNA methylation system in the pea aphid, Acyrthosiphon pisum. Insect Mol. Biol. 19, 215–228.
Wang, Y., Leung, F.C.C., 2009. In silico prediction of two classes of honeybee genes with CpG deficiency or CpG enrichment and sorting according to gene ontology classes. J. Mol. Evol. 68, 700–705.
Wang, Y., Jorda, M., Jones, P.L., Maleszka, R., Ling, X., Robertson, H.M., Mizzen, C.A., Peinado, M.A., Robinson, G.E., 2006. Functional CpG methylation system in a social insect. Science 314, 645–647.
Wang, Y., Mutti, N.S., Ihle, K.E., Siegel, A., Dolezal, A.G., Kaftanoglu, O., Amdam, G.V., 2010. Down-regulation of honey bee IRS gene biases behavior toward food rich in protein. PLoS Genet. 6, e1000896.
Wang, Y., Kocher, S.D., Linksvayer, T.A., Grozinger, C.M., Page, R.E., Amdam, G.V., 2012. Regulation of behaviorally associated gene networks in worker honey bee ovaries. J. Exp. Biol. 215, 124–134.
Wang, X., Wheeler, D., Avery, A., Rago, A., Choi, J.H., Colbourne, J.K., Clark, A.G., Werren, J.H., 2013. Function and evolution of DNA methylation in Nasonia vitripennis. PLoS Genet. 9, ArtNo.: e1003872.
Wang, X.H., Fang, X.D., Yang, P.C., Jiang, X.T., Jiang, F., Zhao, D.J., Li, B.L., Cui, F., Wei, J.N., Ma, C.A., Wang, Y.D., He, J., Luo, Y., Wang, Z.F., Guo, X.J., Guo, W., Wang, X.S., Zhang, Y., Yang, M.L., Hao, S.G., Chen, B., Ma, Z.Y., Yu, D., Xiong, Z.Q., Zhu, Y.B., Fan, D.D., Han, L.J., Wang, B., Chen, Y.X., Wang, J.W., Yang, L., Zhao, W., Feng, Y., Chen, G.X., Lian, J.M., Li, Q.Y., Huang, Z.Y., Yao, X.M., Lv, N., Zhang, G.J., Li, Y.R., Wang, J., Wang, J., Zhu, B.L., Kang, L., 2014. The locust genome provides insight into swarm formation and long-distance flight. Nat. Commun. 5, 1–9.
Weaver, D.B., Anzola, J.M., Evans, J.D., Reid, J.G., Reese, J.T., Childs, K.L., Zdobnov, E.M., Samanta, M.P., Miller, J., Elsik, C.G., 2007. Computational and transcriptional evidence for microRNAs in the honey bee genome. Genome Biol. 8. http://dx.doi.org/10.1186/gb-2007-8-6-r97.
Weber, M., Hellmann, I., Stadler, M.B., Ramos, L., Paabo, S., Rebhan, M., Schubeler, D., 2007. Distribution, silencing potential and evolutionary impact of promoter DNA methylation in the human genome. Nat. Genet. 39, 457–466.
Weiner, S.A., Toth, A.L., 2012. Epigenetics in social insects: a new direction for understanding the evolution of castes. Genet. Res. Int. 2012, Article ID 609810.
Weiner, S.A., Galbraith, D.A., Adams, D.C., Valenzuela, N., Noll, F.B., Grozinger, C.M., Toth, A.L., 2013. A survey of DNA methylation across social insect species, life stages, and castes reveals abundant and caste-associated methylation in a primitively social wasp. Naturwissenschaften 100, 795–799.
West-Eberhard, M.J., 2003. Developmental Plasticity and Evolution. Oxford University Press, Oxford.

Wheeler, D.E., 1986. Developmental and physiological determinants of caste in social Hymenoptera: evolutionary implications. Am. Nat. 128, 13–34.

Wheeler, M.M., Ament, S.A., Rodriguez-Zas, S.L., Robinson, G.E., 2013. Brain gene expression changes elicited by peripheral vitellogenin knockdown in the honey bee. Insect Mol. Biol. 22, 562–573.

Wilson, E.O., 1971. The Insect Societies. Harvard University Press, Cambridge, MA.

Wilson, M.J., McKelvey, B.H., van der Heide, S., Dearden, P.K., 2010. Notch signaling does not regulate segmentation in the honeybee, Apis mellifera. Dev. Genes Evol. 220, 179–190.

Wilson, M.J., Kenny, N.J., Dearden, P.K., 2014. Components of the dorsal-ventral pathway also contribute to anterior-posterior patterning in honeybee embryos (Apis mellifera). EvoDevo 5, 11. http://dx.doi.org/10.1186/2041-9139-5-11.

Wolschin, F., Mutti, N.S., Amdam, G.V., 2011. Insulin receptor substrate influences female caste development in honeybees. Biol. Lett. 7, 112–115.

Wu, S.C., Zhang, Y., 2010. Active DNA demethylation: many roads lead to Rome. Nat. Rev. Mol. Cell Biol. 11, 607–620.

Wurm, Y., Wang, J., Riba-Grognuz, O., Corona, M., Nygaard, S., Hunt, B.G., Ingram, K.K., Falquet, L., Nipitwattanaphon, M., Gotzek, D., Dijkstra, M.B., Oettler, J., Comtesse, F., Shih, C.J., Wu, W.J., Yang, C.C., Thomas, J., Beaudoing, E., Pradervand, S., Flegel, V., Cook, E.D., Fabbretti, R., Stockinger, H., Long, L., Farmerie, W.G., Oakey, J., Boomsma, J.J., Pamilo, P., Yi, S.V., Heinze, J., Goodisman, M.A.D., Farinelli, L., Harshman, K., Hulo, N., Cerutti, L., Xenarios, I., Shoemaker, D., Keller, L., 2011. The genome of the fire ant *Solenopsis invicta*. Proc. Natl. Acad. Sci. U.S.A. 108, 5679–5684.

Xiang, H., Zhu, J.D., Chen, Q.A., Dai, F.Y., Li, X., Li, M.W., Zhang, H.Y., Zhang, G.J., Li, D., Dong, Y., Zhao, L., Lin, Y., Cheng, D.J., Yu, J.A., Sun, J.F., Zhou, X.Y., Ma, K.L., He, Y.H., Zhao, Y.X., Guo, S.C., Ye, M.Z., Guo, G.W., Li, Y.R., Li, R.Q., Zhang, X.Q., Ma, L.J., Kristiansen, K., Guo, Q.H., Jiang, J.H., Beck, S., Xia, Q.Y., Wang, W., Wang, J., 2010. Single base-resolution methylome of the silkworm reveals a sparse epigenomic map. Nat. Biotechnol. 28, 516–520.

Yan, H., Simola, D.F., Bonasio, R., Liebig, J., Berger, S.L., Reinberg, D., 2014. Eusocial insects as emerging models for behavioural epigenetics. Nat. Rev. Genet. 15, 677–688.

Yi, S.V., 2012. Birds do it, bees do it, worms and ciliates do it too: DNA methylation from unexpected corners of the tree of life. Genome Biol. 13, ArtNo.: 174.

Yi, S.V., Goodisman, M.A.D., 2009. Computational approaches for understanding the evolution of DNA methylation in animals. Epigenetics 4, 551–556.

Yoder, J.A., Walsh, C.P., Bestor, T.H., 1997. Cytosine methylation and the ecology of intragenomic parasites. Trends Genet. 13, 335–340.

Zemach, A., McDaniel, I.E., Silva, P., Zilberman, D., 2010. Genome-wide evolutionary analysis of eukaryotic DNA methylation. Science 328, 916–919.

Zeng, J., Yi, S.V., 2010. DNA methylation and genome evolution in honeybee: gene length, expression, functional enrichment covary with the evolutionary signature of DNA methylation. Genome Biol. Evol. 2, 770–780.

Zentner, G.E., Henikoff, S., 2013. Regulation of nucleosome dynamics by histone modifications. Nat. Struct. Mol. Biol. 20, 259–266.

Zhou, X.G., Oi, F.M., Scharf, M.E., 2006. Social exploitation of hexamerin: RNAi reveals a major caste-regulatory factor in termites. Proc. Natl. Acad. Sci. U.S.A. 103, 4499–4504.

Zhou, X.G., Wheeler, M.M., Oi, F.M., Scharf, M.E., 2008. RNA interference in the termite Reticulitermes flavipes through ingestion of double-stranded RNA. Insect Biochem. Mol. Biol. 38, 805–815.

Zhou, V.W., Goren, A., Bernstein, B.E., 2011. Charting histone modifications and the functional organization of mammalian genomes. Nat. Rev. Genet. 12, 7–18.

Zilberman, D., Gehring, M., Tran, R.K., Ballinger, T., Henikoff, S., 2007. Genome-wide analysis of Arabidopsis thaliana DNA methylation uncovers an interdependence between methylation and transcription. Nat. Genet. 39, 61–69.

Zilberman, D., Coleman-Derr, D., Ballinger, T., Henikoff, S., 2008. Histone H2A.Z and DNA methylation are mutually antagonistic chromatin marks. Nature 456, 125–129.

Zondag, L., Dearden, P.K., Wilson, M.J., 2012. Deep sequencing and expression of microRNAs from early honeybee (Apis mellifera) embryos reveals a role in regulating early embryonic patterning. BMC Evol. Biol. 12, 211. http://dx.doi.org/10.1186/1471-2148-12-211.

CHAPTER EIGHT

The Molecular and Evolutionary Genetic Implications of Being Truly Social for the Social Insects

Timothy A. Linksvayer[1]
Department of Biology, University of Pennsylvania, Philadelphia, Pennsylvania, USA
[1]Corresponding author: e-mail address: tlinks@sas.upenn.edu

Contents

Abstract

The social complexity that characterizes the eusocial insects strongly affects all aspects of social insect life, including the molecular and evolutionary genetic basis of social insect traits. Quantitative genetic theory and empirical approaches have been developed over the past 60 years specifically to study the genetic implications of social interactions. Surprisingly, given the obvious biological importance of social interactions in social insects, this research tradition has been and continues to be widely overlooked throughout the social insect literature, including in recent sociogenomic studies focused on understanding the molecular underpinnings of social life. Instead, the overwhelming majority of social insect genetic research has relied on conventional genetic approaches developed for solitary organisms focused on the one-to-one association of an individual's genes to its own traits. I survey social insect studies that conclusively demonstrate the importance of indirect genetic effects (IGEs), which arise from social interactions, for social insect trait expression and evolution. I explain why these genetically based social effects are expected to be ubiquitous in social insects and I explain the relevance of the IGE framework, originally developed within quantitative genetics, for molecular genetic studies of social insect traits such as behaviour and caste. I discuss the problems of ignoring IGEs and relying solely on conventional direct genetic effect approaches. Finally, I discuss the strong potential of using the IGE approach and other

Advances in Insect Physiology, Volume 48
ISSN 0065-2806
http://dx.doi.org/10.1016/bs.aiip.2014.12.003

more systems-level-focused approaches to complement conventional reductionist approaches in elucidating the genetic basis of social insect trait expression and evolution.

Eusocial insects are by definition "truly social" (Hölldobler and Wilson, 2009). Even the simplest eusocial insect colonies, consisting of small groups of individuals, are characterized by some degree of division of labour and group co-ordination by means of chemical and physical social signals. The largest and most complex insect societies teem with millions of individuals with highly specialized roles. Individuals within these societies are in almost constant social contact with colony members from a range of different functional groups and life history stages (Hölldobler and Wilson, 2009). An array of social signals acts to functionally integrate individuals within these complex societies to such an extent that it is difficult to conceptualize the physiology and behaviour of individuals in isolation (Hölldobler and Wilson, 2009; Le Conte and Hefetz, 2008; Seeley, 1995). Such social complexity is expected to profoundly affect all aspects of life, including the genetic basis of traits (e.g. Bloch and Grozinger, 2011; Johnson and Linksvayer, 2010; Kent and Zayed, 2013; Linksvayer and Wade, 2005).

However, research into the genetic basis of social insect traits has overwhelmingly relied on conventional reductionist genetic approaches. These approaches were originally developed for solitary organisms, where there is a simple one-to-one mapping of an individual's genes to its traits (Falconer and Mackay, 1996). In contrast, in social organisms—especially in tightly socially integrated organisms such as social insects—traits are the properties of the genomes of multiple interacting individuals (Fig. 1; Linksvayer, 2006; Linksvayer et al., 2009; Moore et al., 1997; Wolf, 2003). Notably, even though sociogenomic studies have used conventional approaches that do not explicitly consider the genetic implications of social interactions, one of the major findings of 10 years of sociogenomic research is that physiology, development, gene expression profile, neurogenomic state and behaviour critically depend on the social environment (Manfredini et al., 2014; Robinson et al., 2005, 2008).

Generally speaking, traits that depend on social context—both traits that are *defined* by social interactions and traits that are *affected* by social interactions—have long been recognized as presenting a special difficulty for conventional genetic analysis (Cheverud and Moore, 1994; Griffing, 1981; Hahn and Schanz, 1996; Moore et al., 2002). Fortunately, theoretical

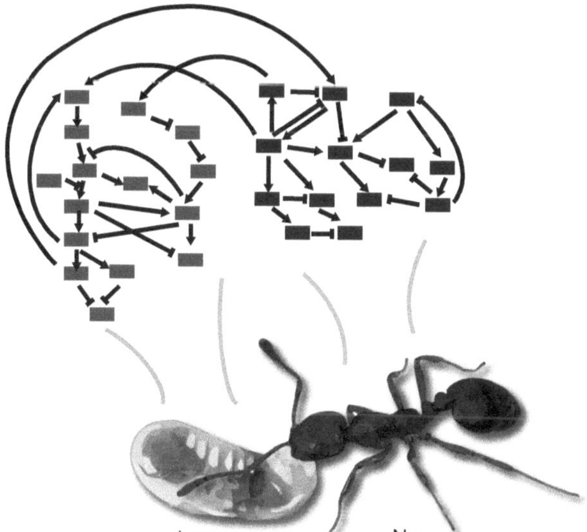

Figure 1 Genes affecting trait expression can be conceptualized as existing within a gene regulatory network. Here, a hypothetical gene regulatory network affecting ant larval traits (behaviour, development, physiology, etc.) is shown. Genes expressed in the larva (red (grey in the print version) boxes) directly influence its traits through *direct genetic effects*. Genes expressed in the nurse worker (blue (black in the print version) boxes) affect the larva's traits indirectly, via social interactions (transfer of proteins, brood care behaviour, etc.), through *indirect genetic effects*. Similarly, genes expressed in the larva feedback to indirectly influence the nurse's expressed traits through social interactions (begging behaviour, brood pheromone, etc.). Edges between boxes of the same colour represent conventional within-organism gene regulation and edges between boxes of different colours represent socially acting, intergenomic regulation. Realistically, genes expressed in foragers, the queen and other brood stages likely also influence the traits expressed by the pictured larva and nurse.

and empirical approaches tailored exactly for this problem have been developed within the field of quantitative genetics over the past 60 years. Indeed, well before W. D. Hamilton developed his famous theory of inclusive fitness for how social interactions affect fitness and expected social evolutionary outcomes (Hamilton, 1964), quantitative geneticists had begun developing models of how social interactions affect trait expression and trait evolution (Dickerson, 1947; Falconer, 1960; Willham, 1963). These models have led to the development of the interacting phenotypes/indirect genetic effect (IGE) framework, which provide formal theory and empirical approaches for studying how an individual's traits and fitness are affected by the genes of social partners (McGlothlin et al., 2010; Moore et al., 1997; Wolf et al.,

1998, 1999). Within this framework, *direct genetic effects*, the focus of conventional genetic approaches, occur when an individual's genotype affects its own traits. In contrast, *IGEs* occur when the traits of an individual depend on social partners' genotypes. These effects are labelled "indirect" because they act indirectly on traits, by shaping the environment (e.g. nutritional, chemical, physical environment) experienced by the focal individual.

Despite this long and rich history of IGE research focused on the genetic implications of social interactions and more recent calls for social insect researchers to consider IGEs (Helantera, 2011; Johnson and Linksvayer, 2010; Keller, 2009; Linksvayer and Wade, 2005; Linksvayer et al., 2009), only a relatively small proportion of studies of the genetic basis of social insects' traits have explicitly used the IGE framework (e.g. Gempe et al., 2012; Linksvayer, 2006, 2007; Linksvayer et al., 2009, 2011; Teseo et al., 2014; van Zweden et al., 2010; Vojvodic et al., 2014; Wang et al., 2008). Surprisingly, the framework has until very recently (Malka et al., 2014) had seemingly little influence on sociogenomic research, the stated goal of which is to understand the molecular basis of social life (Robinson et al., 2005). Most recent reviews of the genetic basis of social behaviour and division of labour in social insects do not even mention IGEs (Bloch and Grozinger, 2011; Lattorff and Moritz, 2013; Libbrecht et al., 2013; Oldroyd and Thompson, 2007; Page et al., 2012; Smith et al., 2008; Yan et al., 2014; Zayed and Robinson, 2012), despite their certain importance.

These reviews typically do explicitly refer to the effect of the "social environment" on trait expression by indicating that an individual's traits, such as its behaviour, neurogenomic state or caste, are influenced by both its own genotype and the social environment it experiences. However, this basic genotype/environment partition is inadequate for social traits. As described further below, the critical missing point is that we know what makes up the social environment: the traits and genes of assorted types of social partners. Instead of simply conceptualizing these as the "social environment" and then moving on to study the direct genetic basis of response to these factors (Fig. 2A), we can explicitly study the genetic contribution of these "extrinsic factors present in the social environment" (Zayed and Robinson, 2012) to trait expression and evolution (Fig. 2B).

The omission of IGE-related theory and empirical studies from much of the literature on the genetic basis of social insect traits may in part be because researchers have often used either quantitative genetic approaches (e.g. Graham et al., 2011; Hunt et al., 1998; Kovacs et al., 2010;

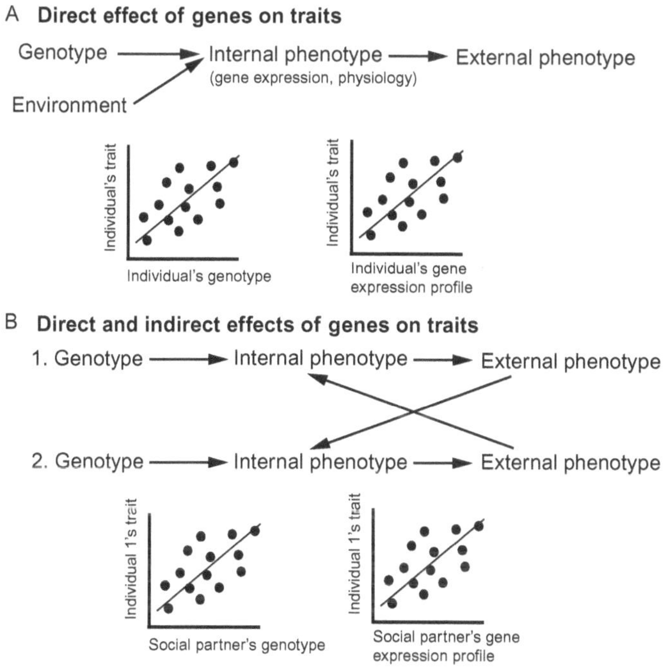

Figure 2 An individual's expressed behavioural and morphological traits (i.e. its external phenotype) depend on its own genotype and the environment it experiences. Here, an internal phenotype that can be quantified by measuring gene expression profiles or physiological traits is modelled as mediating genotypic and environmental effects on trait expression. (A) Direct effects of genes on traits. Conventional quantitative genetic approaches can elucidate the relationship between an individual's genotype and its traits (left graph); alternatively, conventional transcriptomic approaches can elucidate the relationship between an individual's gene expression profile and its traits (right graph). Note that with these approaches, it is possible to study the effects of the social environment on an individual's gene expression profile or other traits, but it is not possible to determine which genes and traits of social partners that make up the social environment are playing a role. (B) Direct and indirect effects of genes on traits. Alternatively, it is possible to explicitly study the indirect genetic effects of social partner genes on the traits of focal individuals. Here, the traits of individual 1 are affected by its own genotype and internal phenotype (patterns of gene expression, physiological traits, etc.), and its external phenotypic traits are also affected by the traits of its social partner, individual 2. As in panel (A), an internal phenotype (e.g. gene expression profile) is modelled as mediating genotypic and environmental effects on trait expression, so that the genetic effects of individual 2 on the traits of individual 1 first flow through the external traits of individual 2 and then to the internal and external traits of individual 1, and vice versa for IGEs of individual 1 on 2. Under this scenario, it is possible to determine how an individual's genotype and gene expression profile influence its own traits (left and right graphs in panel A), but it is also possible to study how the genotype and gene expression profiles of social partners influence the traits of each individual (left and right graphs in panel B).

Oldroyd et al., 1991; Page et al., 2000; Rueppell et al., 2004) or molecular genetic/transcriptomic approaches (e.g. Grozinger et al., 2003; Manfredini et al., 2014; Toth et al., 2007; Whitfield et al., 2006), with seemingly relatively little cross-talk. This is likely at least in part because the approaches are typically used to answer different but complementary questions about the genetic basis of trait variation versus trait expression, respectively. As a result, some researchers may be relatively unfamiliar with the existing quantitative genetic theoretical and empirical literature. Thus, one goal of this chapter is to introduce unfamiliar readers with the IGE literature specifically relevant to social insects.

Most importantly, I seek to explain how the IGE framework, originally developed within a quantitative genetic context, has strong potential to complement conventional approaches to elucidate the complex molecular and evolutionary genetic basis of trait expression and evolution in social insects (Linksvayer et al., 2012). I begin by briefly surveying the IGE literature and social insect studies that explicitly or implicitly studied IGEs. Next, I briefly survey studies of the molecular genetic basis of social insect traits, which have increasingly sought to study how the "social environment" affected trait expression but typically have done so without a formal framework. I explain why a formal theoretical and empirical framework such as the IGE framework is useful and often is necessary for both motivating and interpreting studies of the molecular and evolutionary genetic basis of social traits. Finally, I describe how the IGE framework is currently being extended in concert with emerging transcriptomic and genomic approaches to study the molecular basis of social interactions (Vojvodic et al., 2014), and how these approaches can lead to exciting social systems-level insight into the molecular and evolutionary genetic basis of social insect traits.

1. INTRODUCTION TO IGEs

The first models to explicitly consider the genetic implications of social interactions were developed by animal and plant breeders and focused on maternal effects, which occur when offspring traits depend on maternal traits, for example, as a result of maternal care (Cheverud, 1984; Dickerson, 1947; Falconer, 1960; Kirkpatrick and Lande, 1989; Willham, 1963). For example, in mammals, offspring size is strongly influenced by the quantity and quality of milk produced and provided by the mother (Bijma, 2006; Bouwman et al., 2010). Such maternal effects typically are highly heritable (i.e. affected by maternal genotype), so that the maternal effect can evolve. Maternal genetic effects have been shown to strongly contribute to total heritable variation for many offspring traits and hence strongly affect evolutionary responses to selection (Wade, 1998). As a result, estimating maternal genetic effects has long been considered crucial and also routine for plant and animal breeders (Bijma, 2006; Bouwman et al., 2010).

Maternal effects models were extended to include other types of genetically based social effects (Cheverud and Moore, 1994; Griffing, 1977, 1981; Lynch, 1987), in particular within the interacting phenotypes/IGE framework (McGlothlin et al., 2010; Moore et al., 1997; Wolf et al., 1998, 1999). The "social environment" provided by conspecifics (e.g. in social insect colonies, the mother queen, sibling adults, sibling brood) depends on the traits and genes of social partners. In this way, the social environment contains genes and can itself evolve (Wolf, 2003). This is not the case for abiotic environmental factors, explaining why the social environment is fundamentally different than the abiotic environment. As discussed further below, the simple fact of "genes in the environment" can have profound implications for the molecular and evolutionary genetic basis of traits that are influenced or defined by social interactions as well as for approaches required to study these traits (Fig. 2). Note that IGEs are perhaps most intuitive when considering different classes or life history stages of conspecifics (e.g. parents and offspring; adult nurse worker and immature sibling; Fig. 1) but clearly can occur even when interacting individuals are homogeneous in terms of age, caste, etc. For more detailed reviews of the IGE theory and the IGE empirical literature, see Bleakley et al. (2010), McGlothlin et al. (2010), Wolf et al. (1998) and Wolf and Moore (2010).

2. SURVEY OF STUDIES OF IGEs IN SOCIAL INSECTS

The maternal effect quantitative genetic models described above were first extended to study the genetic basis of honey bee colony level traits such as honey production that are likely primarily influenced by queen and worker genotypes (Bienefeld and Pirchner, 1991; Bienefeld et al., 2007; Chevalet and Cornuet, 1982; Cornuet and Chevalet, 1987).

IGE approaches have also been developed and applied to social insect traits that can be measured on individuals, such as body size and caste (Linksvayer, 2006). Linksvayer and Wade (2005) described the theoretical implications of using an explicit IGE perspective for understanding the genetic basis of social insect traits and the evolution of eusociality (see also

Kent and Zayed, 2013; Linksvayer and Wade, 2009), and several subsequent empirical studies have explicitly used an IGE perspective to study the genetic basis of social insect traits (Gempe et al., 2012; Linksvayer, 2006, 2007; Linksvayer et al., 2009, 2011; Teseo et al., 2014; van Zweden et al., 2010; Vojvodic et al., 2014; Wang et al., 2008). Linksvayer (2006) estimated heritability and genetic correlations for direct and IGEs on body size and caste in an ant population, and Linksvayer (2007) and Linksvayer et al. (2009, 2011) studied the contribution of IGEs to divergence between ant species and honey bee lineages, respectively. Similar to these studies, Teseo et al. (2014) recently showed that reproductive traits in a clonal army ant depend on the interaction between direct and IGEs. Van Zweden et al. (2010) demonstrated that both direct and IGEs contribute to heritable variation in ant hydrocarbon profiles, with the most heritable hydrocarbons being those that were readily transferred among nestmates. Notably, Wang et al. (2008) was the first study to show that gene expression profiles depend strongly on IGEs (see also Gempe et al., 2012): in fact, the gene expression profiles of ant workers depended more strongly on the genotype of worker nestmates at the Gp-9 locus than the worker's own genotype.

These studies explicitly using the IGE framework build on a long series of social insect studies that implicitly considered IGEs. These studies recognized the special difficulty of heritable effects of the social environment (i.e. IGEs) for studying the genetic basis of social insect traits, and also demonstrated that these effects strongly contribute to heritable variation for a range of social insect traits (e.g. Allsopp et al., 2003; Beekman and Oldroyd, 2003; Beekman et al., 2000; Calderone and Page, 1992; Calis et al., 2002; Gotzek and Ross, 2008; Guzmán-Novoa and Page, 1994; Hunt et al., 2003; Keller and Ross, 1995; Melnichenko and Burmistrova, 1963; Moritz and Southwick, 1987; Moritz et al., 1987; Oldroyd et al., 1991; Osborne and Oldroyd, 1999; Pankiw and Page, 2001; Pankiw et al., 2002; Rinderer et al., 1986; Ross and Keller, 2002; Rüppell et al., 2001). These studies have often used a cross-fostering approach to experimentally disentangle the contribution of an individual's own genotype (direct genetic effects) versus the genotype of social partners (IGEs) to its traits.

Altogether these empirical studies which explicitly or implicitly studied IGEs provide strong empirical support for the notion that IGEs are very widespread and strongly contribute to heritable differences within and between social insect populations. Below, I explain why IGEs are actually expected to be ubiquitous in the truly social insects.

3. IGEs ARE EXPECTED TO BE UBIQUITOUS IN SOCIAL INSECTS

Maternal care guarantees maternal effects on offspring trait expression (Wade, 1998). Similarly, cooperative brood care (sib care), a defining feature of eusociality, guarantees sib effects on the traits expressed by developing brood (Linksvayer, 2006; Linksvayer and Wade, 2005). The importance of such sib effects on social insect traits is perhaps best illustrated by studies of honey bee caste development and caste-related traits. It has long been known that honey bee nurse workers regulate the development and caste fate of larvae by controlling whether larvae receive qualitatively and quantitatively distinct queen or worker diets (Haydak, 1970). When larvae are reared in the lab so that this social control is not possible, strict queen–worker dimorphism disappears (Linksvayer et al., 2011). Similarly, when social regulation of the larval nutritional environment is artificially manipulated, caste-related characters are affected (Dedej et al., 1998; Wang et al., 2014). This also occurs when provisioning is more naturally manipulated, by cross-fostering larvae between honey bee strains (Allsopp et al., 2003; Linksvayer et al., 2009, 2011; Osborne and Oldroyd, 1999). These studies demonstrate that the caste developmental programme includes nurse worker genes with IGEs on caste, so that focusing solely on direct genetic effects expressed in developing larvae yields an incomplete picture of the genetic basis of caste. Some of the putative honey bee genes with IGEs on caste have been identified, including two major royal jelly protein genes (Huang et al., 2012; Kamakura, 2011), and hundreds of additional genes with putative IGEs on caste have recently been identified through RNA sequencing of the heads and royal jelly producing glands of nurse bees feeding queen- versus worker-destined larvae (Vojvodic et al., 2014). These nurse worker genes presumably affect larval development by affecting the quality and quantity of provisioned food as well as by affecting the provisioning behaviour of nurse workers.

The social mechanisms by which colony members regulate the development of brood are also known in a range of other social insects, and these mechanisms provide straightforward means by which IGEs act on the traits expressed by developing brood (Fig. 1). For example, in several ant species, nurse workers manipulate larval caste fate through biting developing larvae (Brian, 1973; Penick and Liebig, 2012); nurse *Polistes* wasps influence larval caste fate through antennal drumming (Suryanarayanan et al., 2011); and *Melipona* stingless bee workers influence caste fate via a glandular secretion that is fed to larvae (Jarau et al., 2010). At the same time, larvae can indirectly affect the provisioning behaviour of nurse workers (Fig. 1) and hence affect their own caste fate through begging behaviour (Kaptein et al., 2005) and brood pheromone (Le Conte et al., 1995; Sagili and Pankiw, 2009).

Other types of social signalling mechanisms used in social insect colonies provide additional mechanisms for IGEs on a wide range of social insect traits. By definition, pheromones are chemical signals produced by one individual that affects the behaviour or physiology of the individuals who receive and respond to the signal (Le Conte and Hefetz, 2008; Slessor et al., 2005). Thus, genes underlying pheromone production and signalling can have IGEs on the traits of individuals receiving and responding to the pheromone. As ants are considered to be walking batteries of exocrine glands (Hölldobler and Wilson, 1990, 2009), producing an array of pheromones, there is enormous potential for pheromone-mediated IGEs on the expression of ant traits. Many social insect pheromones have been identified that originate from queen, worker or brood nestmates and affect the physiology, behaviour or development of recipients (Hölldobler and Wilson, 1990; Le Conte and Hefetz, 2008; Slessor et al., 2005).

This range of specific social mechanisms by which social insect individuals communicate—exchange of food, biting, exchange of pheromones, etc.—all provide mechanisms by which genes expressed in nestmate workers, brood and queen(s) have IGEs on development, physiology and behaviour. Indeed, the social mechanisms described above that characterize insect societies virtually guarantee that most, if not all social insect traits, are affected to some degree by IGEs originating in various types of nestmates. Despite the clear implication of the ubiquity of IGEs for social insect traits, the functional and evolutionary importance of IGEs has only rarely been acknowledged (Helantera, 2011; Hunt, 2012; Johnson and Linksvayer, 2010; Keller, 2009; Kent and Zayed, 2013; Linksvayer and Wade, 2005; Linksvayer et al., 2009). In the following sections, I discuss why this omission is problematic.

4. IMPLICATIONS FOR THE MOLECULAR GENETIC UNDERPINNINGS OF SOCIAL INSECT TRAIT EXPRESSION

While past quantitative genetic research in the social insects sought to identify the causal allelic variants underlying trait variation, an alternative molecular genetic approach is to identify and functionally describe the full set of genes and molecular interactions that influence trait expression, regardless whether any particular loci harbour allelic variation (Fig. 1). Increasingly, researchers study the transcriptomic profiles of individuals expressing certain traits (e.g. behaviours, developmental trajectories, etc.) as a first step to identify the set of genes and molecular mechanisms associated with trait expression (Chandrasekaran et al., 2011; Evans and Wheeler, 1999; Feldmeyer et al., 2014; Ferreira et al., 2013; Whitfield et al., 2006).

The first such social insect sociogenomic studies searched for associations between an individual's behaviour or caste and its genome-wide transcriptional profile (Fig. 2A), often while experimentally controlling the social environment (e.g. using colonies containing only a single age cohort of workers; Barchuk et al., 2007; Evans and Wheeler, 1999; Whitfield et al., 2003, 2006). These studies, many of which have focused on the genetic basis of social behaviour, have identified a large number of candidate genes expressed in individual's brains that are associated with their behavioural state (Chandrasekaran et al., 2011). As an aside, an IGE-motivated version of these studies would ask how well an individual's behaviour is predicted by its own versus its social partners' gene expression profiles in order to identify genes with putative direct versus IGEs on behaviour (Fig. 2). Subsequent studies have shown that the neuro-genomic state of workers is highly responsive to social signals, such as honey bee queen mandibular pheromone and forager-produced pheromones, which are known to affect colony-level division of labour (Alaux et al., 2009; Grozinger et al., 2003; Manfredini et al., 2014). These studies provide exciting first steps to elucidating the molecular genetic basis of individual's response to social signals. Other studies have begun to elucidate the molecular genetic basis of pheromone production (Malka et al., 2009, 2014).

Social insect colonies typically contain multiple types of individuals (e.g. various brood stages, queens, workers of different functional groups) and social signal production, reception and response occurs dynamically and simultaneously in these various interacting groups (Fig. 1). Studying signal production, reception and response separately, and in one group at a time, is likely to miss biologically important complexity. An alternative way to study this dynamic process involving signalling and response among multiple life history stages and functional groups within a colony is to use more systems-level approaches (Linksvayer et al., 2012). The IGE approach takes a step in this direction by considering the effect of multiple interacting individuals on each other. While the IGE approach has usually been applied in the context of quantitative genetic studies, the framework can also be readily applied to study the full set of genes expressed in interacting social partners underlying trait expression (Linksvayer et al., 2012; Vojvodic et al., 2014). For example, by simultaneously studying the transcriptomic profiles of interacting nurse workers and developing larvae (Fig. 1), it is possible to identify genes expressed in both larvae and their care-giving nurses that directly, and indirectly, respectively, affect larval development (Vojvodic et al., 2014; Fig. 2).

As RNA sequencing costs continue to drop, it will increasingly be feasible to simultaneously study gene expression profiles of multiple categories of socially interacting individuals within social insect colonies, to provide social systems-level insight into the molecular basis of social regulation of the behaviour, physiology, development and fitness of colony members. Such a more systems-level focused strategy of profiling gene expression of multiple interacting organisms simultaneously has already been used to study the molecular mechanisms underlying host–pathogen interactions (Kollmus et al., 2014; Reid and Berriman, 2012; Tierney et al., 2012; Westermann et al., 2012), and this approach is a conceptually straightforward next step to study the molecular mechanisms underlying social interactions (Linksvayer et al., 2012; Vojvodic et al., 2014). Once candidate genes and gene networks are identified in the various classes of interacting individuals, gene-level approaches can be used to experimentally manipulate expression and quantify effects at the individual- and colony-levels.

Furthermore, there is increasingly exciting potential to link molecular-genetic and evolutionary-genetic approaches to identify how allelic variation within the colony-level network of genes underlying trait expression leads to phenotypic variation for both individual- and colony-level traits. These approaches have been described as "systems genetics" or "genetical genomics" (Ayroles et al., 2009; Mackay et al., 2009) and fit within the rapidly growing field of evolutionary systems biology (O'Malley, 2012; Rockman, 2011). For truly social organisms, like social insects, the colony represents an additional biologically important level of organization, expanding the complex system to be studied from the individual organism level to the colony level. Integrating molecular genetic, functional genomic, quantitative genetic/genomic and emerging evolutionary systems biology approaches will provide insight into the genetic basis and evolution of social insect traits.

5. WHY SHOULD SOCIAL INSECT RESEARCHERS CONSIDER THE IGE FRAMEWORK?

1. In theory, IGEs can strongly affect both the magnitude and direction of evolutionary responses to selection as well as expected patterns of

sequence variation within and between lineages (Linksvayer and Wade, 2009; McGlothlin et al., 2010; Moore et al., 1997; Wolf and Moore, 2010; Wolf et al., 1998). In practice, IGEs have often been found to be as important as direct genetic effects in a wide range of organisms (Bleakley et al., 2010; Wolf and Moore, 2010). As discussed throughout this chapter, IGEs are expected to be especially important in social insects as a result of long evolutionary history of being truly social (Linksvayer and Wade, 2005). In order to properly account for the contribution of socially expressed genes to trait variation and trait evolution, formal evolutionary genetic models must be used. The IGE framework is a natural extension of well-established quantitative genetic models (McGlothlin et al., 2010; Moore et al., 1997; Wolf and Moore, 2010). Correct estimation of IGE parameters and incorporation into evolutionary models is especially tricky in complex social systems such as the eusocial insects, where IGEs can originate simultaneously from multiple sources, including the queen, different functional categories of workers and brood (see Linksvayer, 2006). Furthermore, the contribution of IGEs to trait variation and the evolutionary process remains hidden to conventional genetic approaches that do not explicitly account for genetic components arising from social interactions (Wolf, 2003).

2. IGEs are also expected to make up a large proportion of the full network of genes affecting social insect trait expression (Linksvayer et al., 2009, 2012). Indeed, as described above, all of the mechanisms of social communication (pheromones, exchange of food, physical interactions, etc.) that are apparently so important to the functioning of social insect colonies provide straightforward mechanisms for IGEs originating from various functional categories of nestmates to affect trait expression. The contribution of these social regulatory mechanisms to the molecular genetic basis of trait expression is opaque to conventional approaches and has so far, remained largely hidden. Unless research programmes seeking to describe the molecular mechanisms underlying social insect trait expression explicitly consider IGEs, they will often simply overlook functionally important socially acting mechanisms—the very mechanisms that characterize eusocial colonies (Holldobler and Wilson, 2009; Seeley, 1997). For example, as discussed in detail above, caste development across social insect lineages is well understood to depend on the socially regulated nutritional environment. While a large number of studies have sought to describe the molecular mechanisms of caste development (e.g. Barchuk et al., 2007; Evans and Wheeler, 1999; Foret et al., 2012; Mutti et al., 2011; Wang et al., 2013; Wheeler

et al., 2014), the vast majority of these studies have only considered genes that are differentially expressed in developing larvae in response to the nutritional environment, but have not considered the genes expressed in care-giving nurses that produce alternate nutritional environments, and thus have IGEs on caste (but see Huang et al., 2012; Kamakura, 2011; Vojvodic et al., 2014). An explicit IGE framework ensures that these molecular components that contribute to social regulation are actively studied (Linksvayer et al., 2009).

3. A major roadblock to the increasing number of transcriptomic and genomic studies of social insects is to characterize the function of genes identified as playing important roles in social insect trait expression or evolution. Gene function is typically inferred by trying to identify functionally annotated orthologs from model systems such as *Drosophila*. Notwithstanding the fact that the function of orthologs in *Drosophila* and social insects may differ, a number of recent transcriptomic and genomic studies in social insects have emphasized the large number of taxonomically restricted genes (Feldmeyer et al., 2014; Ferreira et al., 2013; Harpur et al., 2014; Jasper et al., 2014; Johnson and Tsustui, 2011; Simola et al., 2013; Sumner, 2014), which by definition do not have identifiable orthologs, so usually have completely unknown function. Such novel genes and genes with novel functions may be expected to be especially important in the evolution of the elaborate social communication and regulatory mechanisms that characterize insect societies (Johnson and Linksvayer, 2010). Indeed, a recent honey bee study found that the transcriptomic profiles of tissues with novel social function were dominated by a small number of taxonomically restricted genes, indicating that novel genes play important roles in traits associated with social communication and regulation (Jasper et al., 2014). As the IGE framework explicitly focuses on genes that act through these social mechanisms, this approach can help to elucidate the function of novel genes that may often otherwise be overlooked.
4. Finally, social evolution theory is frequently invoked to motivate studies of the genetic basis and evolution of insect societies (Abbot et al., 2011; Bourke and Franks, 1995; Queller and Strassmann, 1998). Social evolution theory pre-supposes the existence of socially acting genes (Hamilton, 1964). The IGE/interacting phenotypes framework provides approaches specifically designed to detect these genes with social effects on trait expression and fitness.

6. CONCLUSIONS

Maternal effects in mammals are well understood to be a biological fact that must be considered by any genetic analysis of mammalian traits (McAdam and Boutin, 2004; Wade, 1998; Wade et al., 2008). Similarly, IGEs are a biological fact in social insects and should be explicitly considered in any genetic analysis of social insect trait expression or trait evolution. This is not only true for quantitative genetic studies but also includes sociogenomic studies focused on the molecular mechanisms underpinning trait expression. The social communication and regulatory systems that characterize eusocial insect colonies guarantee an important role of IGEs in the molecular basis of traits.

I have argued that IGEs are ubiquitous, affecting trait expression for most or all social insect traits. It is reasonable to ask if there is likely any gene in the genome of social insects that does not have indirect effects on some trait. Given the functional integration of individuals within insect societies, it seems likely that a majority of genes may have both direct and indirect effects on trait expression and fitness, perhaps acting through diverse and circuitous social routes. These precise mechanisms are what we need to describe to understand the functional significance of specific genes in social insect genomes. Describing these mechanisms is indeed a daunting task—seemingly much more so than in solitary organisms where there is necessarily only a one-to-one mapping of an individual's genes to its own traits, and as a result perhaps much less pleiotropy than in highly social organisms (Mullen and Thompson, "Understanding Honey Bee Worker Self-Sacrifice: A Conceptual–Empirical Framework"; Thompson et al., 2013). On the other hand, this added social complexity is precisely the characteristic that compels both social insect researchers and casual observers to marvel at insect societies—and is the reason social insects are established as models for social evolution, collective behaviour and the genetic basis of behaviour.

REFERENCES

Abbot, P., Abe, J., Alcock, J., Alizon, S., Alpedrinha, J.A.C., Andersson, M., Andre, J.B., van Baalen, M., Balloux, F., Balshine, S., Barton, N., Beukeboom, L.W., Biernaskie, J.M., Bilde, T., Borgia, G., Breed, M., Brown, S., Bshary, R., Buckling, A., Burley, N.T., Burton-Chellew, M.N., Cant, M.A., Chapuisat, M., Charnov, E.L., Clutton-Brock, T., Cockburn, A., Cole, B.J., Colegrave, N., Cosmides, L., Couzin, I.D., Coyne, J.A., Creel, S., Crespi, B., Curry, R.L., Dall, S.R.X., Day, T., Dickinson, J.L., Dugatkin, L.A.,

El Mouden, C., Emlen, S.T., Evans, J., Ferriere, R., Field, J., Foitzik, S., Foster, K., Foster, W.A., Fox, C.W., Gadau, J., Gandon, S., Gardner, A., Gardner, M.G., Getty, T., Goodisman, M.A.D., Grafen, A., Grosberg, R., Grozinger, C.M., Gouyon, P.H., Gwynne, D., Harvey, P.H., Hatchwell, B.J., Heinze, J., Helantera, H., Helms, K.R., Hill, K., Jiricny, N., Johnstone, R.A., Kacelnik, A., Kiers, E.T., Kokko, H., Komdeur, J., Korb, J., Kronauer, D., Kummerli, R., Lehmann, L., Linksvayer, T.A., Lion, S., Lyon, B., Marshall, J.A.R., McElreath, R., Michalakis, Y., Michod, R.E., Mock, D., Monnin, T., Montgomerie, R., Moore, A.J., Mueller, U.G., Noe, R., Okasha, S., Pamilo, P., Parker, G.A., Pedersen, J.S., Pen, I., Pfennig, D., Queller, D.C., Rankin, D.J., Reece, S.E., Reeve, H.K., Reuter, M., Roberts, G., Robson, S.K.A., Roze, D., Rousset, F., Rueppell, O., Sachs, J.L., Santorelli, L., Schmid-Hempel, P., Schwarz, M.P., Scott-Phillips, T., Shellmann-Sherman, J., Sherman, P.W., Shuker, D.M., Smith, J., Spagna, J.C., Strassmann, B., Suarez, A.V., Sundstrom, L., Taborsky, M., Taylor, P., Thompson, G., Tooby, J., Tsutsui, N.D., Tsuji, K., Turillazzi, S., Ubeda, F., Vargo, E.L., Voelkl, B., Wenseleers, T., West, S.A., West-Eberhard, M.J., Westneat, D.F., Wiernasz, D.C., Wild, G., Wrangham, R., Young, A.J., Zeh, D.W., Zeh, J.A., Zink, A., 2011. Inclusive fitness theory and eusociality. Nature 471, E1–E4.

Alaux, C., Le Conte, Y., Adams, H.A., Rodriguez-Zas, S., Grozinger, C.M., Sinha, S., Robinson, G.E., 2009. Regulation of brain gene expression in honey bees by brood pheromone. Genes Brain Behav. 8, 309–319.

Allsopp, M.H., Calis, J.N.M., Boot, W.J., 2003. Differential feeding of worker larvae affects caste characters in the Cape honeybee, Apis mellifera capensis. Behav. Ecol. Sociobiol. 54, 555–561.

Ayroles, J.F., Carbone, M.A., Stone, E.A., Jordan, K.W., Lyman, R.F., Magwire, M.M., Rollmann, S.M., Duncan, L.H., Lawrence, F., Anholt, R.R.H., MacKay, T.F.C., 2009. Systems genetics of complex traits in *Drosophila melanogaster*. Nat. Genet. 41, 299–307.

Barchuk, A.R., Cristino, A.S., Kucharski, R., Costa, L.F., Simoes, Z.L.P., Maleszka, R., 2007. Molecular determinants of caste differentiation in the highly eusocial honeybee Apis mellifera. BMC Dev. Biol. 7, 70.

Beekman, M., Oldroyd, B.P., 2003. Effects of cross-feeding anarchistic and wild type honey bees: anarchistic workers are not queen-like. Naturwissenschaften 90, 189–192.

Beekman, M., Calis, J.N.M., Boot, W.J., 2000. Parasitic honeybees get royal treatment. Nature 404, 723.

Bienefeld, K., Pirchner, F., 1991. Genetic correlations among several colony characters in the honey-bee (Hymenoptera, Apidae) taking queen and worker effects into account. Ann. Entomol. Soc. Am. 84, 324–331.

Bienefeld, K., Ehrhardt, K., Reinhardt, F., 2007. Genetic evaluation in the honey bee considering queen and worker effects—a BLUP-animal model approach. Apidologie 38, 77–85.

Bijma, P., 2006. Estimating maternal genetic effects in livestock. J. Anim. Sci. 84, 800–806.

Bleakley, B.H., Wolf, J.B., Moore, A.J., 2010. Evolutionary quantitative genetics of social behaviour. In: Szekely, T., Moore, A.J., Komdeur, J., Griffiths, M. (Eds.), Social Behaviour: Genes, Ecology, and Evolution. Cambridge University Press, Cambridge, MA, pp. 29–54.

Bloch, G., Grozinger, C.M., 2011. Social molecular pathways and the evolution of bee societies. Philos. Trans. R. Soc. B Biol. Sci. 366, 2155–2170.

Bourke, A.F.G., Franks, N.R., 1995. Social Evolution in Ants. Princeton University Press, Princeton, NJ.

Bouwman, A.C., Bergsma, R., Duijvesteijn, N., Bijma, P., 2010. Maternal and social genetic effects on average daily gain of piglets from birth until weaning. J. Anim. Sci. 88, 2883–2892.

Brian, M.V., 1973. Caste control through worker attack in the ant *Myrmica*. Insectes Soc. 20, 87–102.

Calderone, N.W., Page, R.E., 1992. Effects of interactions among genotypically diverse nestmates on task specialization by foraging honey bees (*Apis mellifera*). Behav. Ecol. Sociobiol. 30, 219–226.

Calis, J.N.M., Boot, W.J., Allsopp, M.H., Beekman, M., 2002. Getting more than a fair share: nutrition of worker larvae related to social parasitism in the Cape honey bee *Apis mellifera capensis*. Apidologie 33, 193–202.

Chandrasekaran, S., Ament, S.A., Eddy, J.A., Rodriguez-Zas, S.L., Schatz, B.R., Price, N.D., Robinson, G.E., 2011. Behavior-specific changes in transcriptional modules lead to distinct and predictable neurogenomic states. Proc. Natl. Acad. Sci. U.S.A. 108, 18020–18025.

Chevalet, C., Cornuet, J.M., 1982. Theoretical-study on the selection of the honey yield character. 1. Genetic and statistical-model. Apidologie 13, 39–65.

Cheverud, J.M., 1984. Evolution by kin selection: a quantitative genetic model illustrated by maternal performance in mice. Evolution 38, 766–777.

Cheverud, J.M., Moore, A.J., 1994. Quantitative genetics and the role of the environment provided by relatives in behavioral evolution. In: Boake, C.R.B. (Ed.), Quantitative Genetic Studies of Behavioral Evolution. University of Chicago Press, Chicago, IL, pp. 67–100.

Cornuet, J.M., Chevalet, C., 1987. Theoretical-study of the selection of the honey yield character. 2. Combined selection scheme of honey-bee queens in natural mating. Apidologie 18, 253–266.

Dedej, S., Hartfelder, K., Aumeier, P., Rosenkranz, P., Engels, W., 1998. Caste determination is a sequential process: effect of larval age at grafting on ovariole number, hind leg size and cephalic volatiles in the honey bee (Apis mellifera carnica). J. Apicult. Res. 37, 183–190.

Dickerson, G.E., 1947. Composition of hog carcasses as influenced by heritable differences in rate and economy of gain. Iowa Agric. Exp. Stn. Res. Bull. 354, 489–524.

Evans, J.D., Wheeler, D.E., 1999. Differential gene expression between developing queens and workers in the honey bee, Apis mellifera. Proc. Natl. Acad. Sci. U.S.A. 96, 5575–5580.

Falconer, D.S., 1960. Introduction to Quantitative Genetics. The Ronald Press Company, New York.

Falconer, D.S., Mackay, T.F.C., 1996. Introduction to Quantitative Genetics, fourth ed. Longman, Essex, England.

Feldmeyer, B., Elsner, D., Foitzik, S., 2014. Gene expression patterns associated with caste and reproductive status in ants: worker-specific genes are more derived than queen-specific ones. Mol. Ecol. 23, 151–161.

Ferreira, P.G., Patalano, S., Chauhan, R., Ffrench-Constant, R., Gabaldon, T., Guigo, R., Sumner, S., 2013. Transcriptome analyses of primitively eusocial wasps reveal novel insights into the evolution of sociality and the origin of alternative phenotypes. Genome Biol. 14, R20.

Foret, S., Kucharski, R., Pellegrini, M., Feng, S., Jacobsen, S.E., Robinson, G.E., Maleszka, R., 2012. DNA methylation dynamics, metabolic fluxes, gene splicing, and alternative phenotypes in honey bees. Proc. Natl. Acad. Sci. U.S.A. 109, 4968–4973.

Gempe, T., Stach, S., Bienefeld, K., Beye, M., 2012. Mixing of honeybees with different genotypes affects individual worker behavior and transcription of genes in the neuronal substrate. PLoS One 7, e31653.

Gotzek, D., Ross, K.G., 2008. Experimental conversion of colony social organization in fire ants (Solenopsis invicta): worker genotype manipulation in the absence of queen effects. J. Insect Behav. 21, 337–350.

Graham, A.M., Munday, M.D., Kaftanoglu, O., Page, R.E., Amdam, G.V., Rueppell, O., 2011. Support for the reproductive ground plan hypothesis of social evolution and major QTL for ovary traits of Africanized worker honey bees (Apis mellifera L.). BMC Evol. Biol. 11, 95.
Griffing, B., 1977. Selection for populations of interacting phenotypes. In: Pollak, E., Kempthorne, O., Bailey, T.B. (Eds.), Proceedings of the International Conference on Quantitative Genetics. Iowa State University Press, Ames, IA, pp. 413–434.
Griffing, B., 1981. A theory of natural-selection incorporating interaction among individuals. 1. The modeling process. J. Theor. Biol. 89, 635–658.
Grozinger, C.M., Sharabash, N.M., Whitfield, C.W., Robinson, G.E., 2003. Pheromone-mediated gene expression in the honey bee brain. Proc. Natl. Acad. Sci. U.S.A. 100, 14519–14525.
Guzmán-Novoa, E., Page, R.E., 1994. Genetic dominance and worker interactions affect honeybee colony defense. Behav. Ecol. 5, 91–97.
Hahn, M.E., Schanz, N., 1996. Issues in the genetics of behavior: revisited. Behav. Genet. 26, 463–470.
Hamilton, W.D., 1964. The genetical evolution of social behaviour, I. J. Theor. Biol. 7, 1–16.
Harpur, B.A., Kent, C.F., Molodtsova, D., Lebon, J.M.D., Alqarni, A.S., Owayss, A.A., Zayed, A., 2014. Population genomics of the honey bee reveals strong signatures of positive selection on worker traits. Proc. Natl. Acad. Sci. U.S.A. 111, 2614–2619.
Haydak, M.H., 1970. Honey bee nutrition. Annu. Rev. Entomol. 15, 143–156.
Helantera, H., 2011. Ant encounters: interaction networks and colony behavior. Biol. Philos. 26, 935–944.
Hölldobler, B., Wilson, E.O., 1990. The Ants. Harvard University Press, Cambridge, MA.
Hölldobler, B., Wilson, E.O., 2009. The Superorganism: The Beauty, Elegance, and Strangeness of Insect Societies. Norton, New York, NY.
Huang, C.Y., Chi, L.L., Huang, W.J., Chen, Y.W., Chen, W.J., Kuo, Y.C., Yuan, C.M., Chen, C.N., 2012. Growth stimulating effect on queen bee larvae of histone deacetylase inhibitors. J. Agric. Food Chem. 60, 6139–6149.
Hunt, J., 2012. A conceptual model for the origin of worker behaviour and adaptation of eusociality. J. Evol. Biol. 25, 1–19.
Hunt, G.J., Guzman-Novoa, E., Fondrk, M.K., Page, R.E., 1998. Quantitative trait loci for honey bee stinging behavior and body size. Genetics 148, 1203–1213.
Hunt, G.J., Guzman-Novoa, E., Uribe-Rubio, J.L., Prieto-Merlos, D., 2003. Genotype–environment interactions in honeybee guarding behaviour. Anim. Behav. 66, 459–467.
Jarau, S., Van Veen, J.W., Twele, R., Reichle, C., Herrera Gonzales, E., Aguilar, I., Francke, W., Ayasse, M., 2010. Workers make the queens in *Melipona* bees: identification of geraniol as a caste determining compound from labial glands of nurse bees. J. Chem. Ecol. 36, 565–569.
Jasper, W.C., Linksvayer, T.A., Atallah, J., Friedman, D., Chiu, J.C., Johnson, B.R., 2014. Large scale coding sequence change underlies the evolution of post-developmental novelty in honey bees. Mol. Biol. Evol. http://dx.doi.org/10.1093/molbev/msu292.
Johnson, B.R., Linksvayer, T.A., 2010. Deconstructing the superorganism: social physiology, groundplans, and sociogenomics. Q. Rev. Biol. 85, 57–79.
Johnson, B.R., Tsustui, N.D., 2011. Taxonomically restricted genes are associated with the evolution of sociality in the honey bee. BMC Genomics 12, 164.
Kamakura, M., 2011. Royalactin induces queen differentiation in honeybees. Nature 473, 478–483.
Kaptein, N., Billen, J., Gobin, B., 2005. Larval begging for food enhances reproductive options in the ponerine ant *Gnamptogenys striatula*. Anim. Behav. 69, 293–299.

Keller, L., 2009. Adaptation and the genetics of social behaviour. Philos. Trans. R. Soc. B Biol. Sci. 364, 3209–3216.
Keller, L., Ross, K.G., 1995. Gene by environment interaction: effects of a single gene and social environment on reproductive phenotypes of fire ant queens. Funct. Ecol. 9, 667–676.
Kent, C.F., Zayed, A., 2013. Evolution of recombination and genome structure in eusocial insects. Commun. Integr. Biol. 6, e22919.
Kirkpatrick, M., Lande, R., 1989. The evolution of maternal characters. Evolution 43, 485–503.
Kollmus, H., Wilk, E., Schughart, K., 2014. Systems biology and systems genetics—novel innovative approaches to study host–pathogen interactions during influenza infection. Curr. Opin. Virol. 6, 47–54.
Kovacs, J.L., Hoffman, E.A., Marriner, S.M., Rekau, J.A., Goodisman, M.A.D., 2010. Environmental and genetic influences on queen and worker body size in the social wasp Vespula maculifrons. Insectes Soc. 57, 53–65.
Lattorff, H.M.G., Moritz, R.F.A., 2013. Genetic underpinnings of division of labor in the honeybee (Apis mellifera). Trends Genet. 29, 641–648.
Le Conte, Y., Hefetz, A., 2008. Primer pheromones in social hymenoptera. Annu. Rev. Entomol. 53, 523–542.
Le Conte, Y., Sreng, L., Poitout, S.H., 1995. Brood pheromone can modulate the feeding behavior of *Apis mellifera* workers (Hymenoptera, Apidae). J. Econ. Entomol. 88, 798–804.
Libbrecht, R., Oxley, P.R., Kronauer, D.J.C., Keller, L., 2013. Ant genomics sheds light on the molecular regulation of social organization. Genome Biol. 14, 212.
Linksvayer, T.A., 2006. Direct, maternal, and sibsocial genetic effects on individual and colony traits in an ant. Evolution 60, 2552–2561.
Linksvayer, T.A., 2007. Ant species differences determined by epistasis between brood and worker genomes. PLoS One 2, e994.
Linksvayer, T.A., Wade, M.J., 2005. The evolutionary origin and elaboration of sociality in the aculeate Hymenoptera: maternal effects, sib-social effects, and heterochrony. Q. Rev. Biol. 80, 317–336.
Linksvayer, T.A., Wade, M.J., 2009. Genes with social effects are expected to harbor more sequence variation within and between species. Evolution 63, 1685–1696.
Linksvayer, T.A., Fondrk, M.K., Page Jr., R.E., 2009. Honey bee social regulatory networks are shaped by colony-level selection. Am. Nat. 173, E99–E107.
Linksvayer, T.A., Kaftanoglu, O., Akyol, E., Blatch, S., Amdam, G.V., Page Jr., R.E., 2011. Larval and nurse worker control of developmental plasticity and the evolution of honey bee queen–worker dimorphism. J. Evol. Biol. 24, 1939–1948.
Linksvayer, T.A., Fewell, J.H., Gadau, J., Laubichler, M., 2012. Developmental evolution in social insects: regulatory networks from genes to societies. J. Exp. Zool. B Mol. Dev. Evol. 318, 159–169.
Lynch, M., 1987. Evolution of intrafamilial interactions. Proc. Natl. Acad. Sci. U.S.A. 84, 8507–8511.
Mackay, T.F.C., Stone, E.A., Ayroles, J.F., 2009. The genetics of quantitative traits: challenges and prospects. Nat. Rev. Genet. 10, 565–577.
Malka, O., Karunker, I., Yeheskel, A., Morin, S., Hefetz, A., 2009. The gene road to royalty—differential expression of hydroxylating genes in the mandibular glands of the honeybee. FEBS J. 276, 5481–5490.
Malka, O., Nino, E.L., Grozinger, C.M., Hefetz, A., 2014. Genomic analysis of the interactions between social environment and social communication systems in honey bees (Apis mellifera). Insect Biochem. Mol. Biol. 47, 36–45.

Manfredini, F., Lucas, C., Nicolas, M., Keller, L., Shoemaker, D., Grozinger, C.M., 2014. Molecular and social regulation of worker division of labour in fire ants. Mol. Ecol. 23, 660–672.

McAdam, A.G., Boutin, S., 2004. Maternal effects and the response to selection in red squirrels. Proc. R. Soc. B Biol. Sci. 271, 75–79.

McGlothlin, J.W., Moore, A.J., Wolf, J.B., Brodie, E.D., 2010. Interacting phenotypes and the evolutionary process. III. Social evolution. Evolution 64, 2558–2574.

Melnichenko, A., Burmistrova, N., 1963. Directed change of heredity in a bee colony and its biochemical basis. Mezhdunar. Kongr. po Pchelovodstvu, 19th, Selkhozizdat, Moscow. pp. 28–42.

Moore, A.J., Brodie III, E.D., Wolf, J.B., 1997. Interacting phenotypes and the evolutionary process. 1. Direct and indirect genetic effects of social interactions. Evolution 51, 1352–1362.

Moore, A.J., Haynes, K.F., Preziosi, R.F., Moore, P.J., 2002. The evolution of interacting phenotypes: genetics and evolution of social dominance. Am. Nat. 160, S186–S197.

Moritz, R.F.A., Southwick, E.E., 1987. Phenotype interactions in group behavior of honey bee workers (*Apis mellifera* L). Behav. Ecol. Sociobiol. 21, 53–57.

Moritz, R.F.A., Southwick, E.E., Harbo, J.R., 1987. Maternal and pre-eclosional factors affecting alarm behavior in adult honey bees (Apis mellifera L). Insectes Soc. 34, 298–307.

Mutti, N.S., Dolezal, A.G., Wolschin, F., Mutti, J.S., Gill, K.S., Amdam, G.V., 2011. IRS and TOR nutrient-signaling pathways act via juvenile hormone to influence honey bee caste fate. J. Exp. Biol. 214, 3977–3984.

Oldroyd, B.P., Thompson, G.J., 2007. Behavioural genetics of the honey bee Apis mellifera. In: Advances in Insect Physiology, vol. 33, pp. 1–49.

Oldroyd, B., Rinderer, T., Buco, S., 1991. Heritability of morphological characters used to distinguish European and Africanized honeybees. Theor. Appl. Genet. 82, 499–504.

O'Malley, M.A., 2012. Evolutionary systems biology: historical and philosophical perspectives on an emerging synthesis. Adv. Exp. Med. Biol. 751, 1–28.

Osborne, K.E., Oldroyd, B.P., 1999. Possible causes of reproductive dominance during emergency queen rearing by honeybees. Anim. Behav. 58, 267–272.

Page, R.E., Fondrk, M.K., Hunt, G.J., Guzman-Novoa, E., Humphries, M.A., Nguyen, K., Greene, A.S., 2000. Genetic dissection of honeybee (Apis mellifera L.) foraging behavior. J. Hered. 91, 474–479.

Page Jr., R.E., Rueppell, O., Amdam, G.V., 2012. Genetics of reproduction and regulation of honeybee (*Apis mellifera* L.) social behavior. Annu. Rev. Genet. 46, 97–119.

Pankiw, T., Page, R.E., 2001. Genotype and colony environment affect honeybee (Apis mellifera L.) development and foraging behavior. Behav. Ecol. Sociobiol. 51, 87–94.

Pankiw, T., Tarpy, D.R., Page, R.E., 2002. Genotype and rearing environment affect honeybee perception and foraging behaviour. Anim. Behav. 64, 663–672.

Penick, C.A., Liebig, J., 2012. Regulation of queen development through worker aggression in a predatory ant. Behav. Ecol. 23, 992–998.

Queller, D.C., Strassmann, J.E., 1998. Kin selection and social insects. Bioscience 48, 165–175.

Reid, A.J., Berriman, M., 2012. Genes involved in host–parasite interactions can be revealed by their correlated expression. Nucleic Acids Res. 41, 1508–1518.

Rinderer, T.E., Sylvester, H.A., Collins, A.M., Pesante, D., 1986. Identification of Africanized and European honey bees: effects of nurse genotype and comb size. Bull. Entomol. Soc. Am. 32, 150–152.

Robinson, G.E., Grozinger, C.M., Whitfield, C.W., 2005. Sociogenomics: social life in molecular terms. Nat. Rev. Genet. 6, 257–270.

Robinson, G.E., Fernald, R.D., Clayton, D.F., 2008. Genes and social behavior. Science 322, 896–900.
Rockman, M., 2011. Evolutionary genomics and systems biology. Q. Rev. Biol. 86, 353–354.
Ross, K.G., Keller, L., 2002. Experimental conversion of colony social organization by manipulation of worker genotype composition in fire ants (*Solenopsis invicta*). Behav. Ecol. Sociobiol. 51, 287–295.
Rueppell, O., Pankiw, T., Nielsen, D.I., Fondrk, M.K., Beye, M., Page, R.E., 2004. The genetic architecture of the behavioral ontogeny of foraging in honeybee workers. Genetics 167, 1767–1779.
Rüppell, O., Heinze, J., Hölldobler, B., 2001. Complex determination of queen body size in the queen size dimorphic ant Leptothorax rugatulus (Formicidae: Hymenoptera). Heredity 87, 33–40.
Sagili, R.R., Pankiw, T., 2009. Effects of brood pheromone modulated brood rearing behaviors on honey bee (Apis mellifera L.) colony growth. J. Insect Behav. 22, 339–349.
Seeley, T.D., 1995. The Wisdom of the Hive. Harvard University Press, Cambridge, MA.
Seeley, T.D., 1997. Honey bee colonies are group-level adaptive units. Am. Nat. 150, S22–S41.
Simola, D.F., Wissler, L., Donahue, G., Waterhouse, R.M., Helmkampf, M., Roux, J., Nygaard, S., Glastad, K.M., Hagen, D.E., Viljakainen, L., Reese, J.T., Hunt, B.G., Graur, D., Elhaik, E., Kriventseva, E.V., Wen, J.Y., Parker, B.J., Cash, E., Privman, E., Childers, C.P., Munoz-Torres, M.C., Boomsma, J.J., Bornberg-Bauer, E., Currie, C.R., Elsik, C.G., Suen, G., Goodisman, M.A.D., Keller, L., Liebig, J., Rawls, A., Reinberg, D., Smith, C.D., Smith, C.R., Tsutsui, N., Wurm, Y., Zdobnov, E.M., Berger, S.L., Gadau, J., 2013. Social insect genomes exhibit dramatic evolution in gene composition and regulation while preserving regulatory features linked to sociality. Genome Res. 23, 1235–1247.
Slessor, K.N., Winston, M.L., Le Conte, Y., 2005. Pheromone communication in the honeybee (Apis mellifera L.). J. Chem. Ecol. 31, 2731–2745.
Smith, C.R., Toth, A.L., Suarez, A.V., Robinson, G.E., 2008. Genetic and genomic analyses of the division of labour in insect societies. Nat. Rev. Genet. 9, 735–748.
Sumner, S., 2014. The importance of genomic novelty in social evolution. Mol. Ecol. 23, 26–28.
Suryanarayanan, S., Hermanson, J.C., Jeanne, R.L., 2011. A mechanical signal biases caste development in a social wasp. Curr. Biol. 21, 1–5.
Teseo, S., Chaline, N., Jaisson, P., Kronauer, D.J.C., 2014. Epistasis between adults and larvae underlies caste fate and fitness in a clonal ant. Nat. Commun. 5, 3363.
Thompson, G.J., Hurd, P.L., Crespi, B.J., 2013. Genes underlying altruism. Biol. Lett. 9, 0395. http://dx.doi.org/10.1098/rsbl.2013.0395.
Tierney, L., Linde, J., Muller, S., Brunke, S., Molina, J.C., Hube, B., Schock, U., Guthke, R., Kuchler, K., 2012. An interspecies regulatory network inferred from simultaneous RNA-seq of *Candida albicans* invading innate immune cells. Front. Microbiol. 3, 1–14.
Toth, A.L., Varala, K., Newman, T.C., Miguez, F.E., Hutchison, S.K., Willoughby, D.A., Simons, J.F., Egholm, M., Hunt, J.H., Hudson, M.E., Robinson, G.E., 2007. Wasp gene expression supports an evolutionary link between maternal behavior and eusociality. Science 318, 441–444.
van Zweden, J.S., Brask, J.B., Christensen, J.H., Boomsma, J.J., Linksvayer, T.A., d'Ettorre, P., 2010. Blending of heritable recognition cues among ant nestmates creates distinct colony gestalt odours but prevents within-colony nepotism. J. Evol. Biol. 23, 1498–1508.

Vojvodic, S., Johnson, B.R., Harpur, B., Kent, C., Zayed, A., Anderson, K.E., Linksvayer, T.A., 2014. The genomic signature of social interactions regulating honey bee caste development. bioRxiv. http://dx.doi.org/10.1101/012385.

Wade, M.J., 1998. The evolutionary genetics of maternal effects. In: Mousseau, T.A., Fox, C.W. (Eds.), Maternal Effects as Adaptations. Oxford University Press, New York, pp. 5–21.

Wade, M.J., Priest, N.K., Cruickshank, T., 2008. A theoretical overview of maternal genetic effects: evolutionary predictions and empirical tests using sequence data within and across mammalian taxa. In: Maternal Effects in Mammals, University of Chicago Press, Chicago.

Wang, J., Ross, K.G., Keller, L., 2008. Genome-wide expression patterns and the genetic architecture of a fundamental social trait. PLoS Genet. 4, e1000127.

Wang, Y., Azevedo, S.V., Hartfelder, K., Amdam, G.V., 2013. Insulin-like peptides (AmILP1 and AmILP2) differentially affect female caste development in the honey bee (Apis mellifera L.). J. Exp. Biol. 216, 4347–4357.

Wang, Y., Kaftanoglu, O., Fondrk, M.K., Page Jr., R.E., 2014. Nurse bee behaviour manipulates worker honeybee (*Apis mellifera* L.) reproductive development. Anim. Behav. 92, 253–261.

Westermann, A.J., Gorski, S.A., Vogel, J., 2012. Dual RNA-seq of pathogen and host. Nat. Rev. Microbiol. 10, 618–630.

Wheeler, D.E., Buck, N.A., Evans, J.D., 2014. Expression of insulin/insulin-like signalling and TOR pathway genes in honey bee caste determination. Insect Mol. Biol. 23, 113–121.

Whitfield, C.W., Cziko, A.M., Robinson, G.E., 2003. Gene expression profiles in the brain predict behavior in individual honey bees. Science 302, 296–299.

Whitfield, C.W., Ben-Shahar, Y., Brillet, C., Leoncini, I., Crauser, D., LeConte, Y., Rodriguez-Zas, S., Robinson, G.E., 2006. Genomic dissection of behavioral maturation in the honey bee. Proc. Natl. Acad. Sci. U.S.A. 103, 16068–16075.

Willham, R.L., 1963. The covariance between relatives for characters composed of components contributed by related individuals. Biometrics 19, 18–27.

Wolf, J.B., 2003. Genetic architecture and evolutionary constraint when the environment contains genes. Proc. Natl. Acad. Sci. U.S.A. 100, 4655–4660.

Wolf, J.B., Moore, A.J., 2010. Interacting phenotypes and indirect genetic effects. In: Westneat, D.F., Fox, C.W. (Eds.), Evolutionary Behavioral Ecology. Oxford University Press, New York, pp. 225–245.

Wolf, J.B., Brodie III, E.D., Cheverud, J.M., Moore, A.J., Wade, M.J., 1998. Evolutionary consequences of indirect genetic effects. Trends Ecol. Evol. 13, 64–69.

Wolf, J.B., Brodie III, E.D., Moore, A.J., 1999. Interacting phenotypes and the evolutionary process. II. Selection resulting from social interactions. Am. Nat. 153, 254–266.

Yan, H., Simola, D.F., Bonasio, R., Liebig, J., Berger, S.L., Reinberg, D., 2014. Eusocial insects as emerging models for behavioural epigenetics. Nat. Rev. Genet. 15, 677–688.

Zayed, A., Robinson, G.E., 2012. Understanding the relationship between brain gene expression and social behavior: lessons from the honey bee. Annu. Rev. Genet. 46 (46), 591–615.

CHAPTER NINE

Population Genomic and Phylogenomic Insights into the Evolution of Physiology and Behaviour in Social Insects

Clement F. Kent*, Amro Zayed†,[1]
*Janelia Research Campus, Howard Hughes Medical Institute, Ashburn, VA, USA
†Department of Biology, York University, Toronto, Ontario, Canada
[1]Corresponding author: e-mail address: zayed@yorku.ca

Contents

Abstract

Genomics revolutionized the field of social insect research by providing powerful tools to understand the relationship between genes, physiology and behaviour of social insects. Notably, analysis of gene expression and methylation patterns in the different castes of insect colonies highlighted many genes that likely underlie caste-specific physiological and behavioural phenotypes. However, earlier studies of social insect genomes lacked an 'evolutionary' context. *Out of the millions of DNA bases found in the genome of a social insect, which pieces were most important to fitness over the timescale of social evolution?* Here, we review a burgeoning body of literature that utilizes between-species or within-species genomic comparisons to highlight the evolutionary forces that have shaped social insect genomes. These pioneering phylogenetic and population genomic

ISSN 0065-2806
http://dx.doi.org/10.1016/bs.aiip.2015.01.002

studies provide a critically needed evolutionary context to social insect genomes and underscore the importance of adaptive changes in physiology and behaviour in social evolution.

1. INTRODUCTION

The reference genome: a means to an end, but not an end in itself

The publication of the honey bee genome in 2006 (Weinstock et al., 2006) was a highly anticipated event. The field was full of excitement because we were about to finally glance at the makings of a 'eusocial' genome. The feverish excitement quickly abated once we realized that the genome of the honey bee was mostly unremarkable! Most of the genes found in the honey bee genome were shared with solitary animals, and there were no obvious genomic features underlying social evolution in the enigmatic honey bee. The reference genome did however revolutionize research on the molecular basis of caste phenotypes by illuminating the relationships between physiology, transcription, methylation, behaviour and caste differences in honey bees (Hunt et al., 2007; Kamakura, 2011; Kucharski et al., 2008; Lyko et al., 2010; Robinson et al., 2008; Zayed and Robinson, 2012) (the chapter 'Insect Reward Systems: Comparing Flies and Bees' by Søvik et al.; the chapter 'Old Threads Make New Tapestry—Rewiring of Signalling Pathways Underlies Caste Phenotypic Plasticity in the Honey Bee, *Apis mellifera* L.' by Hartfleder et al.). The mechanistic studies vastly increased our knowledge of '*social life in molecular terms*' (Robinson et al., 2005) but provided little insight into the process of social evolution. The genomic regions associated with the evolution of sociality remain shrouded in darkness.

Perhaps the biggest challenge in making sense of a single genome is the lack of evolutionary and ecological context. *Among the 236 million bases of DNA that make up the honey bee genome, which sequences were most important to fitness and most relevant to the bee's social biology and evolution?* A well-annotated reference genome only provides information about the genetic potential of a species by highlighting DNA sequences with likely coding or regulatory functions (Fig. 1A). Transcriptomic and proteomic studies provide another layer of functional annotation by highlighting DNA sequences that are transcribed and translated into proteins. Such studies are also critical for establishing links between gene function and biological processes, and for highlighting the relationship between genes and

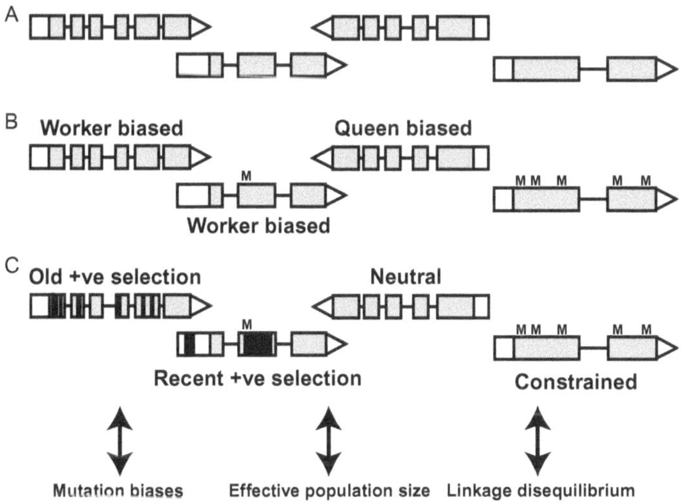

Figure 1 Annotating the evolutionary context of a genome. (A) Reference genomes provide the basic blue prints for social insects by highlighting genomic sequences with coding (grey or yellow boxes) or regulatory (open boxes) capacity. (B) Powered by the reference genome, functional studies can uncover changes in gene expression, protein abundance or methylation status (methylated CpG's are indicated as M in the figure) that influence the expression of caste-specific phenotypes. These studies have greatly enriched our understanding of sociality in molecular terms by cataloguing genes and pathways underpinning division of labour between queens and workers and between the different worker castes. However, not all genes associated with caste phenotypes were directly responsible for the evolution of caste; genes often act in complex networks and adaptive changes in a few key genes can influence the expression patterns of many others. (C) Population genomic studies provide the ultimate tool for highlighting genes and regulatory sequences involved in social evolution. Population genomic studies have the ability to identify DNA sequences (black bars) with adaptive patterns of molecular evolution over multiple time scales, thereby highlighting the loci with the greatest effect of fitness over the course of social evolution. Moreover, population genomic studies allow researchers to infer the 'evolutionary landscape' experienced by social insects—a set of parameters that summarize the relative strength of drift, mutation and selection in social populations. Population genomic studies provide a natural 'glue' that promises to reconcile mechanic and theoretical research on how and why sociality evolved.

transcriptional, proteomic, physiological and behavioural phenotypes (Fig. 1B). But how can we annotate the evolutionarily and ecologically significant parts of a genome? How do we identify the DNA changes that led to the phenotypic innovations found in social insects?

Evolution is caused by changes in the frequency of inherited mutations over time. New mutations can segregate within a species' gene pool for a long time, but can also spread rapidly leading to fixed genetic differences between closely related species. Comparing the reference genomes of closely related animals allows researchers to re-construct the genetic changes that occurred over the course of evolution. These phylogenomic analyses can highlight sequences of DNA with rapid patterns of evolution thereby providing insights into the loci that influenced fitness over the time scale of species divergence. Comparing populations of genomes from a single species—population genomics—allows researchers to directly quantify the relative contribution of natural selection to patterns of within species diversity and between species divergence. Phylogenomic and population genomic analyses thereby allow researchers to annotate genomic regions of 'evolutionary' significance by highlighting DNA sequence changes that were associated with fitness differences in the past or in the present (Fig. 1C). This knowledge will help us understand the genetic basis of adaptive phenotypic changes and provides a natural way to test ideas about how sociality evolved in insects.

Here, we review a small but growing body of literature that strives to understand the evolutionary forces that have shaped the genomes of social insects. We provide a brief overview of how sequence data can be used to infer evidence of ancient and recent natural selection and their advantages and disadvantages. We then summarize some of the case studies on social insect phylogenomics and population genomics and what they have taught us about the evolution and elaboration of sociality. Finally, we discuss future directions and how to best leverage these new approaches to ultimately answer the *How's* and *Why's* of social evolution.

2. A PRIMER ON TESTS OF SELECTION USING SEQUENCE DATA

There are several excellent reviews on how to detect evidence of natural selection using sequence data. Nielsen (2005) provides a great introduction and Vitti et al. (2013) provide a comprehensive list of the latest tools for detecting selection using genomics data. Both reviews tend to focus heavily on methods and examples from humans or traditional genetic model organisms. Here, we provide a brief conceptual overview of some of the most popular methods for inferring selection from sequence data.

2.1 Phylogenomics: Let the tree be your guide!

Creating a reference genome or transcriptome for a new species is exciting and a lot of work. As more genomes of social insects and non-social relatives accumulate, richer and denser phylogenetic trees are being built. Among the low-hanging fruit on this new branch of social insect species trees are genes with signs of accelerated molecular evolution based on comparisons to sequences of the same gene in related species. The conceptual foundations of such studies is the assumption that genes with fast evolving amino-acid sequences are likely experiencing positive selection and that genes with slow evolving amino-acid sequences are likely experiencing negative selection. Imagine a situation where most new amino-acid replacement mutations—also known as non-synonymous mutations—are strongly deleterious. Such mutations will be rapidly removed by natural negative selection thereby conserving the original amino-acid sequence of the protein. Imagine now that orthologous copies of a gene in different species are all experiencing the same level of negative selection. The end result will be minimal changes in the amino-acid sequence across the phylogeny. Now imagine a gene where most non-synonymous mutations are beneficial; such mutations will spread rapidly within the population until they reach 100% frequency—'fixation'—resulting in an amino-acid sequence difference between orthologous copies of a gene between different species. If orthologous copies of a gene in different species are all experiencing a high level of positive selection, we would observe a high rate of amino-acid changes in this protein across the species tree. Of course, the fitness effect of non-synonymous mutations vary in a continuous manner and specific statistical tests are needed to highlight fast evolving genes.

A common tool used to infer selection in molecular phylogenies is d*N*/d*S*, the ratio of non-synonymous (protein changing) to synonymous changes in a coding sequence, normalized to the number of *N* and *S* sites. d*N*/d*S* is sometimes referred to as Ka/Ks or ω, and it can be easily estimated through sequence alignments (Fig. 2). Purely neutral evolution (Kimura, 1985) results in d*N*/d*S* near 1; this occurs when non-synonymous mutations and synonymous mutations have no effect on fitness and fix at rates that are proportional to the rate of mutation. Setting neutral evolution as the bench mark (i.e. d*N*/d*S* near 1), positive selection is thus expected to lead to higher d*N*/d*S* values while negative selection is expected to lead to lower values. In reality, most amino acid changes in genes tend to be deleterious (i.e. d*N*/d*S* near 0, Fig. 2A), so even if a gene has a minority of sites under strong positive selection, d*N*/d*S* may still be much less than 1.

Several factors can distort d*N*/d*S* or lead to high ratios of d*N*/d*S* that are not caused by positive selection. A few of these factors are especially relevant to social insect research, and warrant a brief discussion, below.

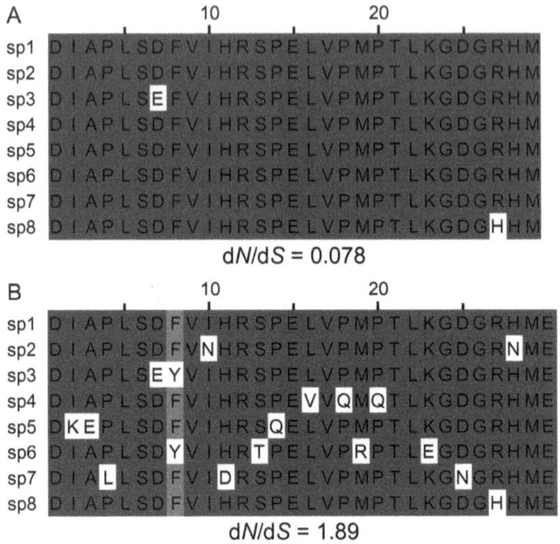

Figure 2 Phylogenomics can highlight genes with fast rates of protein evolution. Alignment of gene sequences from multiple species can be used to estimate the ratio of non-synonymous substitutions per non-synonymous site (d*N*) to the number of synonymous substitutions per synonymous site (d*S*). We present two simulated protein sequences from eight species. In one example (A), the protein is highly conserved with very low d*N*/d*S*. In the other example (B), the protein is rapidly evolving with very high d*N*/d*S*. High d*N*/d*S* can occur because of positive selection but can sometimes also occur because of non-adaptive evolutionary processes.

2.1.1 Demography and changes in population size

There is evidence that eusocial species with large colonies may have smaller effective population sizes (Romiguier et al., 2014b), leading to increased d*N*/d*S* because of a genome-wide relaxation of purifying selection—hereafter called relaxed selection. The power of natural selection is influenced by the fitness effect of mutations and the population's effective size. In small populations, stochastic fluctuations in allele frequencies can easily overpower deterministic forces such as negative selection on weakly deleterious mutations. A large population is likely to harbour many slightly deleterious mutations at very low frequencies, but after a bottleneck, drift can rapidly fix some of these mutations leading to increased d*N*/d*S* that is not caused by adaptive evolution.

2.1.2 DNA chemistry and repair

Mutational processes in the genome are influenced by functional processes such as DNA (de)methylation (Elango et al., 2009; Terrapon et al., 2014) and biased gene conversion (BGC) during meiosis (Kent and Zayed, 2013; Kent et al., 2012). In the first process, methylated C bases mutate to T more often than non-methylated C's. Many widely and strongly expressed (housekeeping) genes in hymenopteran species have gene body methylation, and caste membership modifies methylation levels as well (Foret et al., 2009; Kucharski et al., 2008; Lyko et al., 2010). The concentration of epigenetic changes in gene bodies means that increased methylation-related mutation rates may affect d*N*/d*S*, but the direction of the effect will depend on the proportion of CpG sites which are synonymous or non-synonymous under C→T mutation. Thus the first process affects d*N*/d*S* in proportion to the amount of epigenetic regulation within a gene and the balance of such sites that are *N* or *S*. In genes where the number of CpG sites has been reduced by previous mutations, the loss of a remaining site for epigenetic regulation may have negative fitness consequences. Indeed, in the honey bee genome we showed that C→T mutations at CpG sites may experience negative selection while T→C mutation may experience positive selection (Kent et al., 2012). In the second process, the double-stranded DNA breaks which are essential for meiosis may in some cases be repaired by gene conversion. BGC is slightly more likely to use a G or C allele for repair in the vicinity of a break than to use an A or T allele (Duret and Galtier, 2009; Meunier and Duret, 2004; Pessia et al., 2012). Thus the second process produces mutational change as a side effect of meiosis, at sites near a double-strand break which are heterozygotes for G/C versus A/T alleles. In some Hymenoptera, there are very strong shifts in the GC-content of synonymous sites between AT-rich regions and GC-rich regions. BGC-related mutation disproportionately affects *S* sites in regions with high recombination rates and higher GC-content, where it strongly affects fixation rates (Kent et al., 2012). The net effect of epigenetic-induced mutation and meiosis-induced mutation on d*N*/d*S* will thus depend on the genomic context and functional role of a gene.

2.1.3 Alignments and annotations

The problem of mis-alignments and/or poor gene annotations is very serious even in model organisms with well-annotated genomes (Fletcher and Yang, 2010; Schneider et al., 2009; Wong et al., 2008). Social insect genomes are 'recent'; they have not gone through as many rounds of revision and manual annotation as *Drosophila* for example. Gene predictions may contain errors that influence sequence alignments and divergence estimates. Phylogenomic studies must be very conservative in alignment building and use the best alignment algorithms available (Fletcher and Yang, 2010; Privman et al., 2012). We also recommend the use of evidence of transcription in all species to avoid unknowingly estimating d*N*/d*S* for pseudo-genes.

2.1.4 Saturation

For most of the sequenced ant species, divergence times are on the order of 100 million years (Gadau et al., 2011). Divergence of bees from ants is even older. Over time, higher rates of change at synonymous sites lead to underestimates of d*S*, also known as d*S* saturation (Gharib and Robinson-Rechavi, 2013). d*S* saturation has led to incorrect molecular phylogenies (Praz and Packer, 2014) and warrants special attention when designing phylogenomic studies.

2.1.5 Choice of test

There are several ways of carrying out d*N*/d*S* tests depending on the aims of the study. Branch tests strive to identify fast evolving genes across specific branches of a tree, while site tests identify fast evolving sites within genes; Branch-site tests strive to do both. Roux et al. (2014) recommend the branch-site test based on higher sensitivity shown in simulations (Yang and Dos Reis, 2011), and note that their results differ from previous studies on some of the same species because of the choice of test. This is an important caveat for future phylogenomic studies of social insects.

2.1.6 Shifts in biology and gene-specific patterns of relaxed selection

Sociality involves large-scale changes in life history traits. Specifically, the large number of workers combined with sophisticated division of labour allow social insects to 'solve old problems' with behavioural solutions instead of ancestral cellular, molecular and physiological solutions. For example, social insects have sophisticated behaviours that underlie their social immunity. It has been argued (as summarized later in this review) that the evolution of social immunity resulted in a relaxation of selection on an ancestral set of innate immune genes in some social insects (Evans et al., 2006; Harpur and Zayed, 2013; Viljakainen et al., 2009). This gene or gene-set specific relaxation of selection can increase d*N*/d*S* without involving adaptive evolution.

Two additional limitations of the d*N*/d*S* approach render it uninformative for studying the evolution of regulatory sequences and taxonomically restricted genes. Mutations in regulatory sequences are believed to be very important in phenotypic evolution in general (Carroll, 2008;

Wray, 2007) and in the evolution of social insects in particular (Simola et al., 2013; Toth and Robinson, 2007, 2009). By definition, d*N*/d*S* provides no information on the evolution of regulatory sequences. Although it is conceptually possible to study the divergence of regulatory sequences by benchmarking them to nearby silent sites, such comparisons are often not practical because of difficulties in aligning non-coding sequences even between two related species. Novel taxonomically restricted genes are believed to be very important in adaptive evolution (Chen et al., 2013; Johnson and Tsutsui, 2011; Sumner, 2014). There is evidence that taxonomically restricted genes evolve rapidly relative to conserved genes (Harpur et al., 2014b), and it may be difficult to reliable align fast evolving orthologous novel genes.

Despite the difficulties noted above, the d*N*/d*S* approach is still extremely useful for highlighting groups of genes with fast rates of molecular evolution—candidate genes for subsequent population genetic and molecular studies. Caveats aside, many studies have used phylogenomics to make important insights about the evolution of sociality in ants (Hunt et al., 2011; Roux et al., 2014), termites (Bulmer et al., 2010) and bees (Woodard et al., 2011).

2.2 Population genomics and finding the loci that matter

Population genomics involves sequencing multiple individual genomes within a species, ideally from two or more distinct populations, along with sequencing one or several genomes of a closely related species. By simultaneously quantifying the two phases of molecular evolution (i.e. within species diversity and between species divergence), population genomic datasets offer more powerful approaches for identifying adaptively evolving sequences and for estimating the strength of natural selection on coding and regulatory mutations (Vitti et al., 2013).

The McDonald–Kreitman (MK) test (McDonald and Kreitman, 1991) is a robust and well-characterized test of neutral evolution (Fig. 3). The test compares diversity and divergence at synonymous sites with those of non-synonymous sites. Genes with signs of positive selection are expected to have more divergence relative to diversity at non-synonymous sites when compared to the presumably neutral synonymous sites; beneficial non-synonymous mutations are rapidly fixed by positive selection so they are more likely to be observed in the 'fixed' state versus the 'polymorphic' state relative to neutral mutations. Unlike d*N*/d*S*-type tests, the MK test is mostly

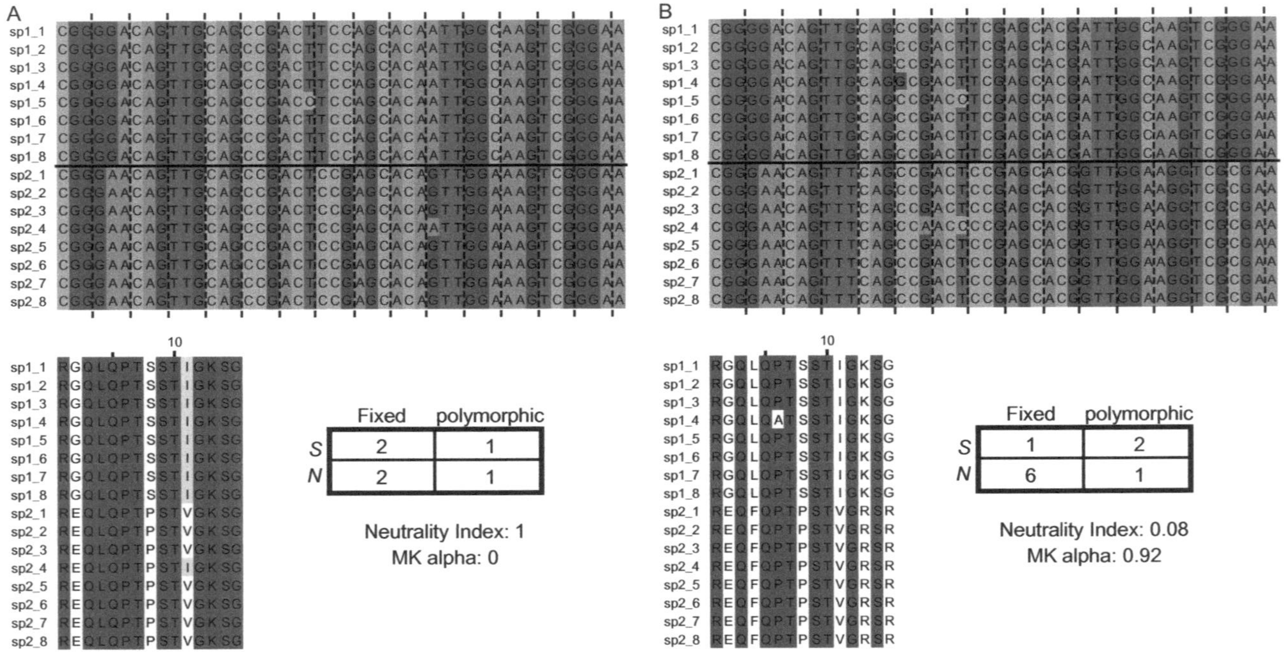

Figure 3 The McDonald–Kreitman (MK) test is the most robust test of natural selection. It involves sequencing multiple haplotypes from two species. In example A, we observe moderate levels of protein divergence at a hypothetical gene between species 1 and 2. However, analysis of nucleotide differences show that non-synonymous (*N*) mutations are fixing at similar rates relative to synonymous (*S*) mutations—the latter serves as a bench mark for neutral evolution. Note that the ratio of fixed to polymorphic mutations is the same for *S* and *N* sites. In example B, we observe high levels of protein divergence between species 1 and 2. We also notice that most *N* mutations were observed as fixed differences between the species, while most *S* mutations were polymorphic—a clear sign of positive selection fixing beneficial *N* mutations at faster rates that neutral *S* mutations over the time scale of species divergence. The mutation counts in MK tables can be easily converted to estimate the neutrality index (where values of 1 indicate neutral evolution) and the MK alpha—the proportion of *N* mutations fixed by positive selection. Bayesian implementations of the MK test can be used to estimate the average selection coefficient acting on *N* mutations, scaled by the effective population size. The MK test can also be used to study selection at intronic or intergenic mutations by benchmarking them against nearby *S* sites. With next-generation sequencing, it is now possible to study the evolution of all genes and putative regulatory sequences in social species using the MK test.

robust to changes in effective population size and mutational biases because such changes affect both diversity and divergence and are thus mostly factored out of the analysis. It is easy to modify MK ratios to estimate statistics that summarize the evolutionary forces acting on a particular sequence, including the proportion of amino-acid substitutions fixed by positive selection (Smith and Eyre-Walker, 2002). It is also possible to conduct MK-type tests within Bayesian population genetic models that can estimate γ, the average selection coefficient on non-synonymous mutations scaled by the effective population size (Eilertson et al., 2012). The MK test can also be modified to study the evolution of regulatory sequences, which can be benchmarked against nearby silent sites (Andolfatto, 2005). The Hudson, Kreitman and Aguade test (Hudson et al., 1987) is less powerful than the MK test when applied to gene coding sequences, but still provides a very useful test for identifying loci under balancing selection (i.e. loci with signs of higher within species diversity and lower within species divergence, relative to a neural locus).

The MK test quantifies adaptive evolution over the timescale of divergence between the two closely related species studied (i.e. typically tens of million years). Population genomic datasets also provide a wealth of information to detect more recent bouts of adaptive evolution. One approach involves estimating statistics that summarize genetic diversity within and between populations across the genome. Common summary statistics include nucleotide diversity, linkage disequilibrium, genetic differentiation between populations (e.g. F_{ST}) and parameters that describe the shape of the histogram of allele frequencies in a given region (e.g. Tajima's D). Positive selection is predicted to reduce genetic diversity, increase linkage disequilibrium and skew the allele frequency spectrum towards more rare variants (Nielsen, 2005). Geographically restricted positive selection can also lead to increased genetic differentiation at a site between two populations (Beaumont and Balding, 2004), often measured by F_{ST} (Fig. 4). Studying the empirical distribution of the above mentioned summary statistics can provide an approximation of which genomic regions experience neutral and nearly neutral evolution (i.e. loci within a reasonable distance from the mean of the empirical distribution), as well as loci with putative signs of selection (i.e. outlier loci) (Kelley et al., 2006). Moreover, the above approaches can be used within a modelling framework to jointly estimate the influence of demography and selection on population genomics data. Tests of selection based on summary statistics often highlight more recent bouts of adaptive evolution (i.e. over the timescale of population

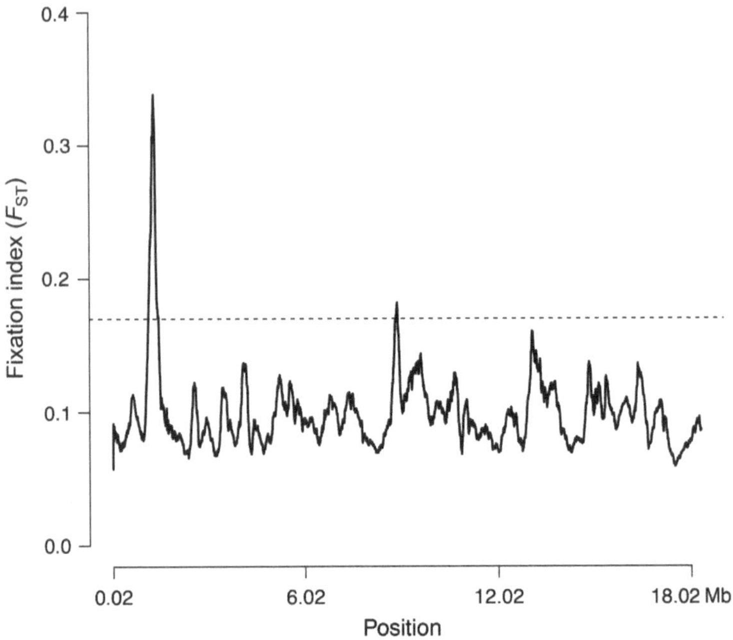

Figure 4 Population genomics reveal recent bouts of local adaptation. When multiple individual genomes from different populations within a species are sequenced, it is possible to identify loci with signs of recent positive selection leading to local adaptation. Comparisons of allele frequencies across the genome can be easily used to estimate levels of genetic differentiation (F_{ST}) between two or more populations. These genome scans can highlight loci or genomic regions with 'outlier' levels of F_{ST}—possible candidate loci underlying local adaptation.

divergence; typically tens to hundreds of thousands of years ago) (Zayed and Whitfield, 2008). The combination of MK and outlier tests provides information about 'old' and 'recent' bouts of adaptive evolution, which can be used to develop and test hypotheses about the traits and genes underlying phenotypic innovation in the studied organisms.

3. A SUMMARY OF MOLECULAR POPULATION GENETICS, PHYLOGENOMICS AND POPULATION GENOMIC STUDIES OF SOCIAL INSECTS

3.1 Molecular population genetic studies of social insects

Several studies have used Sanger sequencing to study the molecular evolution of candidate genes in social insects. These studies provided interesting glimpses into the evolution of immunity, storage proteins and sex determination in social insects.

3.1.1 The evolution of immunity

The evolution of immunity has always been a hot topic in social insects and this interest peaked after the discovery that *Apis mellifera* has a reduced set of immune genes (Evans et al., 2006). Some authors have argued that immune genes in social insects should experience strong positive selection because social living was thought to allow rapid disease spread. Others have argued that social insects don't require many innate immune genes because social behaviour provides a more effective and less costly mechanism to combat pathogens (Evans and Spivak, 2010; Evans et al., 2006; Harpur et al., 2014a). Early studies on termites (Bulmer and Crozier, 2004, 2006) used d*N*/d*S* methods and found some evidence of positive selection in termicin defence compounds and in a transcription factor regulating immune genes. More recently, Bulmer et al. (2010) used population genetics to show the presence of positive selection and a selective sweep in termite antifungal peptide GNBP2, consistent with the hypothesis that sociality promotes the adaptive evolution of innate immunity. d*N*/d*S* approaches also provided evidence of accelerated immune gene evolution both in ants (Viljakainen and Pamilo, 2008) and in ants and honey bees (Viljakainen et al., 2009). Although Viljakainen et al. (2009) noted that high d*N*/d*S* on innate immune genes can be a sign of positive selection or a relaxation of selection on immune genes, they leant towards the more adaptive explanation. Harpur and Zayed (2013) carried out a population genetic study of 13 immune genes, along with 20 randomly chosen genes, in honey bees. The authors were able to recreate Viljakainen et al. (2009)'s observation of accelerated amino-acid evolution of innate immune genes using d*N*/d*S*. However, the within-diversity data was most consistent with relaxed selection; most innate immune genes with high d*N*/d*S* also had more non-synonymous diversity when compared to randomly chosen genes consistent with relaxed selection. A recent ant phylogenomic study (Roux et al., 2014) also documented accelerated amino-acid evolution of innate immune genes using d*N*/d*S* and interpreted this as a sign of positive selection. Population genetic or genomic studies of several ants will be needed to determine if elevated d*N*/d*S* of immune genes in ants are caused by natural selection or by a relaxation of selection as found in honey bees (Harpur and Zayed, 2013). Although most of the honey bee's Toll-pathway genes have signs of relaxed constraint, a few genes do show signs of positive (Harpur and Zayed, 2013; Wallberg et al., 2014) or balancing selection (Harpur and Zayed, 2013). Most studies to date have focused on coding sequences and little attention has been paid to adaptive evolution of immune genes via regulatory mutations; such studies are bound to generate additional insight about the evolution of immunity in social insects.

3.1.2 The evolution of storage proteins

A second group of genes that have attracted considerable interest as possible mediators of social evolution are nutrient transport and storage genes, including vitellogenins (Ihle et al., 2010), royal jelly proteins (Kamakura, 2011) and hexamerins (Hunt et al., 2003; Zhou et al., 2006). Using molecular population genetics and Sanger sequencing, Kent et al. (2011) found that honey bee vitellogenin is under strong positive selection and that adaptive amino acid changes appear to be related to the protein's lipid carrying capacity. The discovery that the vitellogenin gene has been duplicated in some ants (Wurm et al., 2011) was very exciting, especially considering that the paralogs exhibit caste-biased expression (Corona et al., 2013). Corona et al. (2013) used d*N*/d*S* tests to show that one vitellogenin paralog experiences rapid evolution consistent with positive selection, while the others show a great degree of conservation suggestive of negative selection. More recently, researchers have discovered that vitellogenin underwent several ancestral duplications resulting in four paralogs of the gene found in the Hymenoptera, including in the honey bee (Morandin et al., 2014); the different paralogs appear to be functionally distinct and have caste-specific expression, suggesting that these storage proteins have repeatedly specialized to serve roles in caste development and feeding of the young.

3.1.3 The evolution of complementary sex determination

Complementary sex determination is ancestral in the Hymenoptera (Asplen et al., 2009). In this system, heterozygotes at usually one autosomal locus develop into females, while hemizygotes develop into haploid males (Cook and Crozier, 1995; Harpur et al., 2013; Heimpel and De Boer, 2008). Homozyogtes at the sex determining locus typically develop as inviable or sterile diploid males (Cook and Crozier, 1995; Harpur et al., 2013; Heimpel and De Boer, 2008). The production of diploid males represent a significant genetic load that can theoretically lead to rapid extinction in small populations (Zayed and Packer, 2005). Theory indicates that the sex determination locus experiences strong negative frequency dependent selection—a form of balancing selection that maintains a diversity of sex-determining alleles in natural populations. In the honey bee, heterozygosity at the complementary sex determiner gene *csd* provides the primary signal for sex determination (Beye et al., 2003). *csd* arose from a recent duplication from feminizer (*fem*), the honey bee ortholog of *transformer*, which is a conserved sex determination gene in insects (Hasselmann et al., 2008).

Hasselmann and Beye (2004) used population genetic analyses to show that parts of *csd* experience balancing selection, as predicted by theory. Balancing selection at *csd* is so strong that it can maintain more than 87 different sex determining alleles in honey bee populations (Lechner et al., 2014). Balancing selection on *csd* has also been shown to affect the molecular evolution of two neighbouring loci through genetic hitchhiking (Hasselmann et al., 2010). All of our functional knowledge on complementary sex determination is derived from the honey bee. The sex determination locus has not been genetically characterized in other hymenopterans. Recently, phylogenomic studies have identified *fem*- and *csd*-like genes in ants and other hymenopterans (Koch et al., 2014; Privman et al., 2013), but it remains to be demonstrated if these copies are functionally involved in sex determination. Some of the *fem*- and *csd*-like genes show evidence of gene conversion and inter-locus recombination, and phylogenetic signals (i.e. high d*N*/d*S*) of positive selection (Privman et al., 2013), suggesting that these processes are important for generating allelic diversity at the sex locus. Going forward, it will be important to genetically characterize *fem*- and *csd*-like genes in ants and other bees. The identification of the honey bee *csd* involved a massive effort to positionally clone the locus. Population genomic studies can offer a shortcut to identify complementary sex determination loci in other Hymenoptera: the sex determination locus is expected to be in regions of the genome with high within species diversity and low between population divergence as revealed by population genomic scans.

3.2 Phylogenomic studies of social insects

Several studies have used phylogenomics to study rates of molecular evolution of several hundred to several thousand genes in social insects. The first of these studies compared rates of protein sequence evolution in 1511 *Apis mellifera* genes with orthologs in the solitary jewel wasp *Nasonia vitripennis* and at least one other insect species, and found that proteins which are strongly expressed in queens (based on brain cDNA arrays) have faster rates of protein evolution than unbiased genes or worker-biased genes (Hunt et al., 2010). A later study (Hunt et al., 2011) using the newly sequenced *Solenopsis invicta* genome searched for some of the same patterns in ants and in bees. In fire ants, both worker- and queen-biased genes had higher rates of protein evolution than unbiased genes, although the effect was only significant for queen genes. Faster protein sequence evolution can be due to positive selection or reduced purifying selection allowing more slightly deleterious mutations to fix. The authors found that reduced purifying selection seemed to be more important than positive selection for a large number of faster evolving proteins. Further, for over 1000 genes with orthologs in *A. mellifera, S. invicta* and *N. vitripennis*, the rate of protein change in the non-social wasp was significantly associated with rates of change in the social ant and bee. The authors suggested that during the evolution of sociality, genes that were already under relaxed selection in solitary ancestors were differentially co-opted for use in caste-specific functions.

The co-option of already fast evolving proteins during the evolution of sociality is a thought-provoking result. A larger phylogenomic study of 4261 genes in seven ant genomes (Roux et al., 2014) found that '*...genes implicated in neurogenesis and olfaction started to undergo increased positive selection before the evolution of sociality in Hymenoptera*'. While this agrees with Hunt et al. (2011)'s findings in a basic sense (i.e. faster evolving genes in social insects were also faster evolving in solitary ancestors), Roux et al. suggest that positive selection, and not a relaxation of selection, is a precursor for the evolution of phenotypic plasticity associated with sociality. In light of this evidence, it is interesting to consider the results of Farris and Schulmeister (2011), which show that mushroom bodies (a brain structure important for associative learning) are relatively large (compared allometrically to body size) in many predatory and parasitoid Hymenoptera (Farris and Schulmeister, 2011). It is plausible that cognitive and sensory genes that adaptively evolved to facilitate the 'hunting' or 'search for resources' lifestyle in solitary wasps were co-opted during the evolution of sociality to facilitate central-place foraging and/or chemical communication between nestmates.

Roux et al. also found differences between ant and bee genes with putative signs of positive selection that confirmed earlier results (Woodard et al., 2011). Studying the transcriptomes of nine bee species with three independent origins of sociality, Woodard et al. found that advanced social species showed an enrichment of fast evolving genes involved in carbohydrate metabolism. Roux et al. confirmed this result in bees, but found that ants do not show such an enrichment, perhaps due to the loss of flight as the primary foraging lifestyle in ants. Curiously, Woodard et al. found that olfaction and neural development genes tended to show accelerated evolution in the primitively eusocial species in their analysis.

The above studies clearly highlight interesting differences in the rate of molecular evolution of some genes in social insects. The finding that fast evolving genes in social Hymenoptera tend to also be fast evolving in solitary

Hymenoptera is very intriguing for it suggests that genomic or life history features of ancestrally solitary populations enhanced the evolution of sociality. Although social insect phylogenomic studies have provided interesting results, they have also provided some perplexing findings. For example, it is counter intuitive that metabolism genes show accelerated patterns of evolution in highly social bees while genes involved in neuronal development and olfaction have accelerated patterns of evolution in primitively eusocial bees (Woodard et al., 2011) considering the increased complexity of chemical communication and worker behaviour in advanced eusocial bees. Additionally, the finding that queen-biased genes tend to have higher rates of molecular evolution relative to worker-biased genes is also counter intuitive (Hunt et al., 2010, 2011). Although there is evidence that indirect selection on workers is weaker than direct selection on queens (Linksvayer and Wade, 2009, but see Hall and Goodisman (2012)), the ecology and natural history of social insects mostly point to worker traits as the major mechanism allowing colonies to adapt to changes in their environment (Sagili et al., 2011; Wilson, 1985; Wray et al., 2011). It is possible that some of these potential contradictions stem from the inherent biases associated with phylogenomic studies. For example, the need for reliable alignments across multiple species separated by many millions of years precludes analyses of the fastest evolving genes—the genes that are most relevant with respect to adaptive phenotypic evolution. Further, phylogenomic studies typically analyse 10–40% of the genes found in social insects. Finally, it is difficult to use rates of protein evolution as a metric for positive selection because other non-adaptive evolutionary processes can mimic positive selection and inflate d*N*/d*S*. Population genomic studies utilizing both polymorphism and divergence data allow for more robust tests of selection. Below we review some of the early findings of population genomic studies of social insects.

3.3 Population genomic studies of social insects

Social insect population genomics took flight in 2014, with several groundbreaking studies in honey bees (Harpur et al., 2014b; Wallberg et al., 2014), ants, termites and sweat bees (Romiguier et al., 2014a,b). Harpur et al. (2014b) carried out the first population genomic study of social insects by sequencing the individual genomes of 39 *A. mellifera* workers from four different geographic populations along with one *A. cerana* worker, for a total of 80 haploid genomes. Each worker was sequenced using Illumina HiSeq at approximately 38× coverage, for a combined coverage of 1520×. The authors identified approximately 12 million mutations and used genome scans to detect mutations and genomic regions with extreme levels of genetic differentiation between four different honey bee geographic populations; putative loci experiencing local adaptation over short time scales. The authors also used a variant of the MK test to detect selection over the approximately 5–15 MY since *A. mellifera* and *A. cerana* diverged. Molodtsova et al. (2014) later used the Harpur et al. dataset and applied a modified MK test to estimate the signatures of selection acting on putative *cis*-regulatory regions in the honey bee genome. Wallberg et al. (2014) published another landmark paper on honey bee population genomics by sequencing 140 worker genomes from 9 different subspecies of *Apis mellifera*. Wallberg et al. (2014) sequenced a few individuals at high coverage (~20×) but most were sequenced at a lower coverage (~4.4×) resulting in an overall combined coverage of 634× across the study. The large number of individuals and populations sampled in this study allowed powerful use of the F_{ST} statistic to identify mutations with putative signs of positive selection. Rounding off the honey bee population genomic datasets were two studies that used RNA sequencing (RNAseq) to characterize genetic variation in the partial transcriptomes of 6–11 diploid individuals of 3 species of subterranean termites, a harvester ant, a big-headed ant, 2 species of carpenter ants and a sweat bee (Romiguier et al., 2014a,b). The dataset allowed Romiguier et al. to calculate non-synonymous diversity π_N and synonymous diversity π_S for each species. These population genomic studies made important discoveries about social insects, which we summarize below.

3.3.1 Taxonomically restricted genes drive adaptive evolution

Harpur et al. (2014b) showed that taxonomically restricted genes (i.e. genes that are only found in Hymenotpera, Apoida (bees) and *Apis* have higher levels of positive selection than older genes that are conserved across insects. Moreover, bee and *Apis*-restricted genes had significantly higher selection coefficients relative to Hymenoptera-restricted genes. The question of whether major new phenotypic innovations are a result of new genes or old genes is still open. Earlier sociogenomic analyses suggested that sociality evolved through the co-option of conserved genes and gene networks that influence aspects of behaviour and reproductive physiology of solitary insects (Toth and Robinson, 2007, 2009). However, most studies show that novel genes tend to have caste-biased expression patterns suggesting that they underlie caste-specific traits (Feldmeyer et al., 2013; Johnson and

Tsutsui, 2011; Simola et al., 2013; Wissler et al., 2013), but see a recent wasp study that presented a different conclusion (Berens et al., 2014). A recent honey bee study (Jasper et al., 2015) showed that genes with high and tissue-specific expression are predominately taxonomically restricted and are enriched for signatures of positive selection estimated from the Harpur et al. data. The population genomic data support the hypothesis that novel genes play an important role in the adaptive evolution of social insects. It is very likely that novel genes can influence social traits by acting through conserved gene networks. Unfortunately, the molecular and biological functions of novel genes are not well understood because we cannot simply borrow this knowledge from *Drosophila*. Going forward, it will be important to functionally characterize novel genes in social insects and study how they interact with conserved genes to influence social traits.

3.3.2 Worker traits are a primary 'vector' for adaptation in honey bees

Harpur et al. (2014b) leveraged the very rich transcriptomic and proteomic datasets developed for honey bees (Chan et al., 2013; Grozinger et al., 2007; Zayed and Robinson, 2012; Zayed et al., 2012) to show that genes associated with worker traits are enriched for signs of adaptive evolution. For example, genes with higher protein expression in workers than queens had significantly higher levels of positive selection in their coding sequences, relative to genes that show higher expression in queens relative to workers. Additionally, genes associated with worker behaviour through analysis of brain gene expression were often enriched for evidence of local adaptation (i.e. high F_{ST}) especially involving regulatory mutations. Wallberg et al. (2014) also found that single nucleotide polymorphisms (SNPs) with signatures of local adaptation were enriched in genes with worker-biased expression.

3.3.3 Genes sets underlying worker-division of labour experience strong selection

Harpur et al. (2014b) discovered that many genes involved in nutrient and hormonal signalling were under positive selection based on the MK test, including vitellogenin, juvenile hormone-related genes and members of the insulin and TOR signalling pathways. Wallberg et al. (2014) also found SNPs with outlier F_{ST} in key genes in the insulin and vitellogenin signalling pathway, including in *InR* (insulin-like receptor), *Vg* (vitellogenin) and *yl* (vitellogenin receptor). There is a large literature on these pathways and their involvement in the evolution of bee, wasp and ant eusociality

(Corona et al., 2007; Dallacqua et al., 2007; Havukainen et al., 2011; Ihle et al., 2010; Kamakura, 2011; Kent et al., 2011; Martins et al., 2011; Shpigler et al., 2014; Wang et al., 2010; Wurm et al., 2011; Zhou et al., 2006; the chapter 'Old Threads Make New Tapestry—Rewiring of Signalling Pathways Underlies Caste Phenotypic Plasticity in the Honey Bee, *Apis mellifera* L.' by Hartfleder et al.). In particular, it was previously suggested these pathways were co-opted and re-wired to regulate worker division of labour (Amdam and Omholt, 2003; Amdam et al., 2003); the population genomics data lend strong support to this hypothesis in honey bees.

3.3.4 Adaptive evolution of gene regulation

Even in cases where the protein encoded by a gene is under strong purifying selection, adaptive evolutionary change can still occur in regulatory sites. Both Harpur et al. (2014b) and Wallberg et al. (2014) found that many SNPs with outlier levels of genetic differentiation are found outside of gene bodies, suggesting a possible regulatory function. Genes that are associated with worker behaviour, based on analyses of brain gene expression, are much more enriched for outlier SNPs in putative regulatory regions relative to outlier non-synonymous SNPs (Harpur et al., 2014b). Molodtsova et al. (2014) recently estimated the average selection coefficient on mutations in putative *cis*-regulatory regions of most genes in the honey bee genome. The authors used this dataset to show that a conserved transcriptional regulatory network influencing worker behaviour can adaptively evolve via both regulatory and non-synonymous mutations. In most organisms, transcription factors that are more central to regulatory networks (i.e. hub genes) are under stronger purifying selection. This is also true in the honey bee, as evidenced by selection coefficients which are significantly more negative (more purifying) in hub versus non-hub transcription factors (Molodtsova et al., 2014). Genes with adaptive patterns of protein evolution were often restricted to the periphery of the regulatory network. By contrast, the selection coefficients on putative *cis*-regulatory mutations were minimally affected by gene centrality or connectedness.

3.3.5 Recombination and the evolution of worker traits

Wallberg et al. (2014) provide new information about how recent bouts of adaptive evolution—called selective sweeps—influence the evolution of nearby regions. Normally, positive selection is expected to reduce diversity at the focal locus, but also at nearby neutral loci because of linkage disequilibrium. However, social insects in general, and the honey bee in particular, have very high rates of recombination (Wilfert et al., 2007) and this is expected to lead to narrower regions of reduced diversity around selected sites. Indeed, Wallberg et al. (2014) found that regions around mutations with outlier levels of genetic differentiation—suggestive of positive selection—have reduced diversity but only for small regions (20–100 Kb) surrounding the selected site. The relationship between the evolution of recombination and the evolution and elaboration of sociality is a hot topic in sociobiology. Kent et al. (2012) using population genetic analysis of randomly chosen genes, demonstrated a link between recombination, genome structure and the evolution of social behaviour; genes with worker-biased expression were enriched in parts of the honey bee genome with the highest rates of recombination and genetic diversity and where natural selection is most efficient. Kent and Zayed (2013) later suggested that high recombination rates may have been selected for during the evolution of sociality to 'break down' social pleiotropy by allowing natural selection to optimize potentially linked mutations that have antagonistic effects on workers and queens. Wallberg et al. (2014)'s finding that natural selection leaves a 'narrow footprint' in the honey bee genome supports Kent and Zayed (2013)'s proposition. Meanwhile, in ants, it appears that a reduction in recombination was critical for the evolution of 'social' supergenes that influence critical aspects of colony organization (Wang et al., 2013). Although it is still unclear which of the two dichotomous relationships between recombination and sociality (i.e. positive in bees, negative in ants) represents the norm, and which the exception (Linksvayer et al., 2013; the chapter 'Understanding Honey Bee Worker Self-Sacrifice: A Conceptual–Empirical Framework' by Mullen and Thompson), it is clear however that the evolution of recombination appears to be strongly intertwined with the evolution and elaboration of sociality.

3.3.6 Effective population size and the sociality bottleneck

It was previously suggested that social species have reduced effective population size relative to solitary insects because of several reasons including fewer reproductives per unit area, a lower effective carrying capacity because of the large amount of resources required per colony, and/or greater variance in reproductive output between nests (Packer and Owen, 2001). Earlier studies showed lower levels of allozyme variation in social species (Hedrick and Parker, 1997), but these studies seldom corrected for phylogenetic inertia. Romiguier et al. (2014b) discovered a genomic signal of low 'mammalian' like effective population sizes in social insects relative to solitary insects. The authors found the nucleotide diversity at silent sites (π_S) in social insects to be an order of magnitude lower than non-social insects; π_S is a product of the mutation rate and effective population size, so lower diversity suggests lower effective size in social versus solitary insects. Additionally, the authors show that the ratio π_N/π_S is higher in eusocial species. Normally, an excess of non-synonymous mutation suggests that natural selection has been relaxed because of greater genetic drift in populations with small effective size. Romiguier et al. (2014b) estimated the distribution of population-scaled fitness effects of non-synonymous mutations and found that eusocial species have many more mutations with slightly deleterious effects. Again, the parsimonious explanation for this is an order of magnitude lower N_e in the eusocial species. Romiguier et al. (2014b) also used phylogenomic data from seven sequenced ant species to compare dN/dS in ants to non-social insects and to mammals. dN/dS in the eusocial ants is closer to mammals and about double to triple the levels found in fruit flies and mosquitoes with large population sizes; these estimates also suggest lower N_e leading to a relaxation of selection in social insects. However, most of the social species studied by Romiguier et al. (2014b) were *haplodiploid* Hymenoptera while all of the solitary species were *diploid* insects. It is thus possible that haplodiploidy and not sociality *per se* (or other correlated life history trait such as nest building and/or central place foraging) are responsible for the hypothesized reduction in N_e (Packer et al., 2005; Zayed, 2009; Zayed and Packer, 2007; Zayed et al., 2005). Within the seven ants, Romiguier et al. (2014b) show a significant positive correlation between dN/dS and queen/worker dimorphism in size. As this size ratio roughly parallels the degree of eusociality of the ants, it provides some support that sociality, rather than haplodiploidy, may be reducing effective population size. Interestingly, Wallberg et al. (2014) used current levels of genetic diversity to estimate the effective population size of honey bees, which was found to be considerably high. The high N_e estimated by Wallberg et al. (2014) for honey bees challenge Romiguier et al. (2014b)'s evidence for reduced effective population size in social insects. We expect considerable future experimental and theoretical work on this subject.

4. THE FUTURE AND THE NEXT-GEN SYNTHESIS

4.1 Population genomics for the masses

Although sequencing the whole genomes of many individuals was costly until recently, the continuing rapid drop in the cost of sequencing

(Hayden, 2014) will bring population genomics within the reach of individual research programs on most social insects in the immediate future. Challenges to bring the cost of sequencing an individual human genome down to $1,000 (Hayden, 2014) are nearing that goal. It's worth making some back of the envelope calculations about what population genomic studies may cost in the next few years. For example, the honey bee genome is about 1/10 the size of the human genome. With current technology, we are able to sequence individual diploid honey bee genomes at 25 × coverage for approximately $300 (with Illumina HiSeq V.4 chemistry). Sequencing the genome of a single worker or queen bee at that coverage may soon drop to below $100. Genomes for males of haplodiploid species do not require as much coverage since only one allele is present at each site, and will soon cost about $50 per male. Typically one aims to sequence 10 or more unrelated individuals per population, so we may look forward to sequencing costs for a honey bee population in the range of $500–1000 by 2017. RNAseq studies are even more cost effective for a subset of highly expressed genes but have the drawback that many genes are expressed only in certain tissues or developmental stages or are always present at low levels, reducing the number of reads and possibly missing important isoforms. Some social insects such as termites (Terrapon et al., 2014) and the fire ant *Solenopsis invicta* (Wurm et al., 2011) have genomes several times bigger than that of the honey bee; they will certainly be more expensive to sequence than honey bees, but not prohibitively so.

4.2 Population genomics for the *how's* and *why's*

It is clear that sociobiologists are beginning to appreciate the power of phylogenomic and population genomic approaches as a tool for making sense of the genetics and evolution of social insects. To get a forward-looking view of where population genomics and phylogenomics of social insects are heading, we tabulated studies in the published abstracts of the International Union for the Study of Social Insects' 2014 international congress held in Cairns Australia. Fourteen abstracts applied phylogenomic techniques to existing sequenced ant, bee and wasp species as well as a number of new social and solitary hymenopteran species for which genomes may be expected in the next few years. Ten abstracts focused on the genomes of endosymbionts of termites, ants and bees. If we consider the metagenome of a social insect colony to include not only the primary eusocial insect species, but also bacterial, fungal, protist and other symbionts, it is clear that an exciting new frontier in population genomics will involve analysis of selection in multiple co-occurring interdependent species (e.g., when selection on symbiont and host digestive enzyme accompanies nutritional resource shift). Finally, eight abstracts explicitly used one or more population genomic methods to study several species of ants, honey bees, halictid bees and bumblebees. Some hymenoptera are socially polymorphic, and we are eagerly awaiting population genomic studies of the recently sequenced and socially polymorphic sweat bee *Lasioglossum albipes* (Kocher et al., 2013). Although not mentioned at this meeting, the presence of several sequenced termite transcriptomes or genomes suggests population genomics will be fruitful in this large and important group soon as well.

There is a growing interest in understanding how DNA methylation and histone modification are important forms of gene regulation in many organisms including social insects, where DNA methylation can govern developmental commitment to alternative castes (Foret et al., 2009; Kucharski et al., 2008). This topic is covered in detail in this volume (the chapter 'Epigenetics in Social Insects' by Glastad et al.) and is the subject of considerable interest and speculation in the literature (Drewell et al., 2012; Yan et al., 2014). Here, we note that there is a multitude of opportunities for researchers to use population genomics to understand the role of epigenetics in social insects. For instance, studies of social bees implicate metabolic genes as important during social evolution, but little or no research has been done on SNPs in bee non-coding RNA genes such as microRNAs. SNPs in microRNAs are known to have a profound influence on lipid regulation in mammals (Sethupathy, 2013). Studies are beginning to show SNP by environment epigenetic interactions that strongly influence human disorders (Klengel et al., 2013; Ziller et al., 2013); given that the nutritional environment and DNA methylation influence caste determination in bees (Kucharski et al., 2008; Maleszka, 2008), the time is ripe for population genomic analysis of SNPs underling SNP by environment epigenetic interactions in social insects. Further, chromatin structure (nucleosome positioning) is correlated in mammals with SNP positions but in different ways in exons and introns (Liu et al., 2014), suggesting that there may be a deep connection between epigenetic state and mutational processes.

We strongly believe that comparative population genomics will be instrumental in testing and developing hypotheses for why and how sociality evolved, and how sociality affects patterns of molecular evolution. Evolutionary biologists have hypothesized for decades about the 'alleles for sociality'—the causal mutations that facilitate the evolution and maintenance of sociality in insects—but so far, these 'alleles' have mostly remained abstract and ambiguous. Population genomics provide a framework to directly quantify the evolutionary forces shaping patterns of within species diversity and between species divergence and provides tools to detect and quantify the strength of positive selection on genes and regulatory sequences. Most theories for the evolution of sociality assume that social phenotypes are adaptive: social 'alleles' confer a fitness advantage to their bearers, and are driven to fixation by direct positive selection on reproductive members or by indirect kin selection on sterile members of social nests. Loci experiencing positive selection, as revealed by population genomics, are therefore expected to be enriched for genes and regulatory sequences underlying the social traits of insects. Population genomic studies greatly benefit from functional data that connect genes and their products with traits of social insects. Both honey bee population genomic studies greatly benefited from molecular biology, genetic, transcriptomic and behavioural research conduced over the past several decades.

Population genomics naturally links knowledge on the relationship between genes and behaviour with knowledge on the relationship between genes and fitness. Population genomics should greatly facilitate the development of mechanistically explicit theories for the evolution of sociality, by explicitly connecting genes, behaviour and fitness. Instead of the typical 'imagine a mutation that influences sociality', we can now develop and test more explicit hypotheses that make specific predictions about the types and functions of genes involved in the evolution of sociality. For example, ground-plan type hypotheses can now tested by examining if ancestral genes involved in female reproduction experienced positive selection over the course of social evolution. Similarly, we can now test maternal-manipulation hypotheses by examining if genes associated with the chemical or physical manipulation of conspecifics experienced positive selection over the course of social evolution. Population genomic datasets allow researchers to 'time' bursts of positive selection, which will be critical in teasing apart adaptive changes associated with the origin of sociality versus adaptive changes that occurred to elaborate sociality after it evolved. This new level of insight will go a long way in helping the field of sociobiology answer its most fundamental problem—why and how sociality evolved.

ACKNOWLEDGEMENTS

Amro Zayed is funded by a discovery grant from the Natural Sciences and Engineering Research Council of Canada and an Early Researcher Award from the Ontario Ministry of Research and Innovation. We thank Brock Harpur and Alivia Dey for generating diagrams for Figs. 2–4. We also thank Brock Harpur (York University), Dr. Alivia Dey

(York University), Dr. Amy Toth (Iowa State University) and Dr. Timothy Linksvayer (University of Pennsylvania) for providing helpful comments on our manuscript.

REFERENCES

Amdam, G.V., Omholt, S.W., 2003. The hive bee to forager transition in honeybee colonies: the double repressor hypothesis. J. Theor. Biol. 223, 451–464.

Amdam, G.V., Norberg, K., Hagen, A., Omholt, S.W., 2003. Social exploitation of vitellogenin. Proc. Natl. Acad. Sci. U. S. A. 100, 1799–1802.

Andolfatto, P., 2005. Adaptive evolution of non-coding DNA in *Drosophila*. Nature 437, 1149–1152.

Asplen, M.K., Whitfield, J.B., DE Boer, J.G., Heimpel, G.E., 2009. Ancestral state reconstruction analysis of hymenopteran sex determination mechanisms. J. Evol. Biol. 22, 1762–1769.

Beaumont, M.A., Balding, D.J., 2004. Identifying adaptive genetic divergence among populations from genome scans. Mol. Ecol. 13, 969–980.

Berens, A.J., Hunt, J.H., Toth, A.L., 2014. Comparative transcriptomics of convergent evolution: different genes but conserved pathways underlie caste phenotypes across lineages of eusocial insects. Mol. Biol. Evol. http://dx.doi.org/10.1093/molbev/msu330.

Beye, M., Hasselmann, M., Fondrk, M.K., Page, R.E., Omholt, S.W., 2003. The gene *csd* is the primary signal for sexual development in the honey bee and encodes a new SR-type protein. Cell 114, 419–429.

Bulmer, M.S., Crozier, R.H., 2004. Duplication and diversifying selection among termite antifungal peptides. Mol. Biol. Evol. 21, 2256–2264.

Bulmer, M.S., Crozier, R.H., 2006. Variation in positive selection in termite GNBPs and relish. Mol. Biol. Evol. 23, 317–326.

Bulmer, M., Lay, F., Hamilton, C., 2010. Adaptive evolution in subterranean termite antifungal peptides. Insect Mol. Biol. 19, 669–674.

Carroll, S.B., 2008. Evo-devo and an expanding evolutionary synthesis: a genetic theory of morphological evolution. Cell 134, 25–36.

Chan, Q.W., Chan, M.Y., Logan, M., Fang, Y., Higo, H., Foster, L.J., 2013. Honey bee protein atlas at organ-level resolution. Genome Res. 23, 1951–1960.

Chen, S.D., Krinsky, B.H., Long, M.Y., 2013. New genes as drivers of phenotypic evolution. Nat. Rev. Genet. 14, 645–660.

Cook, J.M., Crozier, R.H., 1995. Sex determination and population biology of the Hymenoptera. Trends Ecol. Evol. 10, 281–286.

Corona, M., Velarde, R.A., Remolina, S., Moran-Lauter, A., Wang, Y., Hughes, K.A., Robinson, G.E., 2007. Vitellogenin, juvenile hormone, insulin signaling, and queen honey bee longevity. Proc. Natl. Acad. Sci. U. S. A. 104, 7128–7133.

Corona, M., Libbrecht, R., Wurm, Y., Riba-Grognuz, O., Studer, R.A., Keller, L., 2013. Vitellogenin underwent subfunctionalization to acquire caste and behavioral specific expression in the harvester ant *Pogonomyrmex barbatus*. PLoS Genet. 9, e1003730.

Dallacqua, R.P., Simões, Z.L.P., Bitondi, M.M.G., 2007. Vitellogenin gene expression in stingless bee workers differing in egg-laying behavior. Insect. Soc. 54, 70–76.

Drewell, R.A., Lo, N., Oxley, P.R., Oldroyd, B.P., 2012. Kin conflict in insect societies: a new epigenetic perspective. Trends Ecol. Evol. 27, 367–373.

Duret, L., Galtier, N., 2009. Biased gene conversion and the evolution of mammalian genomic landscapes. Annu. Rev. Genomics Hum. Genet. 10, 285–311.

Eilertson, K.E., Booth, J.G., Bustamante, C.D., 2012. SnIPRE: selection inference using a Poisson random effects model. PLoS Comput. Biol. 8, E1002806.

Elango, N., Hunt, B.G., Goodisman, M.A.D., Yi, S.V., 2009. DNA methylation is widespread and associated with differential gene expression in castes of the honeybee, Apis mellifera. Proc. Natl. Acad. Sci. U. S. A. 106, 11206–11211.

Evans, J.D., Spivak, M., 2010. Socialized medicine: individual and communal disease barriers in honey bees. J. Invertebr. Pathol. 103, S62–S72.

Evans, J.D., Aronstein, K., Chen, Y.P., Hetru, C., Imler, J.L., Jiang, H., Kanost, M., Thompson, G.J., Zou, Z., Hultmark, D., 2006. Immune pathways and defence mechanisms in honey bees, *Apis mellifera*. Insect Mol. Biol. 15, 645–656.

Farris, S.M., Schulmeister, S., 2011. Parasitoidism, not sociality, is associated with the evolution of elaborate mushroom bodies in the brains of hymenopteran insects. Proc. R. Soc. B Biol. Sci. 278, 940–951.

Feldmeyer, B., Elsner, D., Foitzik, S., 2013. Gene expression patterns associated with caste and reproductive status in ants: worker-specific genes are more derived than queen-specific ones. Mol. Ecol. 23, 151–161.

Fletcher, W., Yang, Z., 2010. The effect of insertions, deletions, and alignment errors on the branch-site test of positive selection. Mol. Biol. Evol. 27, 2257–2267.

Foret, S., Kucharski, R., Pittelkow, Y., Lockett, G.A., Maleszka, R., 2009. Epigenetic regulation of the honey bee transcriptome: unravelling the nature of methylated genes. BMC Genomics 10, 472.

Gadau, J., Helmkampf, M., Nygaard, S., Roux, J., Simola, D.F., Smith, C.R., Suen, G., Wurm, Y., Smith, C.D., 2011. The genomic impact of 100 million years of social evolution in seven ant species. Trends Genet. 28, 14–21.

Gharib, W.H., Robinson-Rechavi, M., 2013. The branch-site test of positive selection is surprisingly robust but lacks power under synonymous substitution saturation and variation in GC. Mol. Biol. Evol. 30, 1675–1686.

Grozinger, C.M., Fan, Y., Hoover, S.E., Winston, M.L., 2007. Genome-wide analysis reveals differences in brain gene expression patterns associated with caste and reproductive status in honey bees (*Apis mellifera*). Mol. Ecol. 16, 4837–4848.

Hall, D.W., Goodisman, M.A.D., 2012. The effects of kin selection on rates of molecular evolution in social insects. Evolution 66, 2080–2093.

Harpur, B.A., Zayed, A., 2013. Accelerated evolution of innate immunity proteins in social insects: adaptive evolution or relaxed constraint? Mol. Biol. Evol. 30, 1665–1674.

Harpur, B.A., Sobhani, M., Zayed, A., 2013. A review of the consequences of complementary sex determination and diploid male production on mating failures in the Hymenoptera. Entomol. Exp. Appl. 146, 156–164.

Harpur, B.A., Chernyshova, A., Soltani, A., Tsvetkov, N., Mahjoorighasrodashti, M., Xu, Z., Zayed, A., 2014a. No genetic tradeoffs between hygienic behaviour and individual innate immunity in the honey Bee, *Apis mellifera*. PLoS One 9, e104214.

Harpur, B.A., Kent, C.F., Molodtsova, D., Lebon, J.M., Alqarni, A.S., Owayss, A.A., Zayed, A., 2014b. Population genomics of the honey bee reveals strong signatures of positive selection on worker traits. Proc. Natl. Acad. Sci. U. S. A. 111, 2614–2619.

Hasselmann, M., Beye, M., 2004. Signatures of selection among sex-determining alleles of the honey bee. Proc. Natl. Acad. Sci. U. S. A. 101, 4888–4893.

Hasselmann, M., Gempe, T., Schiott, M., Nunes-Silva, C.G., Otte, M., Beye, M., 2008. Evidence for the evolutionary nascence of a novel sex determination pathway in honeybees. Nature 454, 519–522.

Hasselmann, M., Lechner, S., Schulte, C., Beye, M., 2010. Origin of a function by tandem gene duplication limits the evolutionary capability of its sister copy. Proc. Natl. Acad. Sci. 107, 13378–13383.

Havukainen, H., Halskau, Ø., Amdam, G.V., 2011. Social pleiotropy and the molecular evolution of honey bee vitellogenin. Mol. Ecol. 20, 5111–5113.

Hayden, E.C., 2014. Technology: the $1,000 genome. Nature 507, 294–295.

Hedrick, P.W., Parker, J.D., 1997. Evolutionary genetics and genetic variation of haplodiploids and X-lined genes. Annu. Rev. Ecol. Syst. 28, 55–83.
Heimpel, G.E., De Boer, J.G., 2008. Sex determination in the Hymenoptera. Annu. Rev. Entomol. 53, 209–230.
Hudson, R.R., Kreitman, M., Aguade, M., 1987. A test of neutral molecular evolution based on nucleotide data. Genetics 116, 153–159.
Hunt, J.H., Buck, N.A., Wheeler, D.E., 2003. Storage proteins in vespid wasps: characterization, developmental pattern, and occurrence in adults. J. Insect Physiol. 49, 785–794.
Hunt, G.J., Amdam, G.V., Schlipalius, D., Emore, C., Sardesai, N., Williams, C.E., Rueppell, O., Guzman-Novoa, E., Arechavaleta-Velasco, M., Chandra, S., Fondrk, M.K., Beye, M., Page, R.E.J., 2007. Behavioral genomics of honeybee foraging and nest defense. Naturwissenschaften 94, 247–267.
Hunt, B.G., Wyder, S., Elango, N., Werren, J.H., Zdobnov, E.M., Yi, S.V., Goodisman, M.A., 2010. Sociality is linked to rates of protein evolution in a highly social insect. Mol. Biol. Evol. 27, 497–500.
Hunt, B.G., Ometto, L., Wurm, Y., Shoemaker, D., Yi, S.V., Keller, L., Goodisman, M.A., 2011. Relaxed selection is a precursor to the evolution of phenotypic plasticity. Proc. Natl. Acad. Sci. U. S. A. 108, 15936–15941.
Ihle, K.E., Page, R.E., Frederick, K., Fondrk, M.K., Amdam, G.V., 2010. Genotype effect on regulation of behaviour by vitellogenin supports reproductive origin of honeybee foraging bias. Anim. Behav. 79, 1001–1006.
Jasper, W.C., Linksvayer, T.A., Atallah, J., Friedman, D., Chiu, J.C., Johnson, B.R., 2015. Large scale coding sequence change underlies the evolution of post-developmental novelty in honey bees. Mol. Biol. Evol. 32, 334–346.
Johnson, B.R., Tsutsui, N.D., 2011. Taxonomically restricted genes are associated with the evolution of sociality in the honey bee. BMC Genomics 12, 164.
Kamakura, M., 2011. Royalactin induces queen differentiation in honeybees. Nature 473, 478–483.
Kelley, J.L., Madeoy, J., Calhoun, J.C., Swanson, W., Akey, J.M., 2006. Genomic signatures of positive selection in humans and the limits of outlier approaches. Genome Res. 16, 980–989.
Kent, C.F., Zayed, A., 2013. Evolution of recombination and genome structure in eusocial insects. Commun. Integr. Biol. 6, e22919.
Kent, C.F., Issa, A., Bunting, A.C., Zayed, A., 2011. Adaptive evolution of a key gene affecting queen and worker traits in the honey bee, Apis mellifera. Mol. Ecol. 20, 5226–5235.
Kent, C.F., Minaei, S., Harpur, B.A., Zayed, A., 2012. Recombination is associated with the evolution of genome structure and worker behavior in honey bees. Proc. Natl. Acad. Sci. U. S. A. 109, 18012–18017.
Kimura, M., 1985. The Neutral Theory of Molecular Evolution. Cambridge University Press, Cambridge, UK.
Klengel, T., Mehta, D., Anacker, C., Rex-Haffner, M., Pruessner, J.C., Pariante, C.M., Pace, T.W.W., Mercer, K.B., Mayberg, H.S., Bradley, B., Nemeroff, C.B., Holsboer, F., Heim, C.M., Ressler, K.J., Rein, T., Binder, E.B., 2013. Allele-specific FKBP5 DNA demethylation mediates gene-childhood trauma interactions. Nat. Neurosci. 16, 33–41.
Koch, V., Nissen, I., Schmitt, B.D., Beye, M., 2014. Independent evolutionary origin of fem paralogous genes and complementary sex determination in hymenopteran insects. PLoS One 9, e91883.
Kocher, S.D., Li, C., Yang, W., Tan, H., Yi, S.V., Yang, X., Hoekstra, H.E., Zhang, G., Pierce, N.E., Yu, D.W., 2013. The draft genome of a socially polymorphic halictid bee, *Lasioglossum albipes*. Genome Biol. 14, R142.

Kucharski, R., Maleszka, J., Foret, S., Maleszka, R., 2008. Nutritional control of reproductive status in honeybees via DNA methylation. Science 319, 1827–1830.
Lechner, S., Ferretti, L., Schoning, C., Kinuthia, W., Willemsen, D., Hasselmann, M., 2014. Nucleotide variability at its limit? Insights into the number and evolutionary dynamics of the sex-determining specificities of the honey bee *Apis mellifera*. Mol. Biol. Evol. 31, 272–287.
Linksvayer, T.A., Wade, M.J., 2009. Genes with social effects are expected to harbor more sequence variation within and between species. Evolution 63, 1685–1696.
Linksvayer, T.A., Busch, J.W., Smith, C.R., 2013. Social supergenes of superorganisms: do supergenes play important roles in social evolution? Bioessays 35, 683–689.
Liu, H., Zhai, J., Luo, K., Liu, L., 2014. Chromatin structure is distinct between coding and non-coding single nucleotide polymorphisms. BMC Mol. Biol. 15, 22.
Lyko, F., Foret, S., Kucharski, R., Wolf, S., Falckenhayn, C., Maleszka, R., 2010. The honey bee epigenomes: differential methylation of brain DNA in queens and workers. PLoS Biol. 8, e1000506.
Maleszka, R., 2008. Epigenetic integration of environmental and genomic signals in honey bees. Epigenetics 3, 188–192.
Martins, J.R., Anhezini, L., Dallacqua, R.P., Simões, Z.L.P., Bitondi, M.M.G., 2011. A honey bee Hexamerin, HEX 70a, is likely to play an intranuclear role in developing and mature ovarioles and testioles. PLoS One 6, e29006.
McDonald, J.H., Kreitman, M., 1991. Adaptive protein evolution at the Adh locus in Drosophila. Nature 351, 652–654.
Meunier, J., Duret, L., 2004. Recombination drives the evolution of GC-content in the human genome. Mol. Biol. Evol. 21, 984–990.
Molodtsova, D., Harpur, B.A., Kent, C.F., Seevananthan, K., Zayed, A., 2014. Pleiotropy constrains the evolution of protein but not regulatory sequences in a transcription regulatory network influencing complex social behaviours. Front. Genet. 5, 431.
Morandin, C., Havukainen, H., Kulmuni, J., Dhaygude, K., Trontti, K., Helantera, H., 2014. Not only for egg yolk—functional and evolutionary insights from expression, selection, and structural analyses of Formica ant vitellogenins. Mol. Biol. Evol. 31, 2181–2193.
Nielsen, R., 2005. Molecular signatures of natural selection. Annu. Rev. Genet. 39, 197–218.
Packer, L., Owen, R., 2001. Population genetic aspects of pollinator decline. Conserv. Ecol. 5, 4. http://www.consecol.org/vol5/iss1/art4.
Packer, L., Zayed, A., Grixti, J.C., Ruz, L., Owen, R.E., Vivallo, F., Toro, H., 2005. Conservation genetics of potentially endangered mutualisms: reduced levels of genetic variation in specialist versus generalist bees. Conserv. Biol. 19, 195–202.
Pessia, E., Popa, A., Mousset, S., Rezvoy, C., Duret, L., Marais, G.A.B., 2012. Evidence for widespread GC-biased gene conversion in eukaryotes. Genome Biol. Evol. 4, 787–794.
Praz, C.J., Packer, L., 2014. Phylogenetic position of the bee genera *Ancyla* and *Tarsalia* (Hymenoptera: apidae): a remarkable base compositional bias and an early Paleogene geodispersal from North America to the old world. Mol. Phylogenet. Evol. 81, 258–270.
Privman, E., Penn, O., Pupko, T., 2012. Improving the performance of positive selection inference by filtering unreliable alignment regions. Mol. Biol. Evol. 29, 1–5.
Privman, E., Wurm, Y., Keller, L., 2013. Duplication and concerted evolution in a master sex determiner under balancing selection. Proc. R. Soc. Lond. B Biol. Sci. 280, 20122968.
Robinson, G.E., Grozinger, C.M., Whitfield, C.W., 2005. Sociogenomics: social life in molecular terms. Nat. Rev. Genet. 6, 257–270.

Robinson, G.E., Fernald, R.D., Clayton, D.F., 2008. Genes and social behavior. Science 322, 896–900.
Romiguier, J., Gayral, P., Ballenghien, M., Bernard, A., Cahais, V., Chenuil, A., Chiari, Y., Dernat, R., Duret, L., Faivre, N., Loire, E., Lourenco, J.M., Nabholz, B., Roux, C., Tsagkogeorga, G., Weber, A.A.T., Weinert, L.A., Belkhir, K., Bierne, N., Glemin, S., Galtier, N., 2014a. Comparative population genomics in animals uncovers the determinants of genetic diversity. Nature 515, 261–263.
Romiguier, J., Lourenco, J., Gayral, P., Faivre, N., Weinert, L.A., Ravel, S., Ballenghien, M., Cahais, V., Bernard, A., Loire, E., Keller, L., Galtier, N., 2014b. Population genomics of eusocial insects: the costs of a vertebrate-like effective population size. J. Evol. Biol. 27, 593–603.
Roux, J., Privman, E., Moretti, S., Daub, J.T., Robinson-Rechavi, M., Keller, L., 2014. Patterns of positive selection in seven ant genomes. Mol. Biol. Evol. 31, 1661–1685.
Sagili, R.R., Pankiw, T., Metz, B.N., 2011. Division of labor associated with brood rearing in the honey bee: how does it translate to colony fitness? PLoS One 6, e16785.
Schneider, A., Souvorov, A., Sabath, N., Landan, G., Gonnet, G.H., Graur, D., 2009. Estimates of positive darwinian selection are inflated by errors in sequencing, annotation, and alignment. Genome Biol. Evol. 1, 114–118.
Sethupathy, P., 2013. Needles in the genetic haystack of lipid disorders: single nucleotide polymorphisms in the microRNA regulome. J. Lipid Res. 54, 1168–1173.
Shpigler, H., Amsalem, E., Huang, Z.Y., Cohen, M., Siegel, A.J., Hefetz, A., Bloch, G., 2014. Gonadotropic and physiological functions of juvenile hormone in bumblebee *Bombus terrestris* workers. PLoS One 9, e100650.
Simola, D.F., Wissler, L., Donahue, G., Waterhouse, R.M., Helmkampf, M., Roux, J., Nygaard, S., Glastad, K., Hagen, D.E., Viljakainen, L., Reese, J.T., Hunt, B.G., Graur, D., Elhaik, E., Kriventseva, E., Wen, J., Parker, B.J., Cash, E., Privman, E., Childers, C.P., Munos-Torres, M.C., Boomsma, J.J., Bornberg-Bauer, E., Currie, C., Elsik, C.G., Suen, G., Goodisman, M.A., Keller, L., Liebig, J., Rawls, A., Reinberg, D., Smith, C.D., Smith, C.R., Tsutsui, N., Wurm, Y., Zdobnov, E.M., Berger, S.L., Gadau, J., 2013. Social insect genomes exhibit dramatic evolution in gene composition and regulation while preserving regulatory features linked to sociality. Genome Res. 23, 1235–1247.
Smith, N.G., Eyre-Walker, A., 2002. Adaptive protein evolution in *Drosophila*. Nature 415, 1022–1024.
Sumner, S., 2014. The importance of genomic novelty in social evolution. Mol. Ecol. 23, 26–28.
Terrapon, N., Li, C., Robertson, H.M., Ji, L., Meng, X., Booth, W., Chen, Z., Childers, C.-P., Glastad, K.M., Gokhale, K., Gowin, J., Gronenberg, W., Hermansen, R.A., Hu, H., Hunt, B.G., Huylmans, A.K., Khalil, S.M.S., Mitchell, R.D., Munoz-Torres, M.C., Mustard, J.A., Pan, H., Reese, J.T., Scharf, M.E., Sun, F., Vogel, H., Xiao, J., Yang, W., Yang, Z., Yang, Z., Zhou, J., Zhu, J., Brent, C.S., Elsik, C.G., Goodisman, M.A.D., Liberles, D.A., Roe, R.M., Vargo, E.L., Vilcinskas, A., Wang, J., Bornberg-Bauer, E., Korb, J., Zhang, G., Liebig, J., 2014. Molecular traces of alternative social organization in a termite genome. Nat. Commun. 5, 3636.
Toth, A.L., Robinson, G.E., 2007. Evo-devo and the evolution of social behavior. Trends Genet. 23, 334–341.
Toth, A.L., Robinson, G.E., 2009. Evo-devo and the evolution of social behavior: brain gene expression analyses in social insects. Cold Spring Harb. Symp. Quant. Biol. 74, 419–426.
Viljakainen, L., Pamilo, P., 2008. Selection on an antimicrobial peptide defensin in ants. J. Mol. Evol. 67, 643–652.

Viljakainen, L., Evans, J.D., Hasselmann, M., Rueppell, O., Tingek, S., Pamilo, P., 2009. Rapid evolution of immune proteins in social insects. Mol. Biol. Evol. 26, 1791–1801.

Vitti, J.J., Grossman, S.R., Sabeti, P.C., 2013. Detecting natural selection in genomic data. Annu. Rev. Genet. 47, 97–120.

Wallberg, A., Han, F., Wellhagen, G., Dahle, B., Kawata, M., Haddad, N., Simoes, Z.L.P., Allsopp, M.H., Kandemir, I., De La Rua, P., Pirk, C.W., Webster, M.T., 2014. A worldwide survey of genome sequence variation provides insight into the evolutionary history of the honeybee *Apis mellifera*. Nat. Genet. 46, 1081–1088.

Wang, Y., Kaftanoglu, O., Siegel, A.J., Page Jr., R.E., Amdam, G.V., 2010. Surgically increased ovarian mass in the honey bee confirms link between reproductive physiology and worker behavior. J. Insect Physiol. 56, 1816–1824.

Wang, J., Wurm, Y., Nipitwattanaphon, M., Riba-Grognuz, O., Huang, Y.C., Shoemaker, D., Keller, L., 2013. A Y-like social chromosome causes alternative colony organization in fire ants. Nature 493, 664–668.

Weinstock, G.M., Robinson, G.E., Gibbs, R.A., Worley, K.C., Evans, J.D., Maleszka, R., Robertson, H.M., Weaver, D.B., Beye, M., Bork, P., Elsik, C.G., Hartfelder, K., Hunt, G.J., Zdobnov, E.M., Amdam, G.V., Bitondi, M.M.G., Collins, A.M., Cristino, A.S., Lattorff, H.M.G., Lobo, C.H., Moritz, R.F.A., Nunes, F.M.F., Page, R.E., Simoes, Z.L.P., Wheeler, D., Carninci, P., Fukuda, S., Hayashizaki, Y., Kai, C., Kawai, J., Sakazume, N., Sasaki, D., Tagami, M., Albert, S., Baggerman, G., Beggs, K.T., Bloch, G., Cazzamali, G., Cohen, M., Drapeau, M.D., Eisenhardt, D., Emore, C., Ewing, M.A., Fahrbach, S.E., Foret, S., Grimmelikhuijzen, C.J.P., Hauser, F., Hummon, A.B., Huybrechts, J., Jones, A.K., Kadowaki, T., Kaplan, N., Kucharski, R., Leboulle, G., Linial, M., Littleton, J.T., Mercer, A.R., Richmond, T.A., Rodriguez-Zas, S.L., Rubin, E.B., Sattelle, D.B., Schlipalius, D., Schoofs, L., Shemesh, Y., Sweedler, J.V., Velarde, R., Verleyen, P., Vierstraete, E., Williamson, M.R., Ament, S.A., Brown, S.J., Corona, M., Dearden, P.K., Dunn, W.A., Elekonich, M.M., Fujiyuki, T., Gattermeier, I., Gempe, T., Hasselmann, M., Kage, E., Kamikouchi, A., Kubo, T., Kunieda, T., Lorenzen, M., Milshina, N.V., Morioka, M., Ohashi, K., Overbeek, R., Ross, C.A., Schioett, M., Shippy, T., Takeuchi, H., Toth, A.L., Willis, J.H., Wilson, M.J., Gordon, K.H.J., Letunic, I., Hackett, K., Peterson, J., Felsenfeld, A., et al., 2006. Insights into social insects from the genome of the honeybee *Apis mellifera*. Nature 443, 931–949.

Wilfert, L., Gadau, J., Schmid-Hempel, P., 2007. Variation in genomic recombination rates among animal taxa and the case of social insects. Heredity 98, 189–197.

Wilson, E.O., 1985. The sociogenesis of insect colonies. Science 228, 1489–1495.

Wissler, L., Gadau, J., Simola, D.F., Helmkampf, M., Bornberg-Bauer, E., 2013. Mechanisms and dynamics of orphan gene emergence in insect genomes. Genome Biol. Evol. 5, 439–455.

Wong, K.M., Suchard, M.A., Huelsenbeck, J.P., 2008. Alignment uncertainty and genomic analysis. Science 319, 473–476.

Woodard, S.H., Fischman, B.J., Venkat, A., Hudson, M.E., Varala, K., Cameron, S.A., Clark, A.G., Robinson, G.E., 2011. Genes involved in convergent evolution of eusociality in bees. Proc. Natl. Acad. Sci. U. S. A. 108, 7472–7477.

Wray, G.A., 2007. The evolutionary significance of cis-regulatory mutations. Nat. Rev. Genet. 8, 206–216

Wray, M.K., Mattila, H.R., Seeley, T.D., 2011. Collective personalities in honeybee colonies are linked to colony fitness. Anim. Behav. 81, 559–568.

Wurm, Y., Wang, J., Riba-Grognuz, O., Corona, M., Nygaard, S., Hunt, B.G., Ingram, K.K., Falquet, L., Nipitwattanaphon, M., Gotzek, D., Dijkstra, M.B., Oettler, J., Comtesse, F., Shih, C.J., Wu, W.J., Yang, C.C., Thomas, J., Beaudoing, E., Pradervand, S., Flegel, V., Cook, E.D., Fabbretti, R., Stockinger, H., Long, L., Farmerie, W.G., Oakey, J.,

Boomsma, J.J., Pamilo, P., Yi, S.V., Heinze, J., Goodisman, M.A., Farinelli, L., Harshman, K., Hulo, N., Cerutti, L., Xenarios, I., Shoemaker, D., Keller, L., 2011. The genome of the fire ant Solenopsis invicta. Proc. Natl. Acad. Sci. U. S. A. 108, 5679–5684.

Yan, H., Simola, D.F., Bonasio, R., Liebig, J., Berger, S.L., Reinberg, D., 2014. Eusocial insects as emerging models for behavioural epigenetics. Nat. Rev. Genet. 15, 677–688.

Yang, Z., Dos Reis, M., 2011. Statistical properties of the branch-site test of positive selection. Mol. Biol. Evol. 28, 1217–1228.

Zayed, A., 2009. Bee genetics and conservation. Apidologie 40, 237–262.

Zayed, A., Packer, L., 2005. Complementary sex determination substantially increases extinction proneness of haplodiploid populations. Proc. Natl. Acad. Sci. U.S.A. 102, 10742–10746.

Zayed, A., Packer, L., 2007. The population genetics of a solitary oligolectic sweat bee, *Lasioglossum* (*Sphecodogastra*) *oenotherae* (Hymenoptera: Halictidae). Heredity 99, 397–405.

Zayed, A., Robinson, G.E., 2012. Understanding the relationship between brain gene expression and social behavior: lessons from the honey bee. Annu. Rev. Genet. 46, 591–615.

Zayed, A., Whitfield, C.W., 2008. A genome-wide signature of positive selection in ancient and recent invasive expansions of the honey bee *Apis mellifera*. Proc. Natl. Acad. Sci. U.S.A. 105, 3421–3426.

Zayed, A., Packer, L., Grixti, J.C., Ruz, L., Toro, H., Owen, R.E., 2005. Increased genetic differentiation in a specialist versus a generalist bee: implications for conservation. Conserv. Genet. 6, 1017–1026.

Zayed, A., Naeger, N.L., Rodriguez-Zas, S.L., Robinson, G.E., 2012. Common and novel transcriptional routes to behavioral maturation in worker and male honey bees. Genes Brain Behav. 11, 253–261.

Zhou, X., Oi, F.M., Scharf, M.E., 2006. Social exploitation of hexamerin: RNAi reveals a major caste-regulatory factor in termites. Proc. Natl. Acad. Sci. U. S. A. 103, 4499–4504.

Ziller, M.J., Gu, H., Muller, F., Donaghey, J., Tsai, L.T.Y., Kohlbacher, O., De Jager, P.L., Rosen, E.D., Bennett, D.A., Bernstein, B.E., Gnirke, A., Meissner, A., 2013. Charting a dynamic DNA methylation landscape of the human genome. Nature 500, 477–481.

CHAPTER TEN

Understanding Honey Bee Worker Self-Sacrifice: A Conceptual–Empirical Framework

Emma K. Mullen[1], Graham J. Thompson
Department of Biology, Western University, London, Ontario, Canada
[1]Corresponding author: e-mail address: emullen@alumni.uwo.ca

Contents

Abstract

It is well understood from inclusive fitness theory that genes are important to the evolution of social life. The honey bee *Apis mellifera* has played a central role as a model of gene discovery, but we do not yet have a full understanding of how genes coordinate worker self-sacrifice and reproductive altruism. In this review, we attempt to bring together what we know from theory regarding any 'genes for altruism' that are likely to be present in the honey bee with what we know from practical experiments in molecular biology. First, we highlight an opportunity to bridge the gap between a genetic theory of social evolution that invokes genes in abstraction and the emerging field of sociogenomics that attempts to realize these genes as biochemical molecules that can be characterized, mapped and named. Second, we adopt the kin-theoretic

Advances in Insect Physiology, Volume 48
ISSN 0065-2806
http://dx.doi.org/10.1016/bs.aiip.2014.12.002

framework of Thompson et al. (2013) to help predict the most likely characteristics of genes underlying altruism. Finally, use this framework as a guide to identify what is known, and not known, about genes that specifically underlie reproductive altruism in honey bees. We show that an integration of theory with genomic tools is useful for identifying genes relevant to social evolution and for testing broader ideas about how social genes are expected to evolve. In our view, this integration has only just begun but the honey bee does provide one starting point from which to push inclusive fitness theory into the age of genomics.

1. INTRODUCTION

A central goal of evolutionary biology is to understand the origins of biological complexity. While it has long been recognized that complexity evolves at the level of the gene, only recently has it become more widely appreciated that fundamental increases to complexity result from changes in the way in which genetic information is transmitted between generations (Bourke, 2011; Maynard Smith and Szathmary, 1995). For example, the evolution of eusocial insects, in which individuals pass on their genes only as parts of a whole, is one of the most recent 'major' evolutionary transitions. Though this type of transition to group-living is rare, it is not singular. Eusociality, in one form or another, has evolved at least a dozen times across different spectra of life (Choe and Crespi, 1997). Moreover, eusociality can be evolutionarily labile and lost through reverse transitions (Wcislo and Danforth, 1997) or via extinction of eusocial lineages (Engel, 2001).

The most advanced animal societies are found among humans and insects (Holldobler and Wilson, 2009). Like humans, eusocial ants, bees, wasps and termites show bewildering complexity in how their societies are structured. Yet for insects, this complexity is not associated with culture but rather is derived from a simple division of labour into reproductive and non-reproductive specialists. It is known from kin theory that this reproductive division of labour evolves from selection at the level of the gene (Bourke, 2011; Crozier and Pamilo, 1996; Hamilton, 1964; Keller, 2009; Queller and Strassmann, 1998), and therefore studying divisions of labour from a gene's eye view can provide insights into how complex social systems evolved from more individualistic ancestries.

Insect models provide a good subject to study the genetical evolution of sociality (Beshers and Fewell, 2001; Boomsma and Franks, 2006). For one, social insects are often regarded as pinnacles of social success, whereby the depth of division of labour and degree of interdependence among individuals far exceeds that which is typical of vertebrate social systems (Alexander et al., 1991; Holldobler and Wilson, 2009). Second, owing to their phylogenetic diversity, whereby sociality has arisen repeatedly and to different levels and in different clades (Choe and Crespi, 1997), social insects lend themselves well to comparative analyses of phenotypic (Crespi, 1996; Thompson and Oldroyd, 2004) or genotypic (e.g. Fischman et al., 2011; Woodard et al., 2011) characters. Finally, unlike most vertebrates, some insect species are amenable to behavioural and genetic manipulation, and therefore provide tractable models for studying the interplay between genes, the environment and social versus asocial life (Leboeuf et al., 2013).

2. HONEY BEES AS A MODEL INSECT SOCIETY

Within this sociogenomic theme, the honey bee *Apis mellifera* is a leading model system for understanding the origins of social complexity (Weinstock et al., 2006). Like all eusocial taxa, honey bee societies are characterized by specialized castes. For *A. mellifera*, this specialization is so pronounced that the non-reproductive worker caste is functionally sterile (as it is for other *Apis* spp.). This sterility is maintained by indirect selection that promotes reproductive self-sacrifice and alloparental care from worker individuals, which collectively form an altruistic caste. Worker altruism is stimulated at the proximate level by a collection of social and environmental factors, with a key factor being the presence of a queen (Backx et al., 2012). The queen's pheromones, as well as those from her brood, likely serve as an honest signal to which workers attend and inactivate their ovaries (Kocher et al., 2009). In her presence, workers remain more-or-less sterile, but otherwise some proportion of queenless workers may activate their ovaries and lay eggs.

The precise pheromonal threshold beyond which workers 'switch' their own reproductive machinery *off* in order to assist the queen depends on the social and ecological context, as predicted by response threshold models (Barron and Robinson, 2008), and even from Hamilton's rule: $rb > c$. This rule-of-thumb shows that, despite a cost c to the altruist's direct fitness, genes for reproductive altruism can nonetheless spread if their effect sufficiently benefits (b) relatives of degree r who—by definition—are likely to carry copies of that gene. As stressed by Crozier (2008) and others, the condition under which the inequality is met varies with social, genetic and ecological context. Even for a given set of conditions the threshold response is likely to be genetically variable (Oldroyd and Fewell, 2007); workers with a genetically low threshold will be sensitive to queen signal and a low dose of pheromone is sufficient to suppress ovary development, whereas workers with a genetically high threshold will be relatively insensitive to queen signal. So-called anarchist workers provide a good example of heritable variation in worker sterility. Workers of the anarchist strain have such a high threshold response that they readily activate their ovaries and lay eggs despite the presence of the queen (Barron et al., 2001). The evidence for heritable variation underlying worker ovary activation—essentially, underlying worker sterility—suggests that it may be possible to identify the loci and genes responsible. In effect, genes that explain variation in ovary activation are examples of genes that explain this particular form of reproductive altruism.

With the increasing power and accessibility of tools that facilitate the interrogation of honey bee genomes (Aase et al., 2005; Munoz-Torres et al., 2011; Schulte et al., 2013; Solignac et al., 2007; Weinstock et al., 2006), it has become increasingly feasible to map, isolate and characterize genes associated with social traits. For example, quantitative trait loci that regulate worker sterility in anarchist workers are now mapped to specific linkage groups (Oxley et al., 2008), as are other segregating loci related to other aspects of honey bee social life (e.g. Hunt et al., 1995; Lattorff et al., 2007; Shorter et al., 2012). Building on such mapping studies, molecular biologists have begun to isolate socially relevant loci (e.g. Jarosch et al., 2011; Kiya et al., 2012; Rueppell et al., 2004) and a growing number of genes are becoming carefully characterized (e.g. Fussnecker and Grozinger, 2008; Wang et al., 2009). This progress comes despite idiosyncrasies of honey bee biology that make behavioural genetic studies challenging, at least compared to *Drosophila* (Oldroyd and Thompson, 2007).

Based on these initial findings, there is growing interest in expanding our understanding of the evolutionary genetic basis of honey bee social life (Fischman et al., 2011; Robinson et al., 2008; Smith et al., 2008). Yet despite rapid advances, efforts to date have sought genes associated with myriad bee behavioural traits but have not necessarily focused on genes that underlie reproductive self-sacrifice *per se*. This may be due in part to sociogenomics still emerging from its infancy, but may likewise stem from discord between molecular biologists and bee sociobiologists. If honey bee workers are considered reproductively altruistic, and this altruism has evolved via selection on genes for helping the queen, then it is simply expected that workers carry these genes for altruism (Thompson et al., 2013). Despite this expectation, there remains a considerable gap between social gene theory and our knowledge of the genes themselves.

3. GAP BETWEEN GENE THEORY AND GENES

If Hamilton's inclusive fitness (kin) theory is explicitly gene-centric—literally, *The Genetical Evolution of Social Behaviour*—then why has the theory not been better exploited to uncover socially important genes? We propose that there are two reasons for this gap between sociobiology and molecular biology. First, kin theory enthusiasts—who are most likely to appreciate the genetic basis of sociality—have only truly begun to embrace the tools of molecular biology. On the other hand, molecular specialists—who are best equipped to map and characterize genes of any type—tend not to adopt social species as models of gene discovery. So, despite the massive contributions from some exceptional research groups, there simply has not yet been a field-wide coordinated effort to screen, prioritize and verify candidate genes relevant to honey bee or other social insect biology.

Second, even with the initial success of sociogenomics there has not yet been a robust evolutionary framework developed for guiding gene discovery. After all 'genes for altruism', as Hamilton (1963) would put it, are not just *any* genes, they are genes whose transmission occurs indirectly through non-descendent relatives, and this quality can be exploited to aid in their identification. For example, such genes—'social' genes, if you will—are likely to be differentially expressed (Bourke, 2011; Queller and Strassmann, 1998; Seger, 1981; Thompson et al., 2006), may be more variable (Linksvayer and Wade, 2009) and potentially evolve faster (Bromham and Leys, 2005; Hall and Goodisman, 2012; Kent et al., 2012) than their homologues in non-social lineages, or relative to other genes in the same genome. It should therefore be possible to harness these and other putative qualities of social genes to focus a renewed search for their identification and, beyond that, assemble these genes into functional networks that provide an integrated description of social behaviour, as it is expressed within natural social environments (Chandrasekaran et al., 2011; Mullen et al., 2014).

4. TOWARDS A MOLECULAR UNDERSTANDING OF WORKER SELF-SACRIFICE

In the 50 years since Hamilton's insight, kin theory has become sophisticated and is sometimes misunderstood (cf. Abbot et al., 2011; Crozier, 2008; Dawkins, 1979; Gardner et al., 2011; Queller et al., 2014; Rousset and Lion, 2011; Thompson, 2006)—yet the theory remains simple at its core, and can be interpreted to generate several straightforward predictions regarding the nature of genes for altruism. In a recent review commemorating the semi-centennial anniversary of Hamilton's papers (Herbers, 2013), Thompson et al. (2013) outline what they consider to be some of most likely characteristics of genes underlying altruistic traits, in any organism. Using a kin-theoretic framework, they describe how to identify and analyse genes for altruism in, for example, humans or social insects. This framework is provisional but stands to accelerate social gene discovery by drawing attention from molecular biologists to the abstract 'genes' that are predicted from the theory, but that for the most part have yet to be isolated. Despite the potential utility of this framework, it remains untested against any social genome. In this review, we test this new scheme against what is currently known from the honey bee genome. Our goal is to summarize the molecular evidence for each of Thompson et al.'s predictions and suggests how any data gaps might eventually be filled.

5. PREDICTION 1: GENES FOR ALTRUISM SHOULD SATISFY HAMILTON'S RULE

From an evolutionary perspective, altruism is simply a type of social behaviour defined by its direct fitness consequence to an actor and at least one receiver: the actor (altruist) pays a direct fitness cost, c, and the receiver gains a direct fitness benefit, b. That's it. By comparison, altruism can be distinguished from the other three types of social behaviour—cooperation, selfishness and spite—based on alternative fitness outcomes (reviewed in West et al., 2007). This long-established convention does not imply any 'intent' or even 'choice' on behalf of the actor, nor does it implicate a particular mechanism by which the outcome is achieved. In theory, any functional gene for altruism can evolve if its effect is directed towards other carriers, who will therefore be related by some degree, r—as predicted by the conditions of Hamilton's well-known rule. The notion of a 'gene for altruism' that we use in this chapter follows Thompson et al. (2013) and is no different from its broader use in the field of behavioural genetics. We therefore consider the term a 'gene for altruism', or a gene for *any* behavioural phenotype, to mean a genotypic difference that explains differences in behaviour among individuals (Plomin et al., 2013).

Worker self-sacrifice in honey bees is one of the best-known examples of reproductive altruism (Oldroyd and Thompson, 2007). Even school children appreciate that a worker bee labours on behalf of her colony and dies after she stings in defence of her family. But even if bee biology were unknown to most, the monogamous and genetically cohesive families of ancestral social Hymenoptera (Boomsma, 2009) suggest that the selfless qualities that characterize the worker phenotype evolved via kin-mediated selection on genes that promoted indirect fitness—a key prediction from kin theory (Hughes et al., 2008).

But what are the identities of these genes? By necessity any gene for altruism is under indirect selection, but it is not certain whether any single gene is known where fitness is entirely indirect—that is, with no possibility for direct selection. Even our best examples of genes associated with worker sterility are 'leaky' in the sense that workers still have a direct fitness quotient—for example, when honey bee workers activate their ovaries to lay eggs parthenogenetically. From honey bees, it may therefore be difficult to demonstrate with certainty any genes that are strictly under indirect selection. Still, even with this imperfect experimental model, it should be possible to identify genes that are clearly implicated in sterility (Cardoen et al., 2011; Grozinger et al., 2003, 2007; Thompson et al., 2006, 2008) or other selfless traits—for example, genes that promote sib helping by, say, influencing the timing and target of parental care. Figure 1 shows how such a gene might evolve if it had the simple effect of inducing early onset of parental care, such

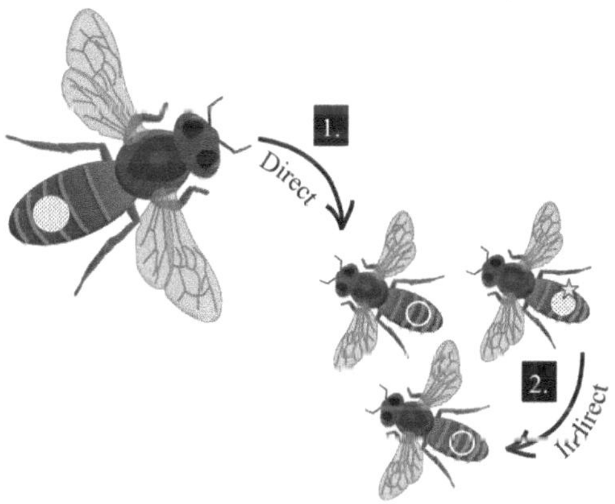

Figure 1 A simple scenario for single-gene effects on the transition from parental-to-alloparental care. (1) A mother directs care towards her offspring under the influence of a single gene that is shown stylistically as a closed circle on her abdomen. The gene is transmitted to her offspring but is typically turned *off* (open circle) until the appropriate age or circumstance. (2) A simple variation on the gene's effect can be seen in the one offspring in which the gene is expressed out of its normal context (star), leading to a form of 'parental' care. If the indirect benefits of directing care towards non-descendent carriers (i.e. *r* is large and positive) outweigh the direct fitness costs (i.e. *c* is small), then this type of gene for altruism (a gene 'for' alloparental care) could evolve.

that care is expressed precociously and towards the production of siblings rather than a female's own offspring (Fig. 1). Again, these genes may prove challenging to identify because honey bees and other advanced eusocial taxa are so far past the parental-to-alloparental evolutionary transition that any genes initially associated with it may have long ago been co-opted to another function. Screens may still prove useful, however, if they focus on the facultative expression of alloparental care among genetically variable workers, or if they map comparable genes found from facultatively eusocial species (e.g. Toth et al., 2007). Finally, other avenues to discover social genes are to infer their indirect transmission through their effect in promoting sterile caste differentiation or their effect in promoting selfless colony defence or stinging (Table 1).

6. PREDICTION 2: GENES FOR ALTRUISM SHOULD BE ENVIRONMENTALLY SENSITIVE

In the solitary or communal ancestors to the corbiculate or other social bee clades, there would be no morphological queen or worker caste. Presumably there was, however, genetic variation among females for when, and towards whom, to direct care. Selection might thus have favoured variants that directed care towards related, but not descendent, brood. If so, the relevant allele associated with this shift from parental-to-alloparental care could increase in frequency (Fig. 1). This scenario assumes, however, that only some of the females make this shift in the timing (tending to early) and direction (tending towards non-descendent kin) of care, while the remaining females (those tending to receive help) continued to be firmly reproductive. To the extent that this, or any other, shift towards alloparental behaviour is mediated by genes, then these genes must have been sensitive to their social environment—that is, the genes switch *on* in altruistic helpers but switch *off* in selfish reproductives (Thompson et al., 2006). As reviewed in Bourke (2011), this conditional expression of genes underlying altruism is a requirement for their evolution—otherwise, any ubiquitous expression of genes costly to personal fitness would surely render them extinct.

One opportunity for identifying these genes is via expression-based screens—for example, microarrays comparing (reproductive) queen and (altruistic) worker transcripts. These screens typically do yield widespread expression differences (Barchuk et al., 2007; Cristino et al., 2006; Evans and Wheeler, 2000; Grozinger et al., 2007). Yet, because queens and workers are differentiated with respect to a large number of traits, it would be difficult

Table 1 Examples of genes implicated in self-sacrificial behaviour in the honey bee *Apis mellifera*

Altruistic trait	Gene name/gene family	Accession	Function	Predictions	References
Caste differentiation	Telomerase	GB52244	Longevity	SE	Grozinger et al. (2007)
	sodh	GB42385	Oxidative process	SE, PE	Grozinger et al. (2007)
	NCC69-like	GB40598	Sodium chloride co-transporter, neural plasticity	SE, PE	Grozinger et al. (2007)
	Hexamerins	GB51697 and others	Storage proteins	SE, PE	Begna et al. (2012), Cristino et al. (2006) and Evans and Wheeler (1999, 2000)
	Fatty acid binding protein	GB49757	Regulates fatty acid transport	SE, PE	Evans and Wheeler (1999)
	Lambda crystallin	GB40673	Eye lens, metabolic enzyme	SE, PE	Evans and Wheeler (1999)
	Oxidoreductases	—	Metabolic enzymes	SE	Cristino et al. (2006) and Evans and Wheeler (1999, 2000)
	CREG	GB52115	Glycoprotein	SE, PE	Barchuk et al. (2007)
	dac	GB44041	Neurogenesis	SE, PE	Barchuk et al. (2007)
	Neurogenesis genes	GB55453, GB44091, GB41629, GB53672	Neurogenesis	SE, PE	Barchuk et al. (2007)

Continued

Table 1 Examples of genes implicated in self-sacrificial behaviour in the honey bee *Apis mellifera*—cont'd

Altruistic trait	Gene name/gene family	Accession	Function	Predictions	References
	Cathepsin D	GB43825	Apoptosis	SE, PE	Barchuk et al. (2007) and Grozinger et al. (2003)
	atx2	GB55453	Adhesion between oocytes and follicular cells	SE, PE	Barchuk et al. (2007) and Grozinger et al. (2003)
	traf 4-like	GB46952	Apoptosis	SE, PE	Barchuk et al. (2007) and Grozinger et al. (2003)
	Mitochondrial genes	GB42024 and others	Metabolism	SE, PE	Corona et al. (1999) and Evans and Wheeler (2000)
	Cuticle proteins	GB48823 and others	Structural constituent of cuticle	SE	Begna et al. (2012) and Evans and Wheeler (2000)
	Unmapped loci	—	—	AV	Tilley and Oldroyd (1997)
Caste differentiation and worker sterility	Histone proteins and DNA methylation genes	GB54085, GB54595, GB55485	DNA binding, methylation	SE	Barchuk et al. (2007), Grozinger et al. (2003), Kucharski et al. (2008) and Mullen et al. (2014)
	Steroid biosynthesis proteins and ecdysteroid regulating proteins	GB40746, GB44718, GB48898, GB55540, GB42142, GB46500, GB42692		SE, PE	Barchuk et al. (2007), Begna et al. (2012), Cardoen et al. (2011, 2012), Grozinger et al. (2003), Hartfelder et al. (1995) and Kocher et al. (2010)

	Insulin/TOR signalling genes	GB43560, GB44905, GB49642, GB47413, GB49911, GB55333 and others	Regulation of growth, metabolism, reproduction and longevity	SE, PE	Cardoen et al. (2011), Corona et al. (2007), De Azevedo and Hartfelder (2008), Grozinger et al. (2007), Grozinger et al. (2003), Mutti et al. (2011), Patel et al. (2007), Wang et al. (2013b), Wheeler et al. (2006) and Wolschin et al. (2011)
	Yolkless	GB40823	Transports vitellogenin into developing ovary, vitellogenin receptor	SE, PE	Cardoen et al. (2011) and Cristino et al. (2006)
Worker sterility	Vitellogenin	GB49544	Egg yolk precursor	SE, SC, PE	Cardoen et al. (2012), Corona et al. (2007), Piulachs et al. (2003) and Thompson et al. (2008)
	Tubulin	GB50238	Structural protein	SE, PE	Cardoen et al. (2012)
	Actin, clone 205-like	GB41310	Structural protein	SE	Cardoen et al. (2012)
	Heat shock proteins	GB40976, GB46339, GB50730 and others	Molecular chaperones, assembly of steroid hormone receptors and response to heat stress	SE, PE	Barchuk et al. (2007), Begna et al. (2012), Cardoen et al. (2011, 2012), Evans and Wheeler (2000), Hartfelder et al. (1995) and Mullen et al. (2014)

Continued

Table 1 Examples of genes implicated in self-sacrificial behaviour in the honey bee *Apis mellifera*—cont'd

Altruistic trait	Gene name/gene family	Accession	Function	Predictions	References
	Niemann–Pick type C2	GB44564, GB42887	Nutrition, binds lipids and cholesterol	SE, PE	Thompson et al. (2006)
	Neuropeptide F	GB50693	Neuropeptide that inhibits JH production	SE, PE	Cardoen et al. (2011)
	Prohormone-2	GB44988	Neuropeptide	SE	Cardoen et al. (2011) and Grozinger et al. (2003)
	myb	GB44791	Transcription factor that regulates eggshell chorion formation	SE, PE	Cardoen et al. (2011) and Grozinger et al. (2003)
	Double parked	GB46452	Regulation of oviposition	SE, PE	Cardoen et al. (2011)
	Piefke	GB46511	Female gonad development	SE	Alaux et al. (2009) and Cardoen et al. (2011)
	Similar to slimfast	GB46448	Regulates growth and TOR pathway	SE, PE	Cardoen et al. (2011) and Grozinger et al. (2007)
	Odorant binding proteins	GB40730, GB50936, GB50151	Specific odorant binding protein	SE, SC	Cardoèn et al. (2011), Grozinger et al. (2007) and Oxley et al. (2008)
	Megator	GB54288	Oogenesis	SE, PE	Barchuk et al. (2007), Cardoen et al. (2011) and Grozinger et al. (2007)

	Kr-h1	GB45427	Transcription factor	SE, PE	Grozinger et al. (2003)
	Push/poe	GB51956	Transmembrane protein involved in glial cell thickness, calmodulin binding, synaptogenesis	SE, PE	Grozinger et al. (2003)
	Klumpfuss	GB42180	Metal ion binding, nucleic acid binding	SE, PE	Grozinger et al. (2003)
	Gemini	GB48238	Transcription factor involved in ovary activation	SE	Jarosch et al. (2011)
	egfr	GB54477	Epidermal growth factor receptor, regulates growth and other functions	SE, PE	Formesyn et al. (2014)
	Unmapped loci	—	—	VA	Backx et al. (2012) and Robinson et al. (1990)
Worker sterility and alloparental care	Major royal jelly proteins	GB55212, GB55213	Nutrition, sex-specific reproductive maturity and other functions	SE, SC, PE	Drapeau et al. (2006), Harpur et al. (2014) and Thompson et al. (2006)
Stinging	14-3-3 Epsilon	GB42560	Neurogenesis	SE, PE	Hunt (2007)
	Tango	GB44259	Transcription factor for neural development	SE, PE	Hunt (2007)

Continued

Table 1 Examples of genes implicated in self-sacrificial behaviour in the honey bee *Apis mellifera*—cont'd

Altruistic trait	Gene name/gene family	Accession	Function	Predictions	References
	Arrestin4-like	GB49099	GPCR signalling, vision and olfaction	SE, PE	Hunt (2007)
	ced-6	GB41603	Mushroom body development	SE	Shorter et al. (2012)
	Disabled	GB42757	SH2 domain binding	SE, PE	Shorter et al. (2012)
	Eph receptor tyrosine kinase	GB41629	Ephrin receptor activity, mushroom body development	SE, PE	Shorter et al. (2012)
	Frazzled	GB41746	Microtubule cytoskeleton organization	SE, PE	Shorter et al. (2012)
	Ankyrin2	GB41601	Axon extension	SE, PE	Shorter et al. (2012)
	Nexin-16	GB55192	Membrane trafficking and synaptic development	SE, PE	Shorter et al. (2012)
	Rab21	GB46645	GTPase activity	SE, PE	Shorter et al. (2012)

We group sets of genes by altruistic trait and provide the gene name, the BeeBase accession, and the gene's basic function as inferred from the original study or from Gene Ontology. For each gene, we indicate its best match to predictions regarding social gene qualities as outlined in Thompson et al.'s (2013). These are: the social gene evolves by Hamilton's rule (HR), is responsive to the social environment (SE), co-evolves with social complexity (SC) and with kin recognition (KR), forms co-adapted gene complexes (GC) with other social genes, is underlain by additive genetic variation (VA) and has pleiotropic effects (PE).

to pinpoint the genes associated with any one physiological or behavioural trait, altruistic or otherwise. The most experimentally relevant studies are therefore those that control for caste and thus compare gene expression differences between reproductive and non-reproductive members of a single caste (e.g. Cardoen et al., 2011; Thompson et al., 2008). These studies reveal more refined sets of genes that are sensitive to social context *per se* and not confounded with genes for caste differentiation. Candidate genes implicated by this type of study in which social cues are manipulated to influence the expression of caste-specific social traits include insulin-signalling genes (Grozinger et al., 2007), pheromone response genes (Cardoen et al., 2011), vitellogenin (Thompson et al., 2008), steroid biosynthesis genes (Cardoen et al., 2011), *Kruppel homolog 1* (Grozinger et al., 2003) and major royal jelly proteins (Thompson et al., 2006), among others (Table 1). These are examples of genes that are conditionally expressed as a function of reproductive altruism (ovary inhibition in workers), but it is not clear whether their expression embodies worker altruism *per se*. To what extent are any of these genes functionally associated with worker reproductive decisions to invest in queen over personal reproduction? Perhaps the best examples come from honey bee major royal jelly proteins (Drapeau et al., 2006) and the genes encoding egg yolk protein vitellogenin (Havukainen et al., 2011). Both classes of genes show nucleotide substitution patterns consistent with bouts of adaptive evolution (Harpur et al., 2014; Kent et al., 2011), and both are directly and functionally implicated in reproductive 'decisions'. First, major royal jelly proteins mediate differentiation into reproductive and non-reproductive castes, while vitellogenin has multiple social functions that include a role in alloparental feeding of larvae by workers. Both vitellogenin and MRJPs are likewise under positive selection (Harpur et al., 2014; Kent et al., 2011), which further suggests that their functional roles are recently under selection.

7. PREDICTION 3: GENES FOR ALTRUISM SHOULD INCREASE IN NUMBER AND COMPLEXITY WITH SOCIAL-BEHAVIORAL SOPHISTICATION

Mutations that predispose essentially selfish individuals towards altruism may initially involve a single gene, such as a conditionally sensitive mutant that reduces reproduction and increases the likelihood to help, or one that increases the likelihood to defend or provision beyond their own brood. If selection promotes this initial gene-based division of labour, then selection may subsequently embellish this division for ever more specialized or efficient divisions that involve larger numbers of downstream and interacting genes. This co-evolutionary scenario between social and genomic complexity is consistent with the original 'genetic release' hypothesis of Gadagkar (1997); because non-social ancestors have no divisions in labour and individuals are thereby constrained to perform a full suite of reproductive and non-reproductive tasks, the genes involved may be similarly constrained via antagonistic pleiotropy (Linksvayer and Wade, 2005). The evolution of task specialization eases this functional constraint and renders genes more responsive to directional selection. If so, then we expect species with well-differentiated castes—that is, advanced eusocial species—to have genomes enriched for genes that are specifically and functionally related to caste and sub-caste specialization, relative to primitively eusocial species that have less well-differentiated castes and only basic divisions in labour (Sumner, 2006).

Even within advanced eusocial genomes, we would further expect that genes specific to queen or worker phenotypes be less constrained than genes that simultaneously contribute to more than one caste phenotype. Figure 2 demonstrates this sociogenomic scenario for a set of five genes. In solitary females, the gene set is constrained, whereas in primitively eusocial species the genes are 'released' from this constraint to take on caste-specific roles. In advanced eusocial females, there is further specialization and gene diversification. In one test of this idea, Hunt et al. (2010) report that rates of molecular evolution are indeed faster at loci associated with queen-biased gene expression relative to loci that are not caste-biased in their expression. Harpur et al. (2014) also report faster rates of evolution at loci where expression is caste-biased, especially in genes that are taxonomically restricted within Apoidea and within *Apis*. More generally, this rapid (released) evolution of genes associated with social task specialization is consistent with the high proportion of novel genes (Sumner, 2014) and non-coding regulatory elements (Simola et al, 2013) that are apparently linked to social diversification in ants. These patterns are consistent with Thompson et al.'s (2013) prediction for co-evolution between gene and task specialization and further suggest that newly acquired genes may be especially involved in this process (Johnson and Tsutsui, 2011; Weinstock et al., 2006).

8. PREDICTION 4: GENES FOR ALTRUISM SHOULD CO-EVOLVE WITH, OR DEPEND ON, THE PREVIOUS EVOLUTION OF GENES FOR KIN RECOGNITION

In order for any gene for altruism to evolve, its effect must be directed towards other carriers in which the gene is typically not expressed. Hence

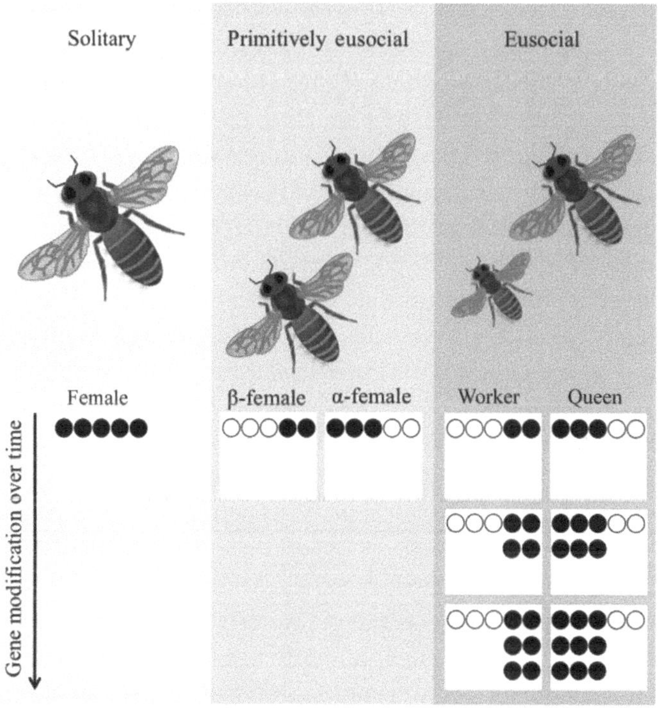

Figure 2 A simple scenario showing the co-evolution between gene and task specialization. A solitary female expresses a set of five genes that enable her full behavioural repertoire. Because the repertoire is diverse, each gene has multiple functions and each is constrained under pleiotropy. In a primitively eusocial population, by contrast, predominantly parental α-females and predominantly alloparental β-females tend towards non-overlapping repertoires. This allows some genes (white) to likewise specialize in reproductive or non-reproductive function, either through prior gene duplication (not depicted) or through caste-specific gene regulation. Finally, a eusocial population with well-differentiated queen and worker castes has yet more genes with caste-specific function. Again, the caste specificity of gene function can again result from prior gene duplication or caste specific gene regulation, or both. This 'genetic release' of previously constrained genes may permit further adaptation and specialization over time, shown here as an increasing number of caste specialized genes.

kin selection should favour kin recognition, which, in social insects, involves perception of shared genetic or environmental cues (Breed, 2014). For honey bees, recognition is mostly a function of nestmates in which guards exclude workers if their identity does not match the colony (Breed, 1998). This Gestalt is likely to have both environmentally acquired and genetically inherited cues (Dani et al., 2005), with the combination varying with species. To be effective, cues need to be variable to distinguish kin from non-kin, but not so individualistic as to reject genuine nestmates with an imperfect match (Agrawal, 2001; Breed, 2014). For heritable cues, this trade-off may be evident in the form of balancing selection at recognition loci (Crozier, 1988; Rousset and Roze, 2007), in which selection favours allelic diversity, but only to a point, beyond which it becomes costly to maintain or becomes overly sensitive and maladaptive (Tsutsui, 2004).

Honey bee nestmate recognition is at least partially a function of variation in cuticular hydrocarbons (e.g. Arnold et al., 2000; Le Conte and Hefetz, 2008; Page et al., 1989), so genes underpinning the variation are in a sense genes 'for' recognition. Mougel et al. (2012) identified QTLs for cuticular hydrocarbon variants, and within these loci they identified seven genes whose homologues in *Drosophila* function in hydrocarbon elongation and desaturation. Although the genes involved in producing the cues are beginning to be uncovered, those involved in detecting and interpreting the cue remain unknown (Breed et al., 2004). Some candidate genes for kin recognition include ASP3c, an antennal chemosensory protein (Calvello et al., 2005; Kamikouchi et al., 2004) and odorant binding proteins (Dani et al., 2010; Danty et al., 1998; Fang et al., 2012). For *A. mellifera* alone, it would be difficult to assess whether loci that underpin nestmate recognition co-evolved with sociality, but this test for correlated evolutionary change between (genes for) recognition and (genes for) sociality may be possible within a broader phylogenetic context (Van Zweden and D'ettore, 2010) (Fig. 3).

9. PREDICTION 5: GENES FOR ALTRUISM RESIDE IN REGIONS OF LOW RECOMBINATION, EXHIBIT CO-EXPRESSION AND SHOW MODULAR GENETIC ARCHITECTURE

If selection promotes task specialization, then it may likewise favour linkage among the genes involved into co-adapted complexes or 'supergenes' that are characterized by low recombination (Schwander et al., 2014). In a manner analogous to the Y-chromosome in humans, co-adapted gene complexes might be favoured in the short term because they collectively give rise to discrete phenotypes, like 'male' or 'worker', that enjoy a high fitness, whereas intermediate phenotypes arising from multi-locus recombination would be less fit, as in 'inter-sexuals' or 'inter-castes' (Linksvayer et al., 2013). Early evidence for social supergenes comes from ants, where Wang et al. (2013a) found that the entire linkage group containing the *b* allele for social monogyny is non-recombining. Further

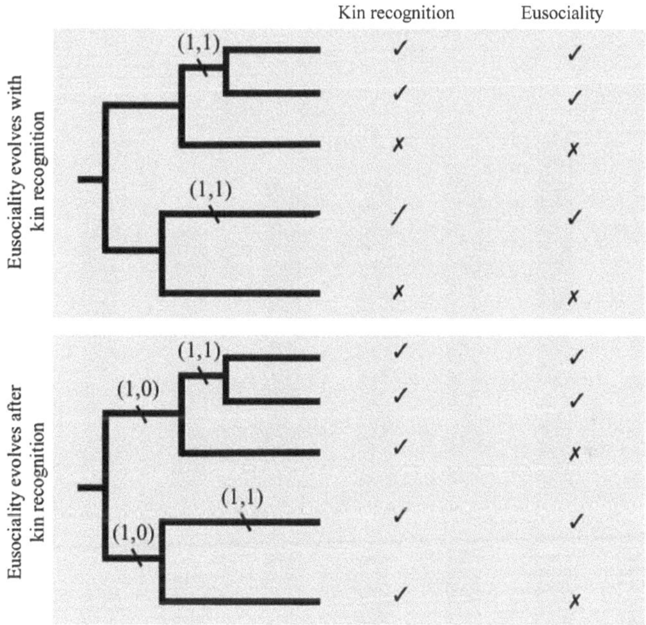

Figure 3 Two alternative phylogenetic scenarios for the co-evolution of eusociality and kin recognition. The top panel shows how state changes (1 = presence, 0 = absence) between two characters—recognition and eusociality—are tightly linked over evolutionary time, suggesting strict co-evolution. The bottom panel, by contrast, shows state changes to be somewhat decoupled, with recognition preceding eusociality. This latter scenario would suggest that kin recognition is an evolutionary precursor to eusociality.

evidence from ants suggests that social chromosomes have evolved repeatedly (Purcell et al., 2014), but as yet there is no clear evidence from honey bees.

Mapping genes for worker sterility (Oxley et al., 2008, Thompson et al., 2008) or worker stinging behaviour (Hunt et al., 1998) suggests that loci are scattered across multiple linkage groups, as opposed to their co-localization within a single region (Fig. 4). Regardless, honey bees do have high genome-wide recombination rates (Beye et al., 2006; Weinstock et al., 2006) and therefore may not readily evolve social chromosomes. High rates of genetic recombination may be functionally associated with bee sociality (Kent and Zayed, 2013) or with the correspondingly low effective population sizes of eusocial populations (Kent et al., 2012). Furthermore, Linksvayer et al. (2013) argue that high recombination rates may actually evolve as a consequence of the initiation of supergenes. The degree to which

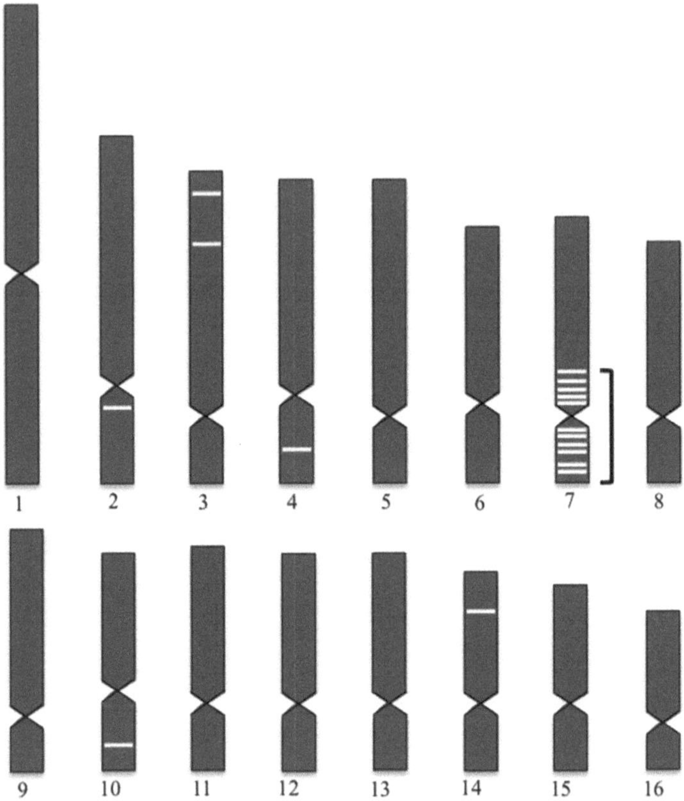

Figure 4 Honey bee karyotype showing hypothetical distribution of loci underlying an altruistic trait. In this hypothetical scenario, we show on chromosome 7 a clustering of genes that are tightly linked, suggesting that selection has promoted their translocation to, or origin within, a common region, and their transmission as a non-recombining unit.

recombination and the formation of any supergenes is relevant to honey bee sociality is unclear, but it should now be possible to fine-map genes that are collectively associated with any number of bee social traits.

10. PREDICTION 6: GENES FOR ALTRUISM SHOULD HAVE AT LEAST PARTIALLY ADDITIVE EFFECTS

Selection is most effective when the traits are underlain by additive genetic variation, and we simply expect kin selection to be most effective when genes for altruism are additive. One way to begin testing for additive effects is to first examine whether the development of a larva into a queen or a worker is influenced by genotype at all. To date, genetic effects on caste differentiation are known from more than 16 species of wasps, ants, bees and termites (Schwander et al., 2010), but in many of these cases the genetic effect may be weak, epistatic (i.e. non-additive), conditional on environmental context (i.e. genetic × environment dependent) or stem from maladaptive interspecific hybridization. In honey bees, experimental evidence for heritable effects on reproductive fate can be seen in queen–worker caste differentiation (Tilley and Oldroyd, 1997), worker egg-laying behaviour (Robinson et al., 1990) and in worker sterility via pheromone-mediated ovary de-activation (Backx et al., 2012), although the exact genes have yet to be identified.

To directly estimate additive effects on aspects of worker sterility, it is possible to employ breeding designs that partition the genetic variance into additive and non-additive components (Oldroyd and Moran, 1983). Worker sterility is an additive trait in several subspecies of honey bees (Jordan et al., 2008), and the narrow sense heritability of ovariole number ranges from 19% to 57% in cape honey bees (*A. mellifera capensis*) (Goudie et al., 2012), and only 6% in anarchistic honey bees (Oxley et al., 2008). Breed et al. (1990) showed that individuals from different subfamilies are more likely to sting than are others, suggesting that there is also a genetic component to this highly social and selfless behaviour. Moritz et al. (1987) estimate the narrow sense heritability for stinging behaviour to be ~81%. This evidence suggests at least partially additive effects to this self-sacrificing behaviour.

11. PREDICTION 7: GENES FOR ALTRUISM SHOULD SHOW STRONG PLEIOTROPY

Social behaviours are complex in that they are the outcome of perceiving, interpreting and responding to cues from conspecifics, and as such can involve several processes, including neuronal, hormonal, motor, etc. Genes that underlie altruism, among other social traits, may therefore be under selection for multiple functions. Examples of social genetic pleiotropy are known from ants, social slime moulds and social vertebrates (Foster et al., 2007; Keller, 2009). For example, a gene responsible for stalk and spore formation in *Dyctiostelium discoideum* may also regulate co-operative behaviour (Foster et al., 2004).

The evidence from honey bees is scarce. In *A. mellifera*, the workers are noticeably different from queens in their physiology, morphology and behaviour, implying a high level of molecular coordination within each caste. For example, workers with inactive ovaries may work faster (Dampney et al., 2004), forage earlier (Oldroyd and Beekman, 2008) and accelerate their metabolism relative to ovary activated workers of the same age. One of the best examples of social pleiotropy is the yolk protein vitellogenin that has evolved multiple social roles in the honey bee (Nelson et al., 2007). In queens, the gene is expressed in the ovaries as a food source to the developing embryo, but in barren workers, vitellogenin is expressed in the hypopharyngeal gland (head) where it is incorporated into royal jelly and fed to larvae as a form of alloparental care (Amdam et al., 2003). This dual social role for vitellogenin—one in parental and the other in alloparental care—demonstrates how a single gene can affect more than one social trait. Other pleiotropic genes from honey bees appear to be involved in insulin signalling (Corona et al., 2007), worker sterility (Cardoen et al., 2011), worker division of labour (Ament et al., 2008), and caste differentiation (Wheeler et al., 2006). Lastly, there is preliminary evidence to suggest that kin recognition loci also have several functions (Holman et al., 2013). It is clear that many social genes will have multiple functions (Table 1).

12. SUMMARY

In this review, we have outlined the need and potential to localize and otherwise identify genes associated with honey bee worker altruism and related self-sacrificial behaviours. We have employed the conceptual framework of Thompson et al. (2013), and it likewise could be similarly deployed against any social taxon for which there were sufficient genomic information to permit the tests. The honey bee provides a good, though not ideal, model to study the social transmission of genes, simply because its biology is so well understood, and its genome is accessible to molecular and bioinformatics screens.

Table 1 shows the nominal support that we were able to glean for each of the gene-finding criteria. Our survey of published studies does yield support for socio-environmental sensitivity (Prediction 2), socio-genetic complexity (Prediction 3) and socio-genetic pleiotropy (Prediction 7). These criteria may therefore be collectively useful evolutionary genetic criteria for helping to identify candidate genes for altruism, at least in honey bees. It may be possible, for example, to combine these search criteria in molecular or bioinformatics screens of social insect genomes to reveal candidate regions or loci that are socially important. Other criteria, including how social genes should satisfy Hamilton's rule (Prediction 1), be strongly additive (Prediction 6), and co-evolve with kin recognition (Prediction 4) are logically sound, but are either difficult to test or have simply not been widely tested for honey bees yet. Finally, evidence for linkage of social genes into co-adapted regions of low recombination (Prediction 5) has been provisionally tested in reference to the honey bee genome and does not appear to be upheld, as it is for other social taxa for which there is better evidence for so-called 'social chromosomes'.

The framework for social gene discovery tested here is provisional and can be modified (Thompson et al., 2013). It is possible that some predictions will have limited practical utility for finding new genes, or show poor or taxon-limited support, or even be falsified, depending on future data and tests. On the other hand, some predictions do seem useful, alone or in combination, for honing in on loci with clear social effects. In combination, Predictions 1 (Hamilton's rule), 3 (sociogenetic complexity) and 6 (additivity) suggest that social genes with altruistic effects can potentially evolve fast via positive indirect selection. Predictions 5 (linkage and low recombination) and 7 (pleiotropy) suggest genetic properties that would make these genes less responsive to positive indirect selection. These contradictions imply that no one social gene will match all seven predictions, but rather that different sets of predictions (e.g. 1, 3, 6 vs. 5, 7) may help identify different sets of social genes. We hope that the conceptual–empirical framework presented here for the honey bee will promote further gene thinking on genes that underlie altruism and other complex social behaviours that Hamilton so long ago predicted.

REFERENCES

Aase, A.L., Amdam, G.V., Hagen, A., Omholt, S.W., 2005. A new method for rearing genetically manipulated honey bee workers. Apidologie 36, 293–299.

Abbot, P., Abe, J., Alcock, J., Alizon, S., Alpedrinha, J.A.C., Andersson, M., Andre, J.B., Van Baalen, M., Balloux, F., Balshine, S., 2011. Inclusive fitness theory and eusociality. Nature 471, E1–E4.

Agrawal, A.F., 2001. Kin recognition and the evolution of altruism. Proc. R. Soc. B Biol. Sci. 268, 1099–1104.

Alaux, C., Le Conte, Y., Adams, H.A., Rodriguez-Zas, S., Grozinger, C.M., Sinha, S., Robinson, G.E., 2009. Regulation of brain gene expression in honey bees by brood pheromone. Genes Brain Behav. 8, 309–319.

Alexander, R.D., Noonan, K., Crespi, B.J., 1991. The evolution of eusociality. In: Sherman, P.W., Jarvis, J., Alexander, R.D. (Eds.), The Biology of the Naked Mole Rat. Princeton University Press, Princeton, NJ.

Amdam, G.V., Norberg, K., Hagen, A., Omholt, S.W., 2003. Social exploitation of vitellogenin. Proc. Natl. Acad. Sci. U.S.A. 100, 1799–1802.

Ament, S.A., Corona, M., Pollock, H.S., Robinson, G.E., 2008. Insulin signaling is involved in the regulation of worker division of labor in honey bee colonies. Proc. Natl. Acad. Sci. U.S.A. 105, 4226–4231.

Arnold, G., Quenet, B., Masson, C., 2000. Influence of social environment on genetically based subfamily signature in the honeybee. J. Chem. Ecol. 26, 2321–2333.
Backx, A.G., Guzman-Novoa, E., Thompson, G.J., 2012. Factors affecting ovary activation in honey bee workers: a meta-analysis. Insect. Soc. 59, 381–388.
Barchuk, A.R., Cristino, A.S., Kucharski, R., Costa, L.F., Simoes, Z.L.P., Maleszka, R., 2007. Molecular determinants of caste differentiation in the highly eusocial honeybee *Apis mellifera*. BMC Dev. Biol. 7, 70.
Barron, A.B., Robinson, G.E., 2008. The utility of behavioral models and modules in molecular analyses of social behavior. Genes Brain Behav. 7, 257–265.
Barron, A.B., Oldroyd, B.P., Ratnieks, F.L.W., 2001. Worker reproduction in honey-bees (*Apis*) and the anarchic syndrome: a review. Behav. Ecol. Sociobiol. 50, 199–208.
Begna, D., Han, B., Feng, M., Fang, Y., Li, J.K., 2012. Differential expressions of nuclear proteomes between honeybee (*Apis mellifera* L.) queen and worker larvae: a deep insight into caste pathway decisions. J. Proteome Res. 11, 1317–1329.
Beshers, S.N., Fewell, J.H., 2001. Models of division of labor in social insects. Annu. Rev. Entomol. 46, 413–440.
Beye, M., Gattermeier, I., Hasselmann, M., Gempe, T., Schioett, M., Baines, J.F., Schlipalius, D., Mougel, F., Emore, C., Rueppell, O., et al., 2006. Exceptionally high levels of recombination across the honey bee genome. Genome Res. 16, 1339–1344.
Boomsma, J.J., 2009. Lifetime monogamy and the evolution of eusociality. Philos. Trans. R. Soc. B Biol. Sci. 364, 3191–3207.
Boomsma, J.J., Franks, N.R., 2006. Social insects: from selfish genes to self organisation and beyond. Trends Ecol. Evol. 21, 303–308.
Bourke, A.F.G., 2011. Principles of Social Evolution. Oxford University Press, Oxford.
Breed, M.D., 1998. Recognition pheromones of the honey bee. Bioscience 48, 463–470.
Breed, M.D., 2014. Kin and nestmate recognition: the influence of W.D. Hamilton on 50 years of research. Anim. Behav. 92, 271–279.
Breed, M.D., Robinson, G.E., Page, R.E., 1990. Division of labor during honey bee colony defense. Behav. Ecol. Sociobiol. 27, 395–401.
Breed, M.D., Diaz, P.H., Lucero, K.D., 2004. Olfactory information processing in honeybee, *Apis mellifera*, nestmate recognition. Anim. Behav. 68, 921–928.
Bromham, L., Leys, R., 2005. Sociality and the rate of molecular evolution. Mol. Biol. Evol. 22, 1393–1402.
Calvello, M., Brandazza, A., Navarrini, A., Dani, F.R., Turillazzi, S., Felicioli, A., Pelosi, P., 2005. Expression of odorant-binding proteins and chemosensory proteins in some Hymenoptera. Insect Biochem. Mol. Biol. 35, 297–307.
Cardoen, D., Wenseleers, T., Ernst, U.R., Danneels, E.L., Laget, D., De Graaf, D.C., Schoofs, L., Verleyen, P., 2011. Genome-wide analysis of alternative reproductive phenotypes in honeybee workers. Mol. Ecol. 20, 4070–4084.
Cardoen, D., Ernst, U.R., Boerjan, B., Bogaerts, A., Formesyn, E., De Graaf, D.C., Wenseleers, T., Schoofs, L., Verleyen, P., 2012. Worker honeybee sterility: a proteomic analysis of suppressed ovary activation. J. Proteome Res. 11, 2838–2850.
Chandrasekaran, S., Ament, S.A., Eddy, J.A., Rodriguez-Zas, S.L., Schatz, B.R., Price, N.D., Robinson, G.E., 2011. Behavior-specific changes in transcriptional modules lead to distinct and predictable neurogenomic states. Proc. Natl. Acad. Sci. U.S.A. 108, 18020–18025.
Choe, J.C., Crespi, B.J., 1997. The Evolution of Social Behavior in Insects and Arachnids. Cambridge University Press, Cambridge.
Corona, M., Estrada, E., Zurita, M., 1999. Differential expression of mitochondrial genes between queens and workers during caste determination in the honeybee *Apis mellifera*. J. Exp. Biol. 202, 929–938.

Corona, M., Velarde, R.A., Remolina, S., Moran-Lauter, A., Wang, Y., Hughes, K.A., Robinson, G.E., 2007. Vitellogenin, juvenile hormone, insulin signaling, and queen honey bee longevity. Proc. Natl. Acad. Sci. U.S.A. 104, 7128–7133.

Crespi, B.J., 1996. Comparative analysis of the origins and losses of eusociality: causal mosaics and historical uniqueness. In: Martins, E.P. (Ed.), Phylogenies and the Comparative Method in Animal Behavior. Oxford University Press, New York.

Cristino, A.S., Nunes, F.M.F., Lobo, C.H., Bitondi, M.M.G., Simoes, Z.L.P., Costa, L.D., Lattorff, H.M.G., Moritz, R.F.A., Evans, J.D., Hartfelder, K., 2006. Caste development and reproduction: a genome-wide analysis of hallmarks of insect eusociality. Insect Mol. Biol. 15, 703–714.

Crozier, R.H., 1988. Kin recognition using innate labels: a central role for piggybacking? In: Grosberg, R.K., Hedgecock, D., Nelson, K. (Eds.), Invertebrate Historecognition. Plenum, New York.

Crozier, R.H., 2008. Advanced eusociality, kin selection and male haploidy. Aust. J. Entomol. 47, 2–8.

Crozier, R.H., Pamilo, P., 1996. Evolution of Social Insect Colonies: Sex Allocation and Kin Selection. Oxford University Press, Oxford.

Dampney, J.R., Barron, A.B., Oldroyd, B.P., 2004. Measuring the cost of worker reproduction in honeybees: work tempo in an 'anarchic' line. Apidologie 35, 83–88.

Dani, F.R., Jones, G.R., Corsi, S., Beard, R., Pradella, D., Turillazzi, S., 2005. Nestmate recognition cues in the honey bee: differential importance of cuticular alkanes and alkenes. Chem. Senses 30, 477–489.

Dani, F.R., Iovinella, I., Felicioli, A., Niccolini, A., Calvello, M.A., Carucci, M.G., Qiao, H.L., Pieraccini, G., Turillazzi, S., Moneti, G., et al., 2010. Mapping the expression of soluble olfactory proteins in the honeybee. J. Proteome Res. 9, 1822–1833.

Danty, E., Arnold, G., Huet, J.C., Huet, D., Masson, C., Pernollet, J.C., 1998. Separation, characterization and sexual heterogeneity of multiple putative odorant-binding proteins in the honeybee *Apis mellifera L.* (Hymenoptera: Apidea). Chem. Senses 23, 83–91.

Dawkins, R., 1979. Twelve misunderstandings of kin selection. Z. Tierpsychol. 51, 184–200.

De Azevedo, S.V., Hartfelder, K., 2008. The insulin signaling pathway in honey bee (*Apis mellifera*) caste development—differential expression of insulin-like peptides and insulin receptors in queen and worker larvae. J. Insect Physiol. 54, 1064–1071.

Drapeau, M.D., Albert, S., Kucharski, R., Prusko, C., Maleszka, R., 2006. Evolution of the yellow/major royal jelly protein family and the emergence of social behavior in honey bees. Genome Res. 16, 1385–1394.

Engel, M.S., 2001. Monophyly and extensive extinction of advanced eusocial bees: insights from an unexpected Eocene diversity. Proc. Natl. Acad. Sci. U.S.A. 98, 1661–1664.

Evans, J.D., Wheeler, D.E., 1999. Differential gene expression between developing queens and workers in the honey bee, *Apis mellifera*. Proc. Natl. Acad. Sci. U.S.A. 96, 5575–5580.

Evans, J.D., Wheeler, D.E., 2000. Expression profiles during honeybee caste determination. Genome Biol. 2(1), research0001.1–research0001.6.

Fang, Y., Song, F.F., Zhang, L., Aleku, D.W., Han, B., Feng, M., Li, J.K., 2012. Differential antennal proteome comparison of adult honeybee drone, worker and queen (*Apis mellifera* L.). J. Proteomics 75, 756–773.

Fischman, B.J., Woodard, S.H., Robinson, G.E., 2011. Molecular evolutionary analyses of insect societies. Proc. Natl. Acad. Sci. U.S.A. 108, 10847–10854.

Formesyn, E.M., Cardoen, D., Ernst, U.R., Danneels, E.L., Van Vaerenbergh, M., De Koker, D., Verleyen, P., Wenseleers, T., Schoofs, L., De Graaf, D.C., 2014. Reproduction of honeybee workers is regulated by epidermal growth factor receptor signaling. Gen. Comp. Endocrinol. 197, 1–4.

Foster, K.R., Shaulsky, G., Strassmann, J.E., Queller, D.C., Thompson, C.R.L., 2004. Pleiotropy as a mechanism to stabilize cooperation. Nature 431, 693–696.
Foster, K.R., Parkinson, K., Thompson, C.R.L., 2007. What can microbial genetics teach sociobiology? Trends Genet. 23, 74–80.
Fussnecker, B., Grozinger, C., 2008. Dissecting the role of Kr-h1 brain gene expression in foraging behavior in honey bees (*Apis mellifera*). Insect Mol. Biol. 17, 515–522.
Gadagkar, R., 1997. The evolution of caste polymorphism in social insects: genetic release followed by diversifying evolution. J. Genet. 76, 167–179.
Gardner, A., West, S.A., Wild, G., 2011. The genetical theory of kin selection. J. Evol. Biol. 24, 1020–1043.
Goudie, F., Allsopp, M.H., Beekman, M., Lim, J., Oldroyd, B.P., 2012. Heritability of worker ovariole number in the Cape honey bee *Apis mellifera capensis*. Insect. Soc. 59, 351–359.
Grozinger, C.M., Sharabash, N.M., Whitfield, C.W., Robinson, G.E., 2003. Pheromone-mediated gene expression in the honey bee brain. Proc. Natl. Acad. Sci. U.S.A. 100, 14519–14525.
Grozinger, C.M., Fan, Y.L., Hoover, S.E.R., Winston, M.L., 2007. Genome-wide analysis reveals differences in brain gene expression patterns associated with caste and reproductive status in honey bees (*Apis mellifera*). Mol. Ecol. 16, 4837–4848.
Hall, D.W., Goodisman, M.A., 2012. The effects of kin selection on rates of molecular evolution in social insects. Evolution 66, 2080–2093.
Hamilton, W.D., 1963. The evolution of altruistic behavior. Am. Nat. 97, 354–356.
Hamilton, W.D., 1964. The genetical evolution of social behaviour, I and II. J. Theor. Biol. 7, 1–52.
Harpur, B.A., Kent, C.F., Molodtsova, D., Lebon, J.M., Alqarni, A.S., Owayss, A.A., Zayed, A., 2014. Population genomics of the honey bee reveals strong signatures of positive selection on worker traits. Proc. Natl. Acad. Sci. U.S.A. 111, 2614–2619.
Hartfelder, K., Kostlin, K., Hepperle, C., 1995. Ecdysteroid-dependent protein-synthesis in caste-specific development of the larval honey-bee ovary. Rouxs Arch. Dev. Biol. 205, 73–80.
Havukainen, H., Halskau, O., Amdam, G.V., 2011. Social pleiotropy and the molecular evolution of honey bee vitellogenin. Mol. Ecol. 20, 5111–5113.
Herbers, J.M., 2013. 50 years on: the legacy of William Donald Hamilton. Biol. Lett. 9, 20130792.
Holldobler, B., Wilson, E.O., 2009. The Superorganism: The Beauty, Elegance, and Strangeness of Insect Societies. W.W. Norton, New York; London.
Holman, L., Van Zweden, J.S., Linksvayer, T.A., D'ettorre, P., 2013. Crozier's paradox revisited: maintenance of genetic recognition systems by disassortative mating. BMC Evol. Biol. 13, 211.
Hughes, W.O.H., Oldroyd, B.P., Beekman, M., Ratnieks, F.L.W., 2008. Ancestral monogamy shows kin selection is key to the evolution of eusociality. Science 320, 1213–1216.
Hunt, G.J., 2007. Flight and fight: A comparative view of the neurophysiology and genetics of honey bee defensive behavior. J. Insect Physiol. 53, 399–410.
Hunt, G.J., Page, R.E., Fondrk, M.K., Dullum, C.J., 1995. Major quantitative trait loci affecting honey-bee foraging behavior. Genetics 141, 1537–1545.
Hunt, G.J., Guzman-Novoa, E., Fondrk, M.K., Page, R.E., 1998. Quantitative trait loci for honey bee stinging behavior and body size. Genetics 148, 1203–1213.
Hunt, B.G., Wyder, S., Elango, N., Werren, J.H., Zdobnov, E.M., Yi, S.V., Goodisman, M.A., 2010. Sociality is linked to rates of protein evolution in a highly social insect. Mol. Biol. Evol. 27, 497–500.

Jarosch, A., Stolle, E., Crewe, R.M., Moritz, R.F.A., 2011. Alternative splicing of a single transcription factor drives selfish reproductive behavior in honeybee workers (*Apis mellifera*). Proc. Natl. Acad. Sci. U.S.A. 108, 15282–15287.

Johnson, B.R., Tsutsui, N.D., 2011. Taxonomically restricted genes are associated with the evolution of sociality in the honey bee. BMC Genomics 12, 164.

Jordan, L.A., Allsopp, M.H., Beekman, M., Wossler, T.C., Oldroyd, B.P., 2008. Inheritance of traits associated with reproductive potential in *Apis mellifera* capensis and *Apis mellifera* scutellata workers. J. Hered. 99, 376–381.

Kamikouchi, A., Morioka, M., Kubo, T., 2004. Identification of honeybee antennal proteins/genes expressed in a sex- and/or caste selective manner. Zool. Sci. 21, 53–62.

Keller, L., 2009. Adaptation and the genetics of social behaviour. Philos. Trans. R. Soc. B Biol. Sci. 364, 3209–3216.

Kent, C.F., Zayed, A., 2013. Evolution of recombination and genome structure in eusocial insects. Commun. Integr. Biol. 6, e22919.

Kent, C.F., Issa, A., Bunting, A.C., Zayed, A., 2011. Adaptive evolution of a key gene affecting queen and worker traits in the honey bee, *Apis mellifera*. Mol. Ecol. 20, 5226–5235.

Kent, C.F., Minaei, S., Harpur, B.A., Zayed, A., 2012. Recombination is associated with the evolution of genome structure and worker behavior in honey bees. Proc. Natl. Acad. Sci. U.S.A. 109, 18012–18017.

Kiya, T., Ugajin, A., Kunieda, T., Kubo, T., 2012. Identification of *kakusei*, a nuclear non-coding RNA, as an immediate early gene from the honeybee, and its application for neuroethological study. Int. J. Mol. Sci. 13, 15496–15509.

Kocher, S.D., Richard, F.J., Tarpy, D.R., Grozinger, C.M., 2009. Queen reproductive state modulates pheromone production and queen-worker interactions in honeybees. Behav. Ecol. 20, 1007–1014.

Kocher, S.D., Ayroles, J.F., Stone, E.A., Grozinger, C.M., 2010. Individual variation in pheromone response correlates with reproductive traits and brain gene expression in worker honey bees. PLoS One 5, e9116.

Kucharski, R., Maleszka, J., Foret, S., Maleszka, R., 2008. Nutritional control of reproductive status in honeybees via DNA methylation. Science 319, 1827–1830.

Lattorff, H.M.G., Moritz, R.F.A., Crewe, R.M., Solignac, M., 2007. Control of reproductive dominance by the *thelytoky* gene in honeybees. Biol. Lett. 3, 292–295.

Le Conte, Y., Hefetz, A., 2008. Primer pheromones in social hymenoptera. Annu. Rev. Entomol. 53, 523–542.

Leboeuf, A.C., Benton, R., Keller, L., 2013. The molecular basis of social behavior: models, methods and advances. Curr. Opin. Neurobiol. 23, 3–10.

Linksvayer, T.A., Wade, M.J., 2005. The evolutionary origin and elaboration of sociality in the aculeate hymenoptera: maternal effects, sib-social effects, and heterochrony. Q. Rev. Biol. 80, 317–336.

Linksvayer, T.A., Wade, M.J., 2009. Genes with social effects are expected to harbor more sequence variation within and between species. Evolution 63, 1685–1696.

Linksvayer, T.A., Busch, J.W., Smith, C.R., 2013. Social supergenes of superorganisms: do supergenes play important roles in social evolution? Bioessays 35, 683–689.

Maynard Smith, J., Szathmary, E., 1995. The Major Transitions in Evolution. W.H. Freeman/Spektrum, Oxford.

Moritz, R.F.A., Southwick, E.E., Harbo, J.B., 1987. Genetic-analysis of defensive behavior of honeybee colonies (*Apis mellifera* L.) in a field-test. Apidologie 18, 27–41.

Mougel, F., Solignac, M., Vautrin, D., Baudry, E., Ogden, J., Tchapla, A., Schweitz, H., Gilbert, H., 2012. Quantitative traits loci (QTL) involved in body colour, wing morphometry, cuticular hydrocarbons and venom components in honeybee. Apidologie 43, 162–181.

Mullen, E.K., Daley, M., Backx, A.G., Thompson, G.J., 2014. Gene co-citation networks associated with worker sterility in honey bees. BMC Syst. Biol. 8, 38.

Munoz-Torres, M.C., Reese, J.T., Childers, C.P., Bennett, A.K., Sundaram, J.P., Childs, K.L., Anzola, J.M., Milshina, N., Elsik, C.G., 2011. Hymenoptera Genome Database: integrated community resources for insect species of the order Hymenoptera. Nucleic Acids Res. 39, D658–D662.

Mutti, N.S., Dolezal, A.G., Wolschin, F., Mutti, J.S., Gill, K.S., Amdam, G.V., 2011. IRS and TOR nutrient-signaling pathways act via juvenile hormone to influence honey bee caste fate. J. Exp. Biol. 214, 3977–3984.

Nelson, C.M., Ihle, K.E., Fondrk, M.K., Page, R.E., Amdam, G.V., 2007. The gene vitellogenin has multiple coordinating effects on social organization. PLoS Biol. 5, 673–677.

Oldroyd, B.P., Beekman, M., 2008. Effects of selection for honey bee worker reproduction on foraging traits. PLoS Biol. 6, 463–470.

Oldroyd, B.P., Fewell, J.H., 2007. Genetic diversity promotes homeostasis in insect colonies. Trends Ecol. Evol. 22, 408–413.

Oldroyd, B., Moran, C., 1983. Heritability of worker characters in the honeybee (*Apis mellifera*). Aust. J. Biol. Sci. 36, 323–332.

Oldroyd, B.P., Thompson, G.J., 2007. Behavioural genetics of the honey bee *Apis mellifera*. Adv. Insect Physiol. 33, 1–49.

Oxley, P.R., Thompson, G.J., Oldroyd, B.P., 2008. Four quantitative trait loci that influence worker sterility in the honeybee (*Apis mellifera*). Genetics 179, 1337–1343.

Page, R.E., Robinson, G.E., Fondrk, M.K., 1989. Genetic specialists, kin recognition and nepotism in honeybee colonies. Nature 338, 576–579.

Patel, A., Fondrk, M.K., Kaftanoglu, O., Emore, C., Hunt, G., Frederick, K., Amdam, G.V., 2007. The making of a queen: TOR pathway is a key player in diphenic caste development. PLoS One 2, e509.

Piulachs, M.D., Guidugli, K.R., Barchuk, A.R., Cruz, J., Simoes, Z.L.P., Belles, X., 2003. The vitellogenin of the honey bee, *Apis mellifera*: structural analysis of the cDNA and expression studies. Insect Biochem. Mol. Biol. 33, 459–465.

Plomin, R., DeFries, J.C., Knopik, V.S., Neiderhiser, J.M., 2013. Behavioral Genetics, fourth ed. Worth Publishers, New York.

Purcell, J., Brelsford, A., Wurm, Y., Perrin, N., Chapuisat, M., 2014. Convergent genetic architecture underlies social organization in ants. In: 17th Congress of the International Union for the Study of Social Insects (IUSSI), Cairns, Australia.

Queller, D.C., Strassmann, J.E., 1998. Kin selection and social insects. Bioscience 48, 165–175.

Queller, D.C., Liao, S., Rong, S., 2014. Nowak-style models refute Nowak's conclusions about eusociality. In: 17th Congress of the International Union for the Study of Social Insects (IUSSI), Cairns, Australia, July 13–18, 2014.

Robinson, G.E., Page, R.E., Fondrk, M.K., 1990. Intracolonial behavioral variation in worker oviposition, oophagy, and larval care in queenless honey-bee colonies. Behav. Ecol. Sociobiol. 26, 315–323.

Robinson, G.E., Fernald, R.D., Clayton, D.F., 2008. Genes and social behavior. Science 322, 896–900.

Rousset, F., Lion, S., 2011. Much ado about nothing: Nowak et al.'s charge against inclusive fitness theory. J. Evol. Biol. 24, 1386–1392.

Rousset, F., Roze, D., 2007. Constraints on the origin and maintenance of genetic kin recognition. Evolution 61, 2320–2330.

Rueppell, O., Pankiw, T., Nielsen, D.I., Fondrk, M.K., Beye, M., Page, R.E., 2004. The genetic architecture of the behavioral ontogeny of foraging in honeybee workers. Genetics 167, 1767–1779.

Schulte, C., Leboulle, G., Otte, M., Grunewald, B., Gehne, N., Beye, M., 2013. Honey bee promoter sequences for targeted gene expression. Insect Mol. Biol. 22, 399–410.
Schwander, T., Lo, N., Beekman, M., Oldroyd, B.P., Keller, L., 2010. Nature versus nurture in social insect caste differentiation. Trends Ecol. Evol. 25, 275–282.
Schwander, T., Libbrecht, R., Keller, L., 2014. Supergenes and complex phenotypes. Curr. Biol. 24, R288–R294.
Seger, J., 1981. Kinship and covariance. J. Theor. Biol. 91, 191–213.
Shorter, J.R., Arechavaleta-Velasco, M., Robles-Rios, C., Hunt, G.J., 2012. A genetic analysis of the stinging and guarding behaviors of the honey bee. Behav. Genet. 42, 663–674.
Simola, D.F., Wissler, L., Donahue, G., Waterhouse, R.M., Helmkampf, M., Roux, J., Nygaard, S., Glastad, K.M., Hagen, D.E., Viljakainen, L., Reese, J.T., Hunt, B.G., Graur, D., Elhaik, E., Kriventseva, E.V., Wen, J.Y., Parker, B.J., Cash, E., Privman, E., Childers, C.P., Munoz-Torres, M.C., Boomsma, J.J., Bornberg-Bauer, E., Currie, C.R., Elsik, C.G., Suen, G., Goodisman, M.A.D., Keller, L., Liebig, J., Rawls, A., Reinberg, D., Smith, C.D., Smith, C.R., Tsutsui, N., Wurm, Y., Zdobnov, E.M., Berger, S.L., Gadau, J., 2013. Social insect genomes exhibit dramatic evolution in gene composition and regulation while preserving regulatory features linked to sociality. Genome Res. 23, 1235–1247.
Smith, C.R., Toth, A.L., Suarez, A.V., Robinson, G.E., 2008. Genetic and genomic analyses of the division of labour in insect societies. Nat. Rev. Genet. 9, 735–748.
Solignac, M., Mougel, F., Vautrin, D., Monnerot, M., Cornuet, J.M., 2007. A third-generation microsatellite-based linkage map of the honey bee, *Apis mellifera*, and its comparison with the sequence-based physical map. Genome Biol. 8 (4), R66.
Sumner, S., 2006. Determining the molecular basis of sociality in insects: progress, prospects and potential in sociogenomics. Ann. Zool. Fenn. 43, 423–442.
Sumner, S., 2014. The importance of genomic novelty in social evolution. Mol. Ecol. 23, 26–28.
Thompson, G.J., 2006. Kin selection in disguise? Insect. Soc. 53, 496–497.
Thompson, G.J., Oldroyd, B.P., 2004. Evaluating alternative hypotheses for the origin of eusociality in corbiculate bees. Mol. Phylogenet. Evol. 33, 452–456.
Thompson, G.J., Kucharski, R., Maleszka, R., Oldroyd, B.P., 2006. Towards a molecular definition of worker sterility: differential gene expression and reproductive plasticity in honey bees. Insect Mol. Biol. 15, 637–644.
Thompson, G.J., Kucharski, R., Maleszka, R., Oldroyd, B.P., 2008. Genome-wide analysis of genes related to ovary activation in worker honey bees. Insect Mol. Biol. 17, 657–665.
Thompson, G.J., Hurd, P.L., Crespi, B.J., 2013. Genes underlying altruism. Biol. Lett. 9, 20130395.
Tilley, C.A., Oldroyd, B.P., 1997. Unequal subfamily proportions among honey bee queen and worker brood. Anim. Behav. 54, 1483–1490.
Toth, A.L., Varala, K., Newman, T.C., Miguez, F.E., Hutchison, S.K., Willoughby, D.A., Simons, J.F., Egholm, M., Hunt, J.H., Hudson, M.E., et al., 2007. Wasp gene expression supports an evolutionary link between maternal behavior and eusociality. Science 318, 441–444.
Tsutsui, N.D., 2004. Scents of self: the expression component of self/nonself recognition systems. Ann. Zool. Fenn. 41, 713–727.
Van Zweden, J.S., D'ettore, P., 2010. Nestmate recognition in social insects and the role of hydrocarbons. In: Blomquist, G.J., Bagnères, A. (Eds.), Insect Hydrocarbons: Biology, Biochemistry, and Chemical Ecology. Cambridge University Press, Cambridge.
Wang, Y., Amdam, G.V., Rueppell, O., Wallrichs, M.A., Fondrk, M.K., Kaftanoglu, O., Page, R.E., 2009. PDK1 and HR46 gene homologs tie social behavior to ovary signals. PLoS One 4, e4899.

Wang, J., Wurm, Y., Nipitwattanaphon, M., Riba-Grognuz, O., Huang, Y.C., Shoemaker, D., Keller, L., 2013a. A Y-like social chromosome causes alternative colony organization in fire ants. Nature 493, 664–668.

Wang, Y., Azevedo, S.V., Hartfelder, K., Amdam, G.V., 2013b. Insulin-like peptides (AmILP1 and AmILP2) differentially affect female caste development in the honey bee (*Apis mellifera* L.). J. Exp. Biol. 216, 4347–4357.

Wcislo, W.T., Danforth, B.N., 1997. Secondarily solitary: the evolutionary loss of social behavior. Trends Ecol. Evol. 12, 468–474.

Weinstock, G.M., Robinson, G.E., Gibbs, R.A., Worley, K.C., Evans, J.D., Maleszka, R., Robertson, H.M., Weaver, D.B., Beye, M., Bork, P., et al., 2006. Insights into social insects from the genome of the honeybee *Apis mellifera*. Nature 443, 931–949.

West, S.A., Griffin, A.S., Gardner, A., 2007. Social semantics: altruism, cooperation, mutualism, strong reciprocity and group selection. J. Evol. Biol. 20, 415–432.

Wheeler, D.E., Buck, N., Evans, J.D., 2006. Expression of insulin pathway genes during the period of caste determination in the honey bee, *Apis mellifera*. Insect Mol. Biol. 15, 597–602.

Wolschin, F., Mutti, N.S., Amdam, G.V., 2011. Insulin receptor substrate influences female caste development in honeybees. Biol. Lett. 7, 112–115.

Woodard, S.H., Fischman, B.J., Venkat, A., Hudson, M.E., Varala, K., Cameron, S.A., Clark, A.G., Robinson, G.E., 2011. Genes involved in convergent evolution of eusociality in bees. Proc. Natl. Acad. Sci. U.S.A. 108, 7472–7477.

INDEX

Note: Page numbers followed by "*f*" indicate figures and "*t*" indicate tables.

C

D

E

Q

R

S

T

Kolibri 360 GmbH
Ferdinand-Jühlke-Straße 7
99095 Erfurt, Deutschland
produktsicherheit@kolibri360.de